Advances in Geoarchaeology and Cultural Heritage

Advances in Geoarchaeology and Cultural Heritage

Guest Editor

Ioannis Liritzis

Basel • Beijing • Wuhan • Barcelona • Belgrade • Novi Sad • Cluj • Manchester

Guest Editor
Ioannis Liritzis
Department of Humanities
Alma Mater Europaea University
Maribor
Slovenia

Editorial Office
MDPI AG
Grosspeteranlage 5
4052 Basel, Switzerland

This is a reprint of the Special Issue, published open access by the journal *Quaternary* (ISSN 2571-550X), freely accessible at: www.mdpi.com/journal/quaternary/special_issues/geoarchaeology_cultural_heritage.

For citation purposes, cite each article independently as indicated on the article page online and using the guide below:

Lastname, A.A.; Lastname, B.B. Article Title. *Journal Name* **Year**, *Volume Number*, Page Range.

ISBN 978-3-7258-3020-6 (Hbk)
ISBN 978-3-7258-3019-0 (PDF)
https://doi.org/10.3390/books978-3-7258-3019-0

Contents

About the Editor

Ioannis Liritzis

Ioannis Liritzis is Distinguished Professor of Natural Sciences in Cultural Heritage, Archaeo-Geo-environment, at the Alma Mater Europaea University, Slovenia and Dean of Class IV (Natural Sciences) in the European Academy of Sciences and Arts (Salzburg). During his career, he has been working in public and research establishments (Ministry of Culture Greece, Academy of Athens, University of the Aegean, EU and USA Universities), performing research on a range of multidisciplinary fields (dosimetry, astronomy, geosciences, archaeometry). Current projects include obsidian dating studies, AI in archaeology, STEM in arts and culture, policy works on education, research, and innovation. He has been awarded honorary titles from the Northwestern Polytechnical University (China), the Samara State Institute of Culture, Corresponding Member of Academie des Sciences, Arts et Belles Letters, Dijon; Edinburgh University, Rhodes University, S. Africa and elsewhere.

Preface

The Special Issue *Advances in Geoarchaeology and Cultural Heritage* brings together diversified interdisciplinary scientific approaches of natural sciences studying the past through archaeological remains and cultural heritage proxies. The reasons and motivations for compiling and editing this scientific work are the great interest in the advanced techniques to link material culture, remote past societies, and human-environment interaction.

It is addressed to earth scientists, archaeologists, archaeometrists, senior scientists, and graduate students. The involved authors come from a wide range of subjects focused on geosciences and cultural heritage issues. I thank managing editors of *Quaternary* MDPI for their assistance in making this volume.

Ioannis Liritzis
Guest Editor

quaternary

MDPI

Editorial

Advances in Geoarchaeology and Cultural Heritage: Editorial

Ioannis Liritzis

Faculty of Humanities, Alma Mater Europaea University (AMEU), Slovenska Ulica 17, 2000 Maribor, Slovenia; ioannis.liritzis@almamater.si

1. Introduction

The non-linear trend of evolution of ancient cultures is decisively affected by climatic change, landscape changes, and more. At the same time, natural sciences which are applied to material culture and, simultaneously, delve into the stratigraphic record reveal and document past daily life. The task of modern humanity is to preserve the memories of the past. This Special Issue is based mainly on selected papers presented in the digital 2nd Sino-Hellenic International Conference on Global Issues of Environment and Culture which took place on 17–19 September 2021, and was hosted in Greece (https://iao.henu.edu.cn/info/1257/5437.htm, accessed on 10 December 2024). Beyond the selected papers, the SI touches on other issues chronologically covering the Quaternary, e.g., palaeoanthropology and the environment, climate change and ancient cultures, disaster archaeology, geoarchaeological issues, archaeometry–archaeological sciences, and digital and remote sensing applications, to mention a few.

The application of natural sciences to solve problems in the fields of archaeology, anthropology, physical and human geography, cultural heritage, and art, covering a long-time range during the quaternary, has been very popular. The various terminologies used (archaeological sciences, geoarchaeology, archaeometry, and STEM in arts and culture) essentially focus on the tandem of science and art (https://euro-acad.eu/library?id=17, accessed on 10 December 2024) (Liritzis 2000 [1]; Liritzis et al., 2024 [2]). After all, such research and its application offers societal and educational benefits, addresses sustainability concerns, and aligns with SDG-11. (With Sustainable Development Goal 11 (SDG 11), countries have pledged to "make cities and human settlements inclusive, safe, resilient and sustainable". Within this goal, Target 11.4 aims to "strengthen efforts to protect and safeguard the world's cultural and natural heritage".) (see also relevant articles in: https://www.mdpi.com/journal/sustainability/special_issues/Archaeology_Cultural_Heritage; accessed on 10 December 2024).

2. Brief Summary

The eleven articles cover diversified nature subjects all related to geoarchaeology and cultural archaeological heritage.

Prof. Stoev et al. [3] studied prehistoric astronomical observatories, which include a specific type of rock-cut monument from the mountainous Thrace in Bulgaria that has a specific shape and orientation in space. which is a part of the characteristic representatives of the archaeoastronomical sites on the Balkan Peninsula. They examined 13 prehistoric astronomical observatories using the methods of archaeoastronomy in order to determine the period of their operation. It was shown that changes in climate (and in the astroclimate accordingly) influenced the type of prehistoric astronomical observatories built.

Dr. Kokkaliari et al. [4] referred to recent findings of archaeological research in the Vathy gulf area, Astypalaia Island, Aegean Sea, Greece, which indicate its continuous habitation since prehistoric times, most importantly in the transitional period from the Final Neolithic to the Early Bronze Age (late 4th/early 3rd millennium BC). Using non-invasive analytical methods (Near-Infrared Spectroscopy—NIR) in combination with the

Citation: Liritzis, I. Advances in Geoarchaeology and Cultural Heritage: Editorial. *Quaternary* **2024**, 7, 59. https://doi.org/10.3390/quat7040059

Received: 11 December 2024
Accepted: 17 December 2024
Published: 18 December 2024

mineral–petrographic characterisation of the main lithological formations of the island, they were able to identify allogenic geo-materials including chalcedony, mica schist, bauxite and meta-bauxite, steatite, and paragonite. Based on the mineralogical and petrographic characterisation of the stone artefacts, a first attempt was made to evaluate the possible raw material sources and to identify potential intra-island modes of stone exploitation.

Dr. Westra et al. [5] studied disasters in the past that have happened throughout human existence. Their traces are found in the environmental record, archaeological evidence, and historical chronicles. They present the Shang and Mycenaean cultures as prime examples, among others, of Bronze Age societies with distinctive geographical, environmental, and cultural features and structures that defined their attitudes and responses to dangerous natural phenomena, such as floods, earthquakes, landslides, and drought. The cultural landscapes of the Aegean Sea in Greece and Yellow River in China share many similarities and dissimilarities and both had vast territorial and cultural expansions. They have an apparent contemporaneity, and both receded and collapsed at about the same time. Thus, through the microgeography of a few select Shang and Mycenaean sites and their relevant environmental, archaeological, and historical contexts, and through environmental effects on a global scale, we may understand chain events of scattered human societal changes, collapses, and revolutions on a structural level.

Liritzis et al. [6] focused on the geoarchaeology of a Kastrouli Late Bronze settlement in the province of Phocis, central Greece, which has been proved to have been an important centre in the periphery of the Mycenaean palaces during the 13th–11th c BC. A methodological approach was applied at the foot of the hillock near the Kastrouli plain site, combining the digital elevation model (DEM) and GIS with electrical resistivity tomography (ERT) traverses of around 300 and 500 m. The existence of an ancient lake with the presence of two natural sinkholes was revealed, which was apparently engineered using hydraulic works. It was noted that the mane-made sinkholes were made to carry our drainage and produce a habitable environment, protecting the cultivated land and avoiding a swamp associated with health issues.

Prof. Koskeridou et al. [7] presented the degradation of coastal environments of an ancient coastal lake wetland, the so-called Lake Lerna in NE Peloponnese, Greece; this is an issue that many areas in Europe are facing. Two drill cores in the area of the ancient lake were analysed and three lithological and faunal units were recovered. The usage of sub-fossil mollusc species for the first time in the region enriched the dataset and contributed significantly to the delimitation of the fauna. They concluded that the lake and its included fauna and flora were mostly affected by climatic fluctuations rather than human intervention.

Dr. Saitis et al. [8] studied beachrocks which are well known as significant proxies for paleoenvironmental analysis as they indicate coastal evolution. The combination of geomorphological and archaeological sea-level indicators significantly contributes to coastal paleogeographic reconstruction. In this study, a beachrock from the Diolkos area (West Corinth canal, Greece) and remnants of the Diolkos slipway were used to reconstruct the coastal evolution before Diolkos' construction until today. They were studied using DGPS-GNSS, and mineralogical analysis and OSL dating of beachrock samples were also carried out. The results showed that the beachrock slab was preserved before the construction of Diolkos below it, followed by its submergence by a co-seismic event after the abandonment of Diolkos during 146 B.C. Consequently, a new beachrock was developed on top of the submerged Diolkos around 120 $\pm$ 14 A.D. The RSL was stable until 1596 $\pm$ 57 A.D. when the beachrock developed even closer to the present-day coastline. After 1596 A.D., it was uplifted by 12 cm until it reached today's position.

Dr. Howland et al. [9] applied tests that elaborated on the suitability of automated point cloud classification tools provided by the popular image-based modelling (IBM) software package Agisoft Metashape for the generation of digital terrain models (DTMs) at the moderately vegetated archaeological site of Kastrouli, a Mycenaean site in Greece. Based on this case study, the mostly automated, geometric classification tool "Classify

Ground Points" provides the best results and produces a quality DTM that is sufficient for mapping and analysis. Each of the methods tested in this paper can likely be improved through the manual editing of point cloud classification.

Dr. Muskara et al. [10] analysed ancient obsidian tools from Kendale Hecala located on the Ambar River in the Upper Tigris Basin, which is located in the province of Diyarbakır in Southeast Anatolia. Macro-observations and the characterisation of archaeological samples were performed using a handheld XRF and provenance was determined through statistical clustering and multivariate analysis of the data. The authors indicate obsidian from different resources at the settlement was used, including from Nemrut Dağ, Bingöl B, and Group 3d. The technological analysis revealed that obsidian was brought to the settlement as nodules and chipped into various tools at the settlement. Understanding the operational sequence of the lithic industry, the chaîne opératoire, including the distribution of raw material from source to site, is important to demonstrate the socio-cultural organisation of the settlement in Southeastern Anatolia during the Ubaid period.

Dr. Robert Bednarik [11], a specialist in rock art, reviewed the most extensive corpus of ancient immovable cultural heritage, including his current work on the direct dating of Chinese Immovable Cultural Heritage. The dating of Chinese rock art by 'direct methods' began in the late 1990s in Qinghai Province. Since then, China has acquired the largest body of direct dating information about rock art of any country. The establishment of the International Centre for Rock Art Dating at Hebei Normal University has been the driving force in this development, with its researchers accounting for most of the results. Dating methods referred to include microerosion analysis, OSL, C-14, and uranium–thorium analysis of carbonate precipitates in caves, and results were reported from Heilongjiang, Inner Mongolia, Ningxia, Jiangsu, Hubei, Guangxi, Yunnan, Qinghai, Tibet, and Xinjiang.

Dr. Rossi et al. [12] reported on the strontium isotopes in archaeology and biogeochemical research carried out over the past 40 years, highlighting its limits and risks. They included basic knowledge material and suggestions for the correct use of these isotopes. The isotopic characteristics of bio-assimilable strontium depend not so much on the isotopic characteristics of the bulk rock but, rather, on those of its more soluble minerals. It is warned that before studying human, animal, and plant remains, the state of conservation and any conditions of isotopic pollution should be carefully checked. Samples should be collected according to random sampling rules. The data should be assessed using a statistical approach. To make comparisons between different areas, it should be borne in mind that the study of current soils can be misleading since the mineralogical modification of soil over time can be very rapidly altered.

Finally, Dr. Anja Hellmuth Kramberger [13] reported on the emergence of the Bronze Age Monkodonja, a Castellieri-type settlement located near Rovinj on the west coast of Istria, Croatia, and the possible effects of climatic changes in the 2nd millennium BC in the Adriatic region. Established around 2000 BC, the settlement experienced significant construction phases, particularly in its defensive architecture built with limestone blocks using dry-stone wall techniques. The settlement appears to have met a violent end around the 15th century BC, which is suggested by destruction layers, widespread burning, and the presence of weapons such as a lance tip, bronze axe, and slingstones. Monkodonja's destruction raises questions about broader military conflicts in the Adriatic region during this period, and it was proposed that a migration of people to the Istrian peninsula brought this new settlement form and other influences, leading to a significant population increase. The appearance of the Castellieri settlement form coincides with a period marked by documented climatic changes [14] and two major natural disasters in the form of volcanic eruptions.

Conflicts of Interest: There is no any conflict of interest.

References

1. Liritzis, I.; Laskaris, N.; Vafiadou, A.; Karapanagiotis, I.; Volonakis, P.; Papageorgopoulou, C.; Bratitsi, M. Archaeometry: An overview. *Sci. Cult.* **2020**, *6*, 49–98. [CrossRef]
2. Liritzis, I. EASA Expert Group: Science, Technology, Engineering, Mathematics in Arts and Culture (STEMAC). *Proc. Eur. Acad. Sci. Arts* **2024**, *3*, 27. [CrossRef]
3. Stoev, A.; Shopov, Y.; Maglova, P.; Ognyanov, O.; Raykova, L. Prehistoric Astronomical Observatories and Paleoclimatic Records in Bulgaria Estimate Astroclimate during 4000–4500 BCE: A Critical Assessment. *Quaternary* **2023**, *6*, 6. [CrossRef]
4. Kokkaliari, M.; Adam, E.; Vlachopoulos, A.; Iliopoulos, I. Tracing Raw Material Sources of Prehistoric Stone Artefacts by Non-Invasive Techniques: The Case of the Early Bronze Age (3rd Mill. BCE) Site of Vathy, Astypalaia, Greece. *Quaternary* **2022**, *5*, 42. [CrossRef]
5. Westra, A.J.D.; Miao, C.; Liritzis, I.; Stefanakis, M. Disasters and Society: Comparing the Shang and Mycenaean Response to Natural Phenomena through Text and Archaeology. *Quaternary* **2022**, *5*, 33. [CrossRef]
6. Liritzis, I.; Evelpidou, N.; Fikos, I.; Stambolidis, A.; Diamanti, N.; Roussari, T.; Tzouxanioti, M.; Louvaris, P.; Tsokas, G.N. Novel Combined Approach of GIS and Electrical Tomography to Identify Marsh/Lake at Kastrouli Late Mycenaean Settlement (Desfina, Greece). *Quaternary* **2022**, *5*, 26. [CrossRef]
7. Koskeridou, E.; Thivaiou, D.; Psarras, C.; Rentoumi, E.; Evelpidou, N.; Saitis, G.; Petropoulos, A.; Ioakim, C.; Katopodis, G.; Papaspyropoulos, K.; et al. The Evolution of an Ancient Coastal Lake (Lerna, Peloponnese, Greece). *Quaternary* **2022**, *5*, 22. [CrossRef]
8. Saitis, G.; Karkani, A.; Evelpidou, N.; Maroukian, H. Palaeogeographical Reconstruction of Ancient Diolkos Slipway by Using Beachrocks as Proxies, West Corinth Isthmus, Greece. *Quaternary* **2022**, *5*, 7. [CrossRef]
9. Howland, M.D.; Tamberino, A.; Liritzis, I.; Levy, T.E. Digital Deforestation: Comparing Automated Approaches to the Production of Digital Terrain Models (DTMs) in Agisoft Metashape. *Quaternary* **2022**, *5*, 5. [CrossRef]
10. Muşkara, Ü.; Konak, A. Characterization of the Obsidian Used in the Chipped Stone Industry in Kendale Hecala. *Quaternary* **2022**, *5*, 3. [CrossRef]
11. Bednarik, R.G. Direct Dating of Chinese Immovable Cultural Heritage. *Quaternary* **2021**, *4*, 42. [CrossRef]
12. Rossi, M.; Iacumin, P.; Venturelli, G. $^{87}Sr/^{86}Sr$ Isotope Ratio as a Tool in Archaeological Investigation: Limits and Risks. *Quaternary* **2024**, *7*, 6. [CrossRef]
13. Hellmuth Kramberger, A. On the Emergence of the Castellieri Settlements and Possible Effects of Climatic Changes in the 2nd Millennium BC in the Adriatic Region. *Quaternary* **2024**, *7*, 56. [CrossRef]
14. Liritzis, I. Disasters and climatic phenomena today and in the past. *Proc. Eur. Acad. Sci. Arts* **2023**, *2*, 22. [CrossRef]

quaternary

Article

On the Emergence of the Castellieri Settlements and Possible Effects of Climatic Changes in the 2nd Millennium BC in the Adriatic Region

Anja Hellmuth Kramberger

Inštitut za Interdisciplinarne Raziskave, Univerza Alma Mater Europaea, Slovenska ulica 17, 2000 Maribor, Slovenia; agrath@web.de or anja.hellmuth@almamater.si

Abstract: The fortified hilltop settlement of Monkodonja, located near Rovinj on the west coast of Istria, Croatia, provides insight into Bronze Age occupation and conflict in the Adriatic region. Established around 2000 BC, as evidenced by a series of C14 dates from human and animal bones, the settlement experienced significant construction phases, particularly in its defensive architecture. Its earliest fortifications, built with limestone blocks using dry-stone wall techniques, date to the 19th century BC, with major expansions in the 16th century BC, where the primary wall was doubled in width and reached over 3 m in thickness. Monkodonja's architectural complexity, notably the West Gate and Acropolis fortifications, and certain types of artifacts reveal influences from southern regions such the eastern Aegean. However, the settlement appears to have met a violent end around the 15th century BC, suggested by destruction layers, widespread burning, and the presence of weapons such as a lance tip, bronze axe, and slingstones. Monkodonja's destruction raises questions about broader military conflicts in the Adriatic region during this period. Possible causes could include localized warfare or connections to larger-scale disturbances. Research in Monkodonja is also significant in the context of the debate surrounding the emergence of the so-called Castellieri settlements in Istria at the beginning of the 2nd millennium BC. As early as the beginning of the 20th century, it was proposed that a migration of people to the Istrian peninsula brought this new settlement form and other influences, leading to a significant population increase. The appearance of the Castellieri settlement form coincides with a period marked by documented climatic changes and two major natural disasters in the form of volcanic eruptions.

Keywords: bronze age; Adriatic region; Istria; phenomen of Castelliere settlements; climate changes; changes in society; migrations

check for updates

Citation: Hellmuth Kramberger, A. On the Emergence of the Castellieri Settlements and Possible Effects of Climatic Changes in the 2nd Millennium BC in the Adriatic Region. *Quaternary* **2024**, *7*, 56. https://doi.org/10.3390/quat7040056

Academic Editor: Ioannis Liritzis

Received: 3 October 2024
Revised: 7 November 2024
Accepted: 6 December 2024
Published: 11 December 2024

1. Istria in the 2nd Millennium BC and the "Phenomenon" of the Castellieri-Settlements (Introduction)

Istria is the largest peninsula in the Northern Adriatic, and it owes its name to ancient sources such as the "*Periegesis*" by Hecataeus of Miletus, written between 560 and 480 BC, in which the pre-Roman inhabitants of the region are referred to as *Histri* or *Istri* [1] (p. 45), [2], [3] (pp. 25–27).

Today, the largest part of Istria is located within Croatia, with the northern part largely belonging to Slovenia, and a small area around Muggia in Italy. The sea currents in the eastern Adriatic flow northward and pass Istria along its western coast and it is assumed that they were already used for navigation in prehistoric times (Figure 1) [1] (pp. 49–50, Figure 41), [4] (pp. 13–20). Istria, which geologically consists of limestone, is characterized by sedimented red soil and the typical karst vegetation growing on it, including sub-Mediterranean holm oak and pine forests [1] (pp. 56–58) and [5], (p. 43). The landscape is marked by numerous rounded hills, bowl-shaped karst sinkholes (dolines), fissures, and caves. In the lowlands around the hills, fields for agriculture and livestock farming can be found today.

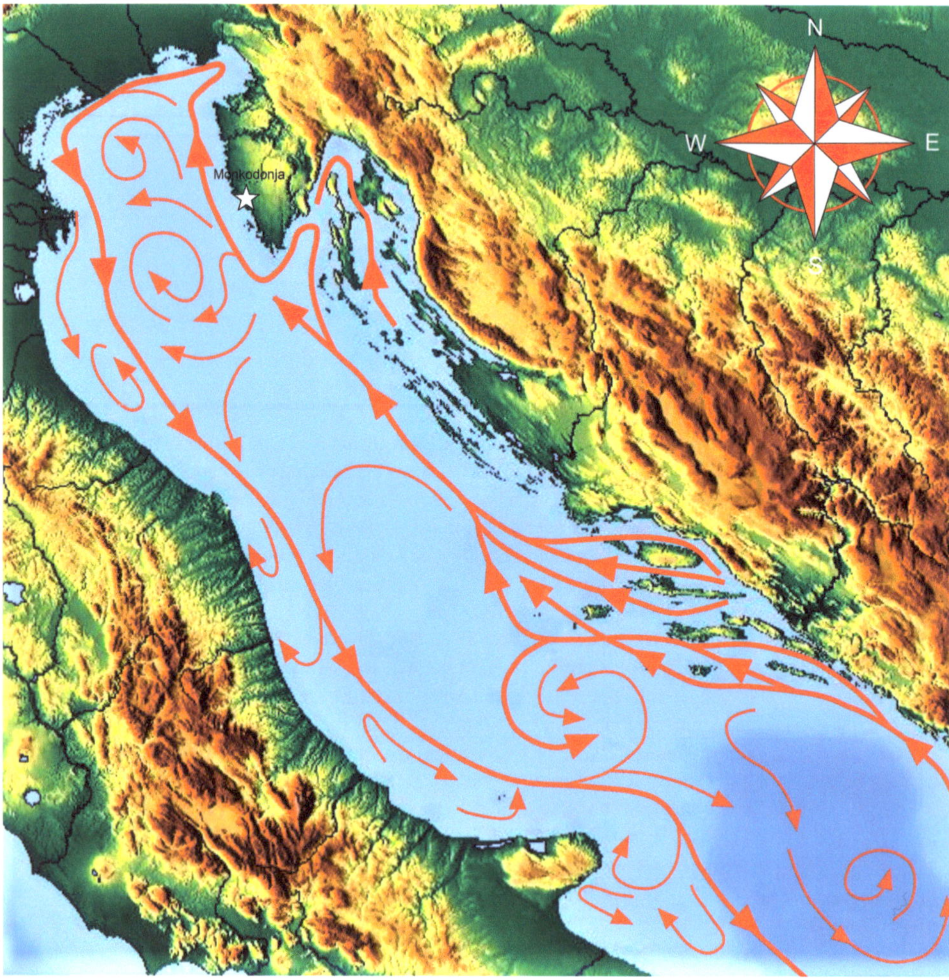

Figure 1. The Adriatic region showing the course of sea currents and the location of the Istrian peninsula (graphic: author, data from [1] (Figure 14)).

From the beginning of the developed Early Bronze Age around 2000 BC, significant changes in the settlement patterns of the Istrian peninsula can be observed: the emergence of a specific type of settlement known as Castellieri, Gradinas, or Kastelliere. These settlements, located on the countless karst hills, are characterized by their thick, often concentric walls built in dry-stone technique—without mortar. Today, these walls are often visible as ring-shaped structures beneath the canopy of the macchia or among the dense vegetation (Figures 2 and 3). By the mid-19th century, over 300 of these hilltop settlements were counted, and today about 500 have been recorded [1] (pp. 27–26, Figure 5), [6,7], [8] (p. 158, Figure 3), [9–12].

On the other hand, it is important to note that in the 19th century, the landscape of Istria looked significantly different than it does today. At that time, the peninsula was less densely covered with vegetation due to extensive livestock grazing [1] (pp. 27, 56–58), [8] (p. 156) and [5] (p. 43). This means that the remains of prehistoric settlements were much more visible, thereby attracting attention and interest.

Figure 2. Aerial view showing the fortified hilltop settlement of Monbrodo, where the concentric walls are covered by the vegetation canopy and clearly stand out in the terrain, Istria, Croatia (Adapted from [13] (Figure 3a)).

The historian and archaeologist Pietro Kandler from Trieste initially speculated in the mid-19th century that the Kastelliere settlements were fortresses to secure the Roman road network [1] (pp. 27–30). This assumed dating to the Roman period was quickly questioned by contemporaries such as Carlo de Franceschi, Tomaso Luciani, and Antonio Covaz, who pointed to the discovery of stone axes and coarse handmade pottery in the settlements. The first scientific study of the fortified hilltop settlements was published in the 1870s by Richard F. Burton, the British consul in Trieste [15]. Burton described in detail the dry-stone construction of the walls and noted that the hilltops on which the settlements are located had been artificially leveled. Burton was accompanied during a visit to the settlement of Kunci near Labin in 1874 by Carlo Marchesetti, a physician and natural scientist, who was the director of the "Museo Civico di Storia Naturale" in Trieste. Marchesetti himself began a systematic study of the hilltop settlements in the 1880s. He registered and mapped over 400 sites and described various settlement types based on their construction, the buildings within the enclosures, and associated necropolises.

The appearance of the new Castellieri settlement type during the Early Bronze Age in Istria, specifically at the end of the Istra I phase according to Čović 1983 [1] (pp. 36, 510, Figure 332), has been recognized by researchers since the late 19th and early 20th centuries as a significant phenomenon, often linked to possible migrations [12] (pp. 123–125). To understand why the emergence of these fortified hilltop settlements in Istria is so remarkable, one must first consider the settlement situation in Istria before 2000 BC. Up until the end of the 3rd millennium BC, Istria appears to have been sparsely populated. Only a few Neolithic and Copper Age coastal stations are known, such as Kargadur (cf. [16]), along with settlement traces on mountain plateaus or cave findings. Most of these consist of finds from karst caves, such as Grotta dei Ciclami/Orehova Pejca [17,18], Laganiši Cave [19], Pupićina Cave [20], Oporovina Cave [21], Garbinovica Cave [22], Pećina Ispod Sela Srbani [23], and Vela Cave [24].

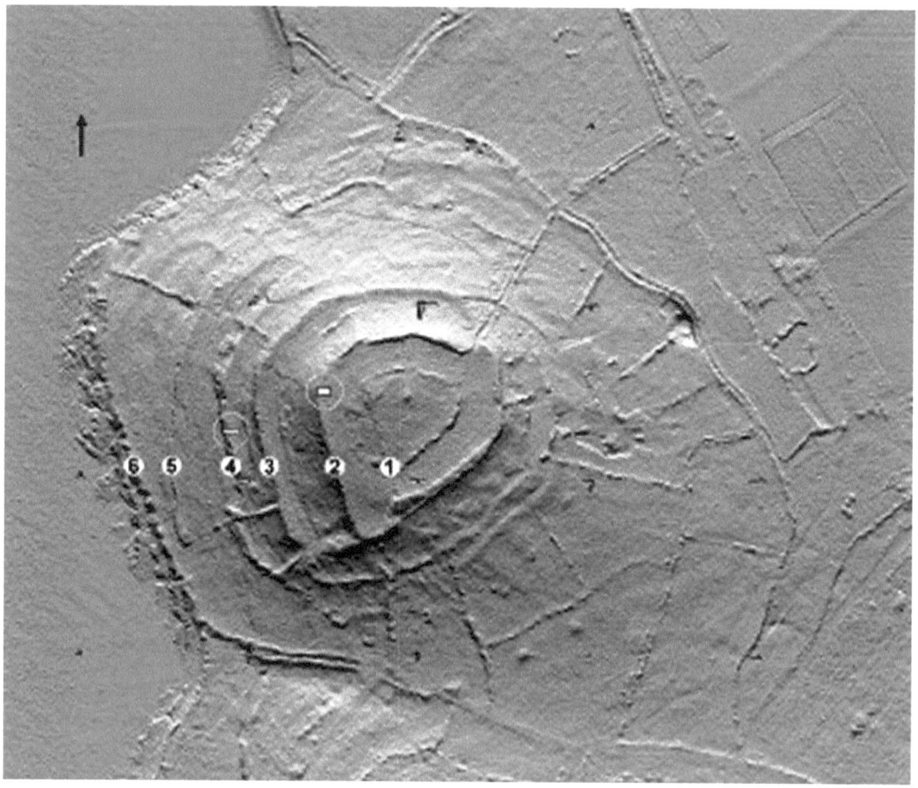

Figure 3. Lidar view of Monbrodo (Adapted from [14] (Figure 2)).

In the Bronze Age, this situation changed dramatically. Not only did the number of known sites increase significantly, but settlements were now frequently found in previously uninhabited mountainous areas [8] (pp. 157–158, Figures 2 and 3). This seemingly "explosive" rise in settlement sites with the onset of the developed Early Bronze Age has repeatedly led researchers to consider that migrations or colonization might have played a role [1] (pp. 30, 494–495), [8] (p. 159), and [12] (pp. 123–125). Although many questions about the Castelliere settlements remain unanswered—for instance, it is still unclear how many settlements were inhabited simultaneously—there are indications that at least some of the founders of the Castellieri in Istria were newcomers. It is suspected that the local population at the transition from the late Copper Age to the Early Bronze Age was too sparse and Istria too thinly populated to carry out such extensive construction projects [1] (p. 494). It is evident that the large-scale quarrying and the construction of meter-thick walls would have required a considerable number of workers, organized and led by specific individuals. Furthermore, the architecture, excavation findings, and artifacts suggest influences from regions as distant as the Carpathian Basin and the Eastern Mediterranean [25,26], [1] (pp. 500–504), and [27] (pp. 337–391). A particularly striking example among the ceramic finds are the tripod plates, which appear in Istria for the first time with the Castelliere settlements and have predecessors in tripods from the Eastern Mediterranean, for example, on Crete and Cyprus [25] and [27] (pp. 215–220) (Figure 4). The use of tripods was likely associated with a specific form of food preparation and culinary traditions that were previously unknown in Istria before the 2nd millennium BC. The Early Bronze Age pottery differs markedly from the characteristic Neolithic and Chalcolithic ceramics of Istria, which were distinguished by striking geometric and curvilinear incised decorations, sometimes with inlays cf e.g., [16] (p. 115). Bronze Age pottery shows distinctly different decorations such as plastic ornaments and fluting, and black polished fine ceramics and barbotine also appear for the first time, to name just a few examples.

Although there is a large number of known sites, only a few have undergone systematic archaeological excavation and much of our knowledge comes from surveys, surface finds—mainly pottery—and occasional rescue excavations (cf 7). Only a very small number of settlements have been investigated through extensive research excavations. In recent years, non-invasive methods, such as geophysical surveys, LiDAR technology, and three-dimensional laser scanning for digital terrain modeling, have been increasingly utilized [28,29] (Figure 3). These methods have provided new insights, such as revealing the frequency of burial mound necropolises near settlements, which were previously obscured by dense vegetation [30] (p. 28). However, due to these limitations, our understanding is still based on incomplete and uneven data.

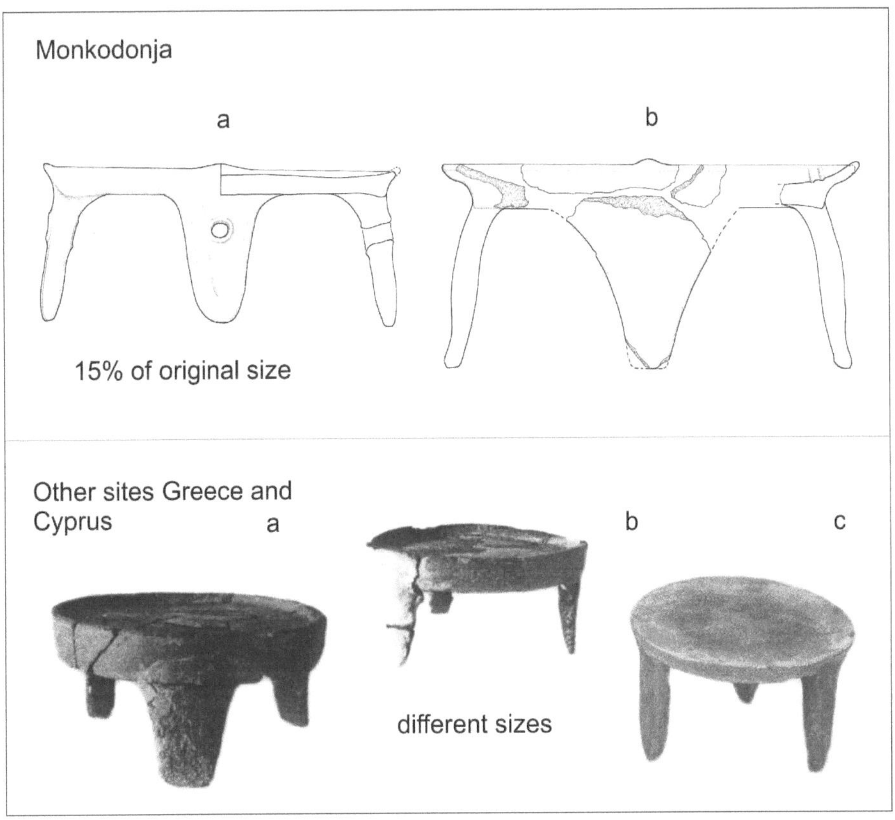

Figure 4. Upper: (**a**,**b**) Ceramic tripods from Monkodonja, Akropolis (Sonda 3), Istria, Croatia (drawings author), lower: tripods from the Eastern Mediterranean region, (**a**,**b**)–Ayia Irini, Kea (Adapted from [31]), (**c**)–Cyprus (Adapted from [32] (p. 50)).

Despite challenges within the existing data, this text seeks to explore the emergence of Castellieri settlements and analyze the factors that might have driven increased settlement activity on the Istrian peninsula from the Early Bronze Age onward. Potential influences on Bronze Age settlement patterns in the late 3rd and 2nd millennia BC include climatic changes and major natural disasters. Additionally, developments throughout the 2nd millennium BC and the question of possible migrations or 'migration waves', as proposed by C. Marchesetti in the early 20th century cf. [2] (p. 117) and [12] (pp. 151–155), will be examined.

The settlement of Monkodonja (Figure 5), the most thoroughly studied archaeological site in the Croatian part of Istria, will serve as a case study for further analysis. Section 2 will provide a brief overview of Monkodonja, following an introduction to the palaeoecology of Istria at the beginning of the 2nd millennium BC. Next, the emergence of Castellieri settlements and the migration-wave theory will be examined, considering the impact of climatic changes and major natural disasters at the transition from the 3rd to the 2nd millennium BC and throughout the 2nd millennium BC.

Figure 5. Aerial view of the fortified hilltop settlement of Monkodonja, with the Adriatic coast visible in the background, Istria, Croatia (Adapted from [33] (Figure 1)).

2. Paleoecology of the Istrian Peninsula and the Gradina Monkodonja in the First Half of the 2nd Millennium BC (Materials and Methods)

2.1. *Paleoecology of the Istrian Peninsula*

The emergence of the new settlement form on the Istrian Peninsula and the associated postulated population increase had an impact on the natural environment.

As far as the Holocene environment and climate of Istria are concerned, our knowledge is still limited, mainly due to a lack of suitable paleo-ecological sites and research [34] (p. 109) and [35] (pp. 514, 519–520). Pollen and organic remains are rarely preserved in Holocene layers, but there are some sites where human influence is evident in the pollen diagram.

Research at the site of Polje Čepić in the east-central part of the Istrian peninsula, at the northernmost tip of the Adriatic, has shown that at the beginning of the 2nd millennium BC (associated with Core Segment Polje Čepić S3), an open mixed forest landscape predominated [34] (p. 119). In comparison to findings from Core Segment S2 at Polje Čepić, which corresponds to a layer from the 5th millennium BC, a decline in oak and other tree pollen species was observed, with trees and shrubs accounting for 70% of the total pollen count. This decline, along with the simultaneous increase in non-arboreal taxa, where herbs make up 7.3% of the total pollen, indicates a more open landscape. Based on sedimentological and palynological data, particularly the increase in non-arboreal pollen in Core Segment S3, it is suggested that humans contributed to these changes through deforestation [34] (pp. 120–122) and [35] (pp. 514, 519). Similar findings have also been made at the site of Prapoče/Čičarija in northern Istria [35] (p. 511). In the 3rd millennium BC, linden forests (Tilia) predominated, and oak (Quercus), hornbeam (*Carpinus betulus*), hazel (Corylus), fir (Abies), beech (Fagus), and alder (Alnus) were also present. At the beginning of the 2nd millennium BC, the landscape became more open, which is reflected in the decrease of tree pollen in the pollen diagram and the increase of herb pollen.

2.2. Gradina Monkodonja–Research and Dating

In the context of the discussion about the appearance of the Castellieri settlements in Istria, attention must be paid to the fortified hillfort settlement Monkodonja near Rovinj, south of the Limski kanal, which represents the best-studied site in the Croatian part of Istria [1,27,33,36] (Figure 5). The research conducted at Monkodonja has led to numerous new insights, for example, regarding the dating of the so-called Castelliere settlements of Istria, the organization of the settlement, or the long-distance contacts.

In the early 1950s, the first systematic excavations at the fortified settlement Monkodonja were conducted under Boris Baćić, then director of the Archaeological Museum of Istria in Pula [5]. In 1997, archaeological investigations were resumed as part of a Croatian-German-Slovenian joint project funded by the DFG (German Research Foundation) and continued until 2009 under the direction of Kristina Mihovilić (Archaeological Museum of Istria, Pula), Bernhard Hänsel, and Biba Teržan (Free University of Berlin).

A prominent feature of the hilltop settlement of Monkodonja, as well as other contemporary settlements in Istria, is, as mentioned, its defensively effective architecture made of broken limestone, constructed using dry-stone wall techniques. The settlement is divided into an acropolis, upper and lower town, and outer fortification by ring-shaped walls. There may have also been an outer settlement, as indicated by traces observed during the clearing of the macchia vegetation at the foot of the hill. The fortified settlement was accessed through several gates, with the main entrance being the large western gate of the main fortification [5] (pp. 111–251). Also noteworthy are the burial sites—stone cist graves—discovered in the area of the gate [37]. Similar graves have been found in other hilltop settlements, such as Vrčin/Monte Orcino or the Gradina Brioni [37] (p. 161). The architecture of the western gate appears highly complex due to various phases of construction and modification (Figure 6). Such a building form is uncommon for Central Europe during the Early Bronze Age and points southward and southeastward. Complex, repeatedly modified gate structures are also found in the southern part of the western Adriatic, in Apulia, at sites like Coppa Nevigata [38,39] (Cazzella, Moscoloni, Recchia 2012; Cassano et al. 1987) and Roca [40]. Excavators have frequently highlighted the parallels between the gate's construction and findings from Aegina, as well as the parallels between the graves near the gate and those in Mycenae [37] (p. 179), [1] (pp. 174–177, Figures 112–123 with reference to [41]) and [42] (pp. 160–165). It was established that the settlement is not only spatially divided into different areas by the various ring-shaped walls but that within these areas, social stratification can be observed. This can be inferred from the building techniques, as small freestanding single buildings have been excavated, as well as large multi-room buildings with courtyards, which were found in the acropolis area. In certain parts of the settlement, storage buildings were also identified, evidenced by the extraordinarily high number of large storage vessels and animal bones [43] and [27] (pp. 305–318). Most likely, large quantities of food produced in the surrounding area of the fortified settlement were centrally stored within the strongly fortified walls and then redistributed from there [44]. This is an economic system well-known from the Early and Middle Bronze Ages in the Near East and the Eastern Mediterranean.

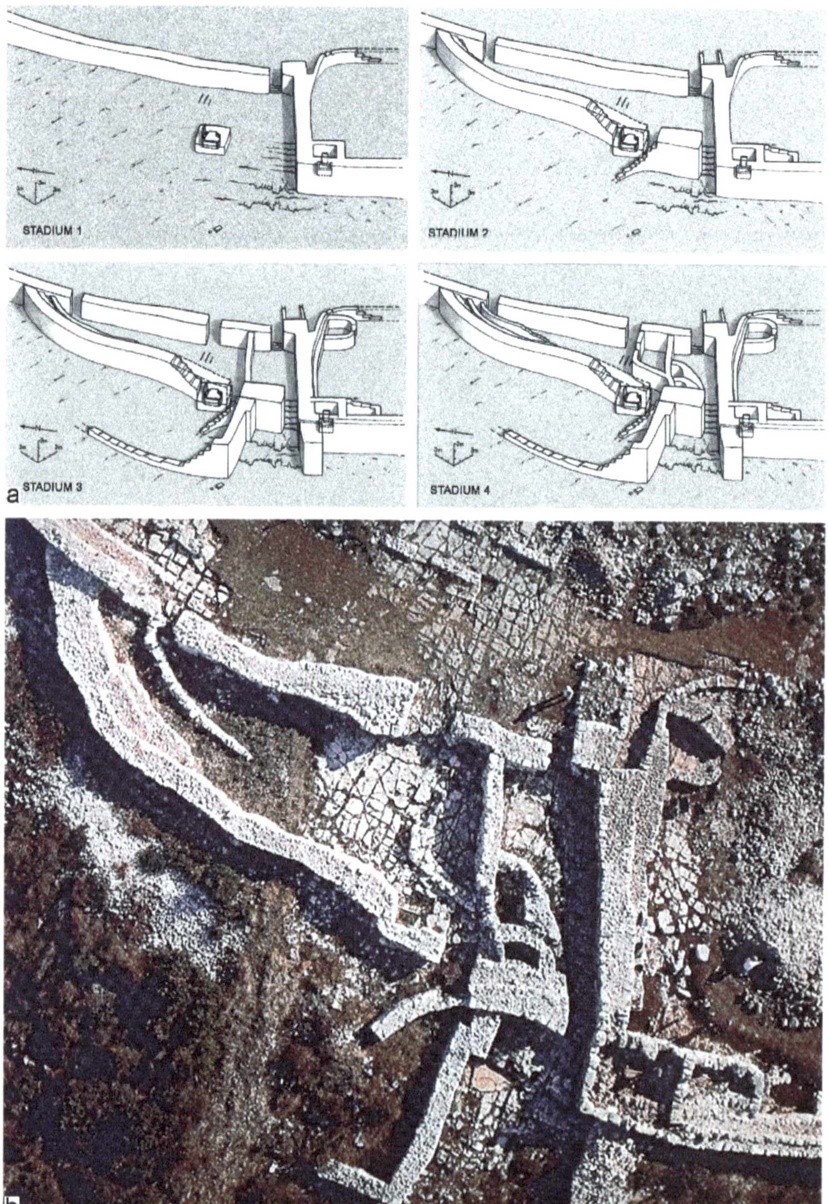

Figure 6. Aerial view showing the excavated area of the main fortification with the west gate of Monkodonja (**b**) and the various expansion phases of the wall and gateway (**a**), Istria, Croatia (Adapted from [33] (Figure 4)).

The gradina of Monkodonja is dated, according to Reinecke's chronological scheme for Central Europe, to the period between the end of Bz A1 and the transition from Bz B1 to Bz B2/C1. This corresponds to the late phase of Istra I to Istra II according to Čović 1983 for Istria and to Bronzo Antico 2 to Bronzo Medio 2 according to Peroni 1994 cf. [1] (p. 510, Figure 332). The Middle Bronze Age, Bronzo Medio 1 and 2 in Italy, corresponds to SH I to SH IIB (Late Helladic I to IIB) in the Aegean cf. [45] (p. 216, Figure 24). A series of 45 C14 dates attest that the settlement was founded around or before 1800 BC, possibly as early as 2000 BC, if we take into account the dating of human bones from the two stone cist graves at the western gate [1] (pp. 141, 146–147, 424–452). These dates were determined under the direction of Pieter M. Grootes at the Leibniz Laboratory for Age Determination and Isotope Research in Kiel, based on animal and human bone finds, and the evaluation of the datings was carried out with the collaboration of Bernhard Weninger (Figure 7a,b).

The calibration of the C14 data was conducted by B. Weninger using the program CALIB rev.5.01 with the INTCAL04 dataset, as well as using the program CalPal and the INTCAL09 dataset [46,47]. Eight additional radiocarbon datings were performed by Tomasz Goslar at the Poznan Radiocarbon Laboratory based on human bones from burial mounds in the Monkodonja necropolis from the neighboring hill Mušego and were calibrated using the program OxCal v4.4.2 [48] (Figure 8).

An issue that was thoroughly discussed in connection with the publication of the C14 dates from Monkodonja and the associated burial mound necropolis Mušego, and therefore does not need to be repeated in detail here, is the potential error due to the marine reservoir effect, as fish consumption has been documented in the settlement [1] (pp. 426–427, 438–440, 450). However, unaffected by the marine reservoir effect are the bones of ruminants, sheep/goat, and cattle, and it was found that the 20 samples from the acropolis and main fortification areas of Monkodonja provide a block of relatively close-set dates, which fall within a time window between the 19th and early 15th centuries BC [1] (pp. 438–440, Figures 320, 446 and 447). These dates define the main phase of settlement, with two significantly younger and one older date from animal bones interpreted as results of sporadic visits to the site before and after the settlement's peak period.

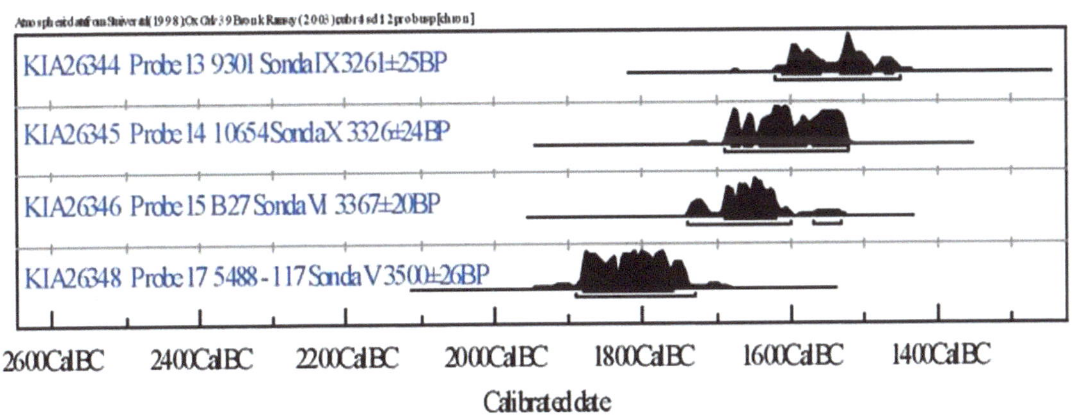

(a)

Figure 7. *Cont.*

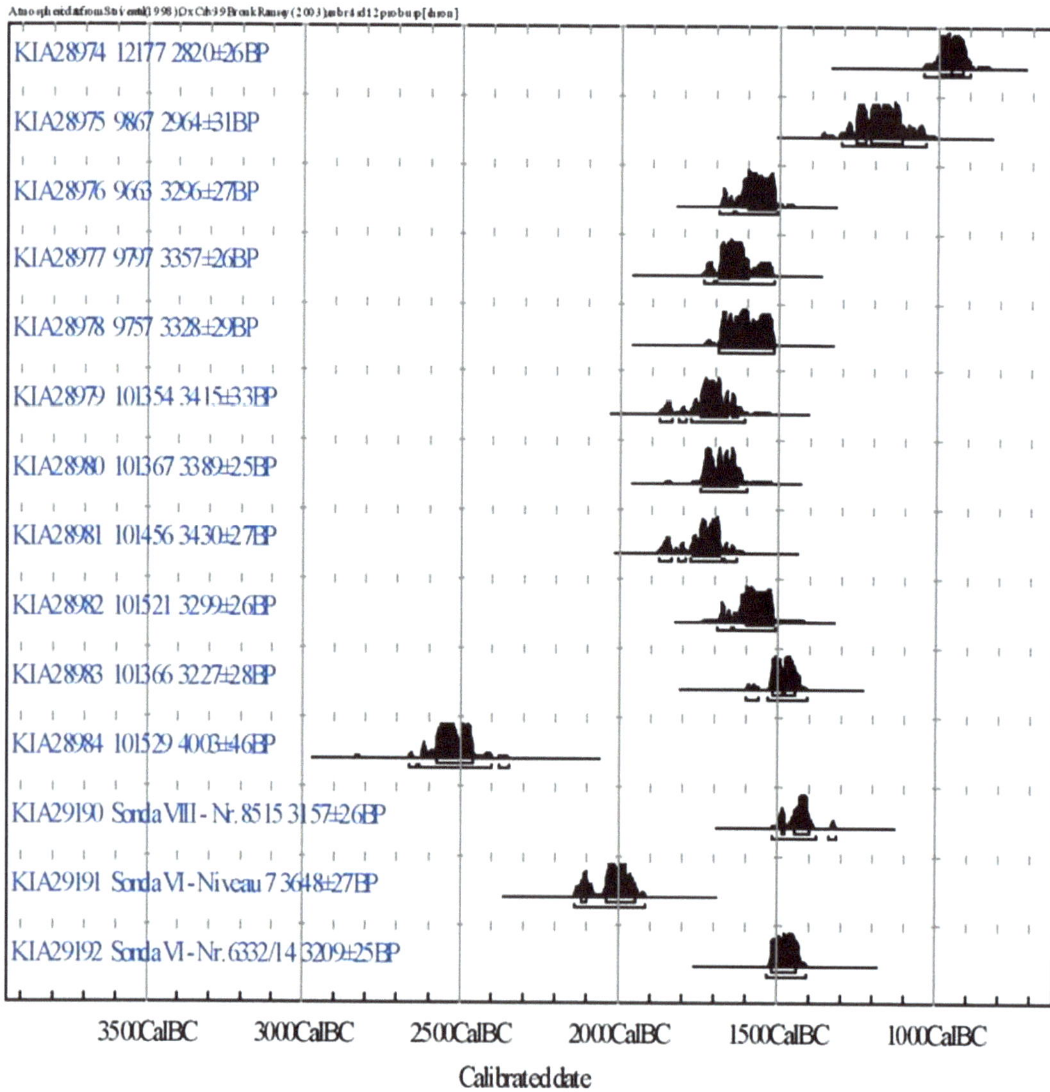

(b)

Figure 7. (**a**,**b**): Calibrated C14 dates from the 2005 and 2006 measurements of human and animal bone finds from the fortified hilltop settlement of Monkodonja, Istria, Croatia, Leibniz Laboratory for Age Determination and Isotope Research Kiel (Adapted from [49] (p. 38)).

Among the human bones from two stone cist graves discovered in the gate area of Monkodonja, there are six dates that align with the animal bone dates, ranging from the late 19th to the 15th centuries BC, as well as others that are distinctly older, dating to the late 3rd century BC [1] (Figure 320). The latter should be viewed with caution. In this context, the excavators have also suggested that these early dates might not solely indicate a reservoir effect but may point to specific burial rituals where human bones from older graves were relocated [1] (p. 448).

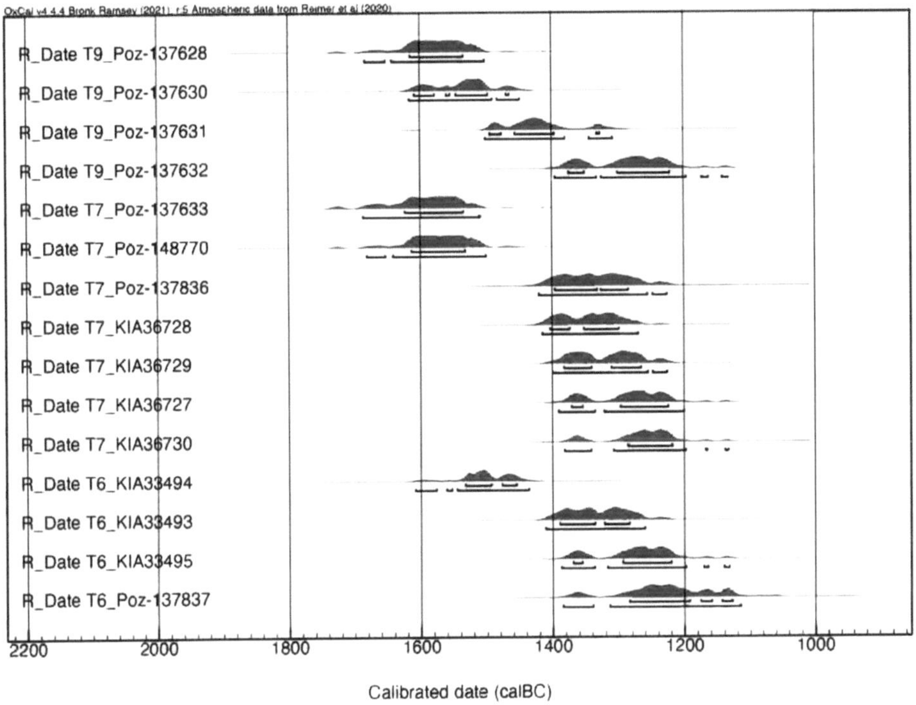

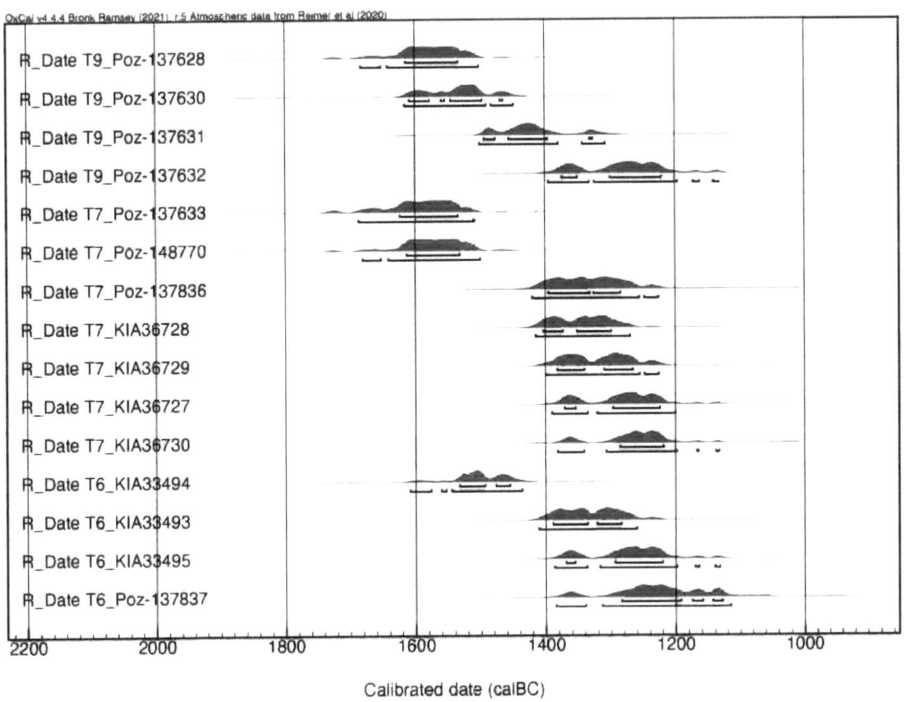

Figure 8. Compilation of all calibrated C14 dates from the burial mounds on Mušego, Istria, Croatia, based on human bones from the Poznan Radiocarbon Laboratory (Poz) and the Leibniz Laboratory for Radiometric Dating and Isotope Research Kiel (KIA) (Adapted from [48] (Figure 4)).

2.3. The Phases of the Construction of the Fortifications

The C14 datings show that the first construction activities on the main fortification wall took place in the 19th century BC, and the earlier phase of the primary use of the site dates to the 18th–17th century BC. The first, oldest wall of the main fortification was built using dry-stone techniques with limestone blocks averaging 60–70 cm in length and 40–50 cm in height. However, in the gate areas, especially at the North Gate, blocks over a meter in length and 60 cm in height were documented, and the wall had an average thickness of 1.50 m [1] (pp. 113–115). In certain excavation areas, the wall was preserved up to a height of 2 m [1] (pp. 120, 122, 127, 137, 143–144, Figures 72 and 86a). In the 16th century BC, there must have been an impulse to significantly strengthen and expand the main fortification [1] (pp. 143–147). The additions on the inner side of the wall doubled its average width, so that in its final phase, the main fortification reached a thickness of over 3 m. The main gate, the large western gate leading into the settlement of Monkodonja, was also extensively reinforced and expanded [1] (pp. 154–155, 158–164, Figures 101 and 102). While in its first, oldest phase the wall only featured a passage through an angled section of the main fortification, in its final construction phase it took the form of a chamber gate with a long gateway flanked by two massive projections or bastions (Figure 6).

Several phases of expansion were also evident in the fortification of the acropolis, during which one gate was closed, and the wall was partially widened to three times its original thickness [1] (pp. 133, 278–301) (Figure 9). Hänsel, Mihovilić, and Teržan point out that the final construction work on the wall in the acropolis area in the 16th century BC appears somewhat hasty due to its slightly varying and less precise construction, suggesting that the inhabitants of the settlement may have reinforced the walls out of necessity and in haste [1] (pp. 304, 446). There is some evidence that the settlement was destroyed in the 15th century BC during a military attack and subsequently abandoned [33] (pp. 111–114) and [1] (p. 452). On the one hand, the C14 data suggest that the settlement was no longer inhabited after the late 15th century BC; on the other hand, the discovery of weapons such as a Middle Bronze Age spearhead, a bronze axe, bone projectiles, sling stones, and extensive burn marks in the acropolis area point to a violent end [1] (pp. 304–305, 349, 506–507) and [27] (p. 308). Likewise, the ceramic spectrum is homogeneous and shows no forms that would date to the Late Bronze Age or Iron Age [27]. It should also be mentioned that numerous human bones were found in all excavated areas of Monkodonja [50]. These include grave finds—burials in the two stone cists near the western gate, destroyed graves, and infant burials—but also human skeletal remains from at least 109 individuals, for which B. Teßmann found evidence that some of these bodies had been exposed to the elements for an extended period and showed signs of animal gnawing [50] (pp. 540–541). This could suggest that some victims of the presumed attack on the settlement might have been left on the surface and under debris after its destruction, without receiving proper burial. It should be noted, however, that this is only one possible interpretation and there may just as well be other reasons for the presence of human bones in the settlement, which are investigated by B. Teßmann.

Figure 9. Aerial view showing the excavated area of the acropolis of Monkodonja, Istria, Croatia (Adapted from [33] (Figure 5)).

3. The Emergence of Castellieri Settlements and Potential Links to Mobility, Climate Change, and Natural Disasters in the Late 3rd and 2nd Millennia BC (Discussion)

Many studies have looked at changes in the environment during the Bronze Age in the Mediterranean and Central Europe, focusing on how these changes affected Bronze Age societies and the links between climate and society (e.g., [35,51–56]).

In the following section, the emergence of the Castellieri settlements in Istria and the theory of migration waves will be discussed against the background of climatic changes during the late 3rd and 2nd millennium BC. For this purpose, a brief overview of climatic changes during the relevant period of time is given, followed by the mention of two major natural disasters in the Mediterranean region in the first half of the 2nd millennium BC, which also had an impact on the Bronze Age population.

3.1. Climate Changes in the Late 3rd and 2nd Millennium BC

In recent decades, there has been increasing research on the occurrence and impact of climate changes and Rapid Climate Change (RCC) events during the Holocene on prehistoric humans, their cultural development, subsistence strategies, and mobility (e.g., [57,58]).

A period during which climate changes are recorded on a global scale includes the late 3rd and the 2nd millennium BC. During the time between 2200 and 1200 BCE, a diverse range of early to late Bronze Age cultures emerged and developed across Central Europe, the Mediterranean, and Western Asia, each contributing uniquely to the broader cultural and historical landscape of the time. Of particular importance and covered extensively in research is a so-called Rapid Climate Change Event, documented for 4.2 ka BP, the 4.2-kiloyear event [51,52,56,59–66]. This refers to a period correlated with the Early Bronze Age in various regions (cf. e.g., [52] (Figure 1), [67] (Figure 3.1), [56], [1] (p. 510, Figure 332), and [68] (Plates 80–83)). Mineralogical and geochemical data from a core (M5–422) taken from the Gulf of Oman document a sudden onset of drought that occurred around 4025 ± 150 BC [65] (p. 382) and [69]. This drought event was evidently short-lived, lasting only a few decades, and may have been the result of large-scale changes in the boundary conditions between ocean, atmosphere, and vegetation. As Cullen and others note, this abrupt climate shift had an unusually large amplitude compared to the rest of the Holocene, almost matching the mineralogical and geochemical amplitudes associated with the Younger Dryas period drought [65]. All available evidence suggests that the 4.2-kiloyear event, marked by cooling and unusual drought, led to dramatic changes in regional climates during the mid-Holocene (e.g., [70–73]). However, it remains uncertain what specific effects these changes had on Early Bronze Age populations in different regions around 2200 BC. As noted above, there is an extensive research apparatus on this topic, discussing the possible impact on the early Bronze age societies in different regions of Europe and Western Asia [56]. Criticism has been directed particularly at oversimplified or generalized interpretations of the connections between changes in the archaeological record and climatic shifts [52] (pp. 333–334), [53] and [35] (p. 519).

For the Late Bronze Age, paleoclimatic data suggest further changes following the cooling phase, pointing to a dry and warm phase associated with the so-called 3.2 kyr cal BP drying event [51], [52] (p. 337), [74] and [57] (pp. 44–48). For the eastern Mediterranean, the destruction of coastal cities such as Ugarit and Tell Tweini around 1200 BC is documented, and these events are connected to socio-economic changes against the backdrop of climatic shifts, which are also discussed in relation to the controverse topic of the so-called "Sea Peoples"(e.g., 50 (p. 4), [74–77]) which, however, cannot be discussed further at this point, as this article focuses on the first half of the 2nd millennium BC. A warmer, drier climate may have led to crop failures and resulting famines in various regions, which in turn could have resulted in increased socio-economic stress and conflicts (cf. [70]). However, the data basis for correlating climatic changes with the disruptions of the Late Bronze Age is sometimes regarded as inconsistent [53] (pp. 11–14) or criticized for assuming that Bronze Age populations lacked adaptability to change [52] (p. 334). Furthermore, high-resolution oxygen and carbon isotope data from a Stalagmite (S1) in the Mavri Trypa Cave in the southwestern Peloponnese, Greece, provide key climate insights surrounding the destruction of the Mycenaean Palace of Nestor at Pylos in the 12th century BC (~3150–3130 BP) [78]. The data suggest that a period of increased aridity following the palace's destruction could have contributed to agricultural challenges and other consequences, such as undermining the ability to rebuild the palace and restore the socio-economic stability of Mycenaean society.

3.2. Natural Disasters During the First Half of the 2nd Millennium BC

Possible effects on Bronze Age societies in the 2nd millennium BC may have been caused also by several documented major natural disasters, such as the Avellino eruption of Mount Vesuvius in Campania, Italy (e.g., [79–81]), and the Minoan eruption (or Thera/Santorini eruption) in the southern Aegean (e.g., [82–85], [42] (pp. 53–62)).

The Plinian eruption of Vesuvius, known as the Pomici di Avellino eruption, ranks among the most explosive Holocene events in the Mediterranean region and had far-reaching impacts on Early Bronze Age populations [79]. The eruption likely dates to the 20th–19th centuries BC, as indicated by a C14 dating of a goat bone from the destruction layer of the Bronze Age settlement Croce del Papa (Nola, Naples) [86] (p. 239) and [81] (p. 814, Figure 7). This dating is supported by an analysis of lake sediments from the Agro Pontino graben, central Italy, which contain a thin, continuous tephra layer composed of lithics, crystals, and a small amount of volcanic glass [80]. The immediate effects of the eruption were catastrophic, and it appears that the devastated areas only became densely populated again in the Middle

Bronze Age, during the Bronzo medio 3 phase, meaning there was significant depopulation or abandonment of the area for several centuries [80,87].

At the end of the 17th century BC, the eruption of the Thera volcano in the volcanic island arc of the southern Aegean marked another volcanic event of enormous magnitude. A Minoan settlement in Akrotiri was covered and preserved by several meters of pumice (e.g., [82–85,88], [42] (pp. 53–62)). The eruption was so extensive that neighboring islands were also blanketed with pumice and ash, and the Minoan ash layer has been detected in numerous sites throughout the eastern Aegean, as far as Anatolia, the Black Sea, and the Nile Delta [83] (p. 43) and [84] (pp. 50, 52, 54, Figure 3). The eruption was accompanied by tsunamis, causing additional destruction over a vast area of the surrounding regions [84] (pp. 54–55). Some researchers, such as W.L. Friedrich [84] (p. 46) and C. Oppenheimer [84] (p. 53), believe that the eruption triggered global climate changes and crop failures, which subsequently led to population migrations in the affected areas.

3.3. Natural Disasters and Climate Change as a Catalysts for Increased Settlement Activity in Istria During the 2nd Millennium BC?

The previous analysis indicates climatic changes occurred during the transition from the 3rd to the 2nd millennium BC, marked by the 4.2-kiloyear event. The impact of this event on Early Bronze Age populations across various regions remains insufficiently understood. Additionally, two significant natural disasters are documented for the first half of the 2nd millennium BC: the Avellino eruption of Vesuvius and the Thera eruption, both of which undoubtedly had immediate and long-term effects on the populations in directly affected areas. These events likely influenced social structures and population movements across the Eastern Mediterranean and Adriatic regions. Notably, these circumstances align with the emergence of the Castellieri—a new settlement form in Istria—and a hypothesized population increase, potentially due to immigration.

However, regarding the rise in settlement activity and the emergence of Castellieri settlements in Istria, a key issue persists: based on current data, we still cannot determine how densely populated Istria was at the beginning of the 2nd millennium BC or the exact number of Castellieri settlements that existed simultaneously.

In the region around Rovinj, it has been established that at least 30 hilltop settlements were occupied during the Late Early and Middle Bronze Age, within an area of approximately 23 square kilometers, with distances between settlements ranging from just 1 to 5 km [37,89] and [1] (pp. 54–55). Excavations conducted between 2016 and 2018 at several coastal sites in this area, as part of a Korean-Croatian joint project, also yielded ceramic finds comparable to those from Monkodonja. These findings indicate at least partial contemporaneous settlement use during the developed Early and Middle Bronze Ages [14,90–92]. This evidence suggests that over the 300–400 years Monkodonja—the best-studied settlement—was occupied, the region indeed had a dense settlement network, with a system of smaller and larger satellite and central settlements positioned within sight of one another (cf. [37] (p. 156, Figure 5)) (Figure 10).

Pollen diagrams currently available indicate that with the onset of the Bronze Age, the rise of the Castellieri settlements, and the suggested population increase, significant changes in the paleoecology of the Istrian peninsula occurred, likely driven by human influence through deforestation. Future research will expand the database, providing deeper insights into these human-induced environmental changes. Nevertheless, these findings already support the assumption of increasing settlement activity on the Istrian peninsula from the beginning of the Bronze Age.

Isotope analyses could provide evidence of mobility or the immigration of newcomers; however, there is very limited data available for the Early and Middle Bronze Age in Istria. C. Gerling and T. Douglas Price analyzed two human teeth from graves near the entrance of the Monkodonja settlement, two human teeth from burial mound 7 at Mušego (the Monkodonja necropolis), and a pig tooth from Monkodonja to examine their strontium (Sr), oxygen (O), and carbon (C) isotope content [93]. The isotope analysis was expected to reveal differences, as it is hypothesized that the settlement's founding generation included immigrants, and the teeth from burial mound 7 at Mušego are younger than those from the graves near the settlement's gate-it is therefore about a "later generation". The pig's tooth was included to provide a baseline for local isotope values, as the pig was assumed to have been born and raised locally in Istria, spending its entire life in the area.

Figure 10. Bronze Age settlement system showing visual communication around the central settlement of Monkodonja, around the middle of the 2nd millennium BC, Istria, Croatia (graphic by the author, Data from [37] (Figure 5) and [1] Figure 17)).

Strontium varies depending on the geographical location or geological substrate, the age, and the composition of the rocks [93] (p. 231). Through weathering processes, strontium enters the groundwater and thus the food chain, being incorporated in place of calcium (Ca) into the hard tissues (like bones or teeth) of living beings. To determine the isotopic signal of biologically available strontium at the location of the first years of life of a human or animal, the first permanent molars (M1) are examined, as the enamel of the first permanent molars mineralizes early in childhood and remains without active metabolism and systematic remodeling throughout life (cf. e.g., [94]). Changes in location can be detected by determining the ratio of stable ^{87}Sr to radiogenic ^{86}Sr. In addition to the $^{87}Sr/^{86}Sr$ analyses, oxygen isotope measurements are also used to determine changes in location [93] (p. 232). It is relevant in this context that the ratio between the stable isotopes ^{18}O to ^{16}O, $\delta^{18}O$, varies depending on factors such as temperature, distance to the sea, altitude, and latitude (cf. e.g., [95] (pp. 80–86)). Thus, information on the oxygen isotope ratios is also incorporated during the time of tooth mineralization, providing clues about the climatic conditions and locations during childhood [93] (p. 232).

The results of the analyses of the strontium and oxygen isotope ratios in the tooth enamel of the four human individuals from Monkodonja and Mušego by Gerling and Price indicate a uniform origin for the investigated individuals [93] (pp. 233, 234–237, Figure 179). Slight differences from the isotope ratios in the pig tooth are attributed to differences in diet. The $^{87}Sr/^{86}Sr$ ratios in the human samples range from 0.70903 to 0.70955, while in the pig, it is 0.70944. $^{87}Sr/^{86}Sr$ ratios for Mesozoic rocks from the Jurassic and Cretaceous periods, which dominate the Monkodonja region, are typically between 0.7080 and 0.7100 [93] (pp. 234, 230, Figure 178). These values align with the geological conditions of the region but also apply to other regions with similar geological backgrounds, making it impossible to deduce different geological origins or mobility of the individuals from Monkodonja and Mušego based on this data alone. However, Gerling and Price also found that the determined $\delta^{18}O$ values of the human individuals, while homogeneous within themselves, deviate from the expected values for the region and the pig tooth [93] (pp. 236–237). The $\delta^{18}O$ values in human teeth range from $-2.43‰$ to $-1.68‰$, with the value for the pig being $-3.93‰$. Converting the $\delta^{18}O$ values to meteoric water yields a range from -3.47 to $-2.24‰$ [93] (p. 236 with reference to [96]), which, according to Gerling and Price, represents a significant difference from the modern oxygen isotope ratio of the region where the archaeological site Monkodonja is situated. The deviations could either be due to inconsistencies in the regression equation used or suggest a southern or western origin of the individuals. To better understand the results, further reference samples are needed, and currently the results do not contribute to the questions about the immigration of the Early Bronze Age inhabitants of Istria.

While ancient DNA analyses are not effective for detecting short-term mobility trends, it is worth noting that studies of Early and Middle Bronze Age populations in Istria have been conducted. As ongoing research examines mitochondrial haplogroups from Monkodonja and other Bronze Age sites in Istria, current findings should be considered preliminary. In collaboration with the Department of Archaeogenetics (DAG) at the Max Planck Institute for the Science of Human History in Jena, researchers V. Villalba-Mouco and W. Haak analyzed samples from 22 individuals across six Bronze Age sites in Istria [97]. These samples include 14 from Monkodonja, two each from Maklavun, Šandalja, and Škicini, and one each from Kavran and Žamnjak. Of the 22 samples, 13 met initial DNA quality criteria and underwent 1240k SNP (single nucleotide polymorphism) capture analysis [97] (p. 166, Table 1). Genetic sex determination identified nine male and four female individuals, and Y-chromosome haplogroups were determined for five males, revealing paternally inherited lineages [97] (p. 168). Notably, male individuals from Monkodonja belonged to haplogroups J2 (MNK011) and J2b (MNK004), from Maklavun to R1b1a2 (MKU001) and J2b2a (MKU002), and from Kavran to R1b1a2 (KVN001). Villalba-Mouco and Haak describe these haplogroups as common in Bronze Age southeastern Europe. It is assumed that haplogroup R may represent a lineage associated with Eurasian expansion into Europe and the origins of prominent Western and Central European haplogroups R1a and R1b, while J2 likely originated in the Middle East [98]. Mitochondrial haplogroups, analyzed for two individuals from Maklavun and one from Monkodonja, included mtDNA haplogroup J1c3 for MKU001, T2b for MKU002 (Maklavun), and K1a18 for MNK007 (Monkodonja) [97] (p. 168). As mentioned above, researches on are still ongoing.

While limited scientific data have yet to provide new insights into mobility and migration in Istria, material culture findings continue to indicate connections with both nearby and distant regions, as discussed extensively by the excavators of the Monkodonja hillfort and the author of this article. As noted, the architecture, excavation findings, and artifacts suggest influences from areas as far-reaching as the Carpathian Basin and the Eastern Mediterranean [25,26], [1] (pp. 500–504) and [27] (pp. 337–391). Regarding ceramics, it is worth to remind once more on the tripod plates, which first appear in Istria with the Castellieri settlements and have predecessors in tripods from the Eastern Mediterranean [25] and [27] (pp. 215–220) (Figure 4). While these findings clearly indicate interaction networks, it remains uncertain whether these connections involved only knowledge exchange or also an exchange of people.

3.4. The Middle of the 2nd Millennium BC-Signs of a Far-Reaching Crisis in the Adriatic Region?

C. Marchesetti postulated as early as the beginning of the 20th century that there were two "waves" of settlement occupation or migrations in Bronze Age Istria (cf. [2] (p. 117), [12] (pp. 151–155)). According to our current knowledge of the appearance of the first Castellieri settlements in Istria, based on the research results from Monkodonja, Marchesetti's postulated first "migration wave" would coincide with the beginning of the 2nd millennium BC. The question arises whether there are indications of other more or less drastic changes in the Bronze Age settlement pattern of Istria, with which Marchesetti's postulated second wave of migration could be linked.

The settlement situation for the late Middle Bronze Age and Late Bronze Age after the 15th century BC—i.e., from Bz C2 according to Reinecke's scheme for Central Europe, or the late phase Istra III according to Čović 1983 for Istria, or Bronzo Medio 3 to Bronzo Recente for Italy according to Peroni 1994 (cf. [1] (p. 510, Figure 332))—is not yet sufficiently clarified. Prior to the extensive research on Monkodonja in the late 1990s and 2000s, scholars assumed that the "phenomenon" of Castellieri settlements was specific to the Middle Bronze Age, and Monkodonja was dated to phases Bz B2-C according to Reinecke's scheme for Central Europe (cf. [5] (p. 117)). However, extensive research and the afore-mentioned series of 45 C14 dates have shown that Monkodonja was already settled in the late Early Bronze Age, Bz A1 according to Reinecke (Figure 7a,b). This late Early Bronze Age and early Middle Bronze Age dating should also be considered for those settlements that yielded identical ceramic finds, such as the Gradina Monbrodo, located about 2 km from Monkodonja and directly on the coast, as well as the settlements of Monvi and Muja [14,90–92], and other sites that have so far been dated based on characteristic surface ceramic finds [8]. Settlement activity at Monkodonja ends in the 15th century BC, but this apparently did not signal the end of all settlement activities in the entire region, as some C14 dates from the burial mounds on the neighboring hill Mušego, which fall between the 14th–13th centuries BC, show [48] (pp. 173–179) (Figure 8). In connection with the C14 datings from the burial mounds of Mušego, which were determined based on human bones, it should also be noted that the first series of 8 datings, conducted at the Leibniz Laboratory for Age Determination and Isotope Research in Kiel and dated to the 14th–13th century BC, was considered problematic [1] (pp. 449–450). They appear too young, especially when a possible reservoir effect is also taken into account. Five additional radiocarbon dates which come from the Poznan Radiocarbon Laboratory are older, they date between the late 17th and 15th century BC which would be contemporaneous with the main settlement phase of Monkodonja [48] (pp. 176–178, Figures 2–4). Two other dates from Poznan fall into a period between the 14th and 13th centuries BC, and are therefore clearly younger than the settlement. Even considering the potential influence of the reservoir effect, some dates from the Mušego necropolis correspond closely to the period of the settlement's existence, at least in its later phase. Additionally, there are dates that are distinctly more recent, suggesting that settlement activity in the area did not entirely cease after the Monkodonja settlement ended in the 15th century BC.

However, no scientific datings and stratified ceramic finds have yet been linked to settlements of the second half of the 2nd millennium, so it is unclear what the characteristic spectrum of ceramic finds at the end of the Middle Bronze Age and during the Late Bronze Age in Istria looks like. Only from the 12th century BC onward is there a broader data base from various necropolises with urn burials, such as from the Limska Gradina [2] (p. 117) and [99]. Based on current data and research, it appears that in the area south of the Limski kanal and between Rovinj and Bale, there was a decline or disruption in settlement activities after the 15th century BC—after the end of Monkodonja.

The question arises whether the presumed violent end of the settlement of Monkodonja involved a military conflict of a regionally limited, local scale, or if it was part of more far-reaching events in the Adriatic region, and what the possible triggers for these events might have been see also [95] (p. 235).

Conflict research and the reconstruction of warfare in prehistoric societies represent a broad field of study, relevant for various time periods and regions (e.g., [100–102]). Destruction layers, burn horizons, human skeletal remains, and the discovery of weapons in settlements can, but do not necessarily, indicate military events. These findings are particularly interesting for the Late Bronze Age and Early Iron Age periods, for which there are not only material cultural records but also written testimonies that point to armed conflicts. For example, during the 13th–12th centuries BC, drastic societal changes occurred in the eastern Mediterranean, and are, as mentioned above, often controversially discussed in the context of the so-called "Sea Peoples". One notable example is the site of Tell Tweini in Syria, where a destruction layer with weapon finds was discovered, dated to this period using scientific methods (e.g., [74,77,103]). For the Adriatic region in the 2nd millennium BC, however, we do not have any written sources that provide evidence of possible armed conflicts, but can only infer such conflicts from material remains.

Of relevance to our considerations are findings from the Adriatic region, which, during a similar timeframe as in Monkodonja, around the middle of the 2nd millennium BC, provide evidence of widespread destruction in settlements that could offer insights into more far-reaching events in this area. In this context, attention should be drawn to the settlements of Roca and Coppa Nevigata in the southern Italian region of Apulia. Roca is located near Otranto, on the Strait of Otranto, the narrow body of water connecting the Adriatic with the Ionian Sea.

The settlement of Roca is situated on a promontory fortified in prehistory and was inhabited from the Middle Bronze Age until the Hellenistic period, up to the 2nd century BC, with a medieval overbuilding also present [40]. The Middle Bronze Age fortification was constructed using monumental dry-stone wall techniques, incorporating post settings, with the wall already having a width of 12 m in its initial construction phase, which was later expanded to 20 m in a subsequent phase. The preserved height of the wall was 3 m, and it is presumed that it was originally at least 2 m higher [40] (Table 7.I), [45] (p. 94). Of particular note is the widespread destruction of the Middle Bronze Age complex around the middle of the 2nd millennium BC, or in the 15th or early 14th century BC during the Bronzo Medio 3 [104] (Figure 3), [40] (pp. 394, 402–403) and [45] (pp. 95, 145). It is highly likely that the destruction of the settlement was the result of an armed attack, during which deliberate fires were set, and the entire complex was burned down [40] (p. 396). In the area of the monumental gate, the skeleton of an armed man was discovered [37] (Figures 3.100–3.101) and [45] (p. 95). Victims of the destruction were found buried under the ruins of the buildings. In the area of the so-called Postern C, the skeletons of seven individuals were found—one man, one woman, and several children and adolescents—who had apparently hidden there and were unable to escape when the fire broke out [40] (pp. 78–86, 399–401, Figure 3.36, 3.54–3.58).

Unlike Monkodonja, however, the settlement of Roca was rebuilt after its destruction, and the fortifications were further expanded. Both for chronology and for indicating long-distance contacts, the finds of Mycenaean pottery are significant [40] (pp. 346–347, 403–404) and [45] (pp. 94–101, 145–148).

Similar to Monkodonja and Roca, a Middle Bronze Age destruction layer has also been documented in the significant prehistoric fortified settlement of Coppa Nevigata, located on the Gargano Peninsula in Apulia. This destruction is dated to around the middle of the 2nd millennium BC or the 15th century BC [104] (Figure 2) and [38] (pp. 459, 462). As in Roca, Coppa Nevigata also shows Late Bronze and Iron Age settlement phases, with the Late Bronze Age settlement, in particular, indicating contacts with the Aegean through Mycenaean pottery [38] (pp. 411–426), [86] (pp. 242–243), [45] (pp. 144–145), [105] (pp. 94–98) and [39] (pp. 131–145). However, based on chemical and petrographic analyses, it has been determined that the Mycenaean-type pottery is not imported from Greece but is instead Italian-made products [45] (p. 144) and [106] (p. 157).

It is overall tempting to link the mentioned examples of destruction of large and likely significant settlements such as Monkodonja, Roca or Coppa Nevigata around the middle of the 2nd millennium BC, both in the western and eastern Adriatic region, and to interpret them as results of more extensive conflicts. However, in this context, the data basis is also severely limited, and it cannot be stated with certainty that all the settlements were destroyed within a closely overlapping time period. However, if we return to the hypothesis of various "waves of migration" to Istria, the data available so far might suggest that after the 15th century BC, there was a certain decline in settlement in Istria, followed by a renewed "upsurge" with the beginning of the Urnfield Culture period.

4. Conclusions

During the 2nd millennium BC, significant changes in settlement activities occurred in the area of the Northern Adriatic on the Istrian Peninsula with the appearance of the so-called Castellieri. As early as the beginning of the 20th century, C. Marchesetti postulated that the occupation took place in the form of two "migration waves", meaning that newcomers migrated into this region, bringing new influences, knowledge, and traditions. It was only with the Early Bronze Age that fortified hilltop settlements with concentric stone walls emerged, which had no precedents in Istria. However, we are familiar with monumental stone architecture from the Early Bronze Age in the fortifications of the Eastern Mediterranean region. No precursors have been found in Istria for the rest of the material culture of the Early and Middle Bronze Age either. For example, if we examine tripod plates in ceramics, these clearly represent a form whose predecessors are found in the Eastern Mediterranean. Furthermore, Bronze Age ceramics in general show no stylistic or technological similarities to the Neolithic and Chalcolithic ceramics of Istria, which suggests the immigration of different ceramic traditions. However, among the ceramic forms, not only are there those that seem stylistically influenced by the Eastern Mediterranean, but also forms and technological features that point in another direction—namely, the Carpathian Basin. Thus, Istria appears as a crossroads, reflecting influences from both the Eastern Mediterranean and Central Europe.

Until the 1990s, research assumed that the spread of Castellieri was primarily a phenomenon of the Middle Bronze Age in Istria. However, extensive modern excavations at Monkodonja, supported by a series of C14 datings, have shown that settlement activities began earlier, in the developed Early Bronze Age, likely around 2000 BC. This shifts the appearance of the Castellieri on the Istrian Peninsula to a time close to the so-called 4.2-kiloyear event, an abrupt climate change event that brought significant cooling and aridity, which may have had more or less noticeable effects in certain regions. The impact of the abrupt climate change on Early Bronze Age societies cannot yet be fully understood, posing a risk of oversimplifying presumed causal relationships. Nevertheless, it seems plausible that the climatic changes were felt more acutely in some regions than in others, possibly leading to droughts and, consequently, migrations in search of better living conditions. The postulated first "migration wave" or the supposed settlement boom on the Istrian Peninsula around 2000 BC could thus be a result of the climatic changes at the turn of the 3rd to the 2nd millennium BC. Based on the current data, it is considered certain that there was a significant increase in settlement activity on the Istrian Peninsula at the beginning of the 2nd millennium BC.

If we accept the dating of the Avellino eruption of Mount Vesuvius in Campania to the 20th–19th centuries BC, this catastrophic event would have also led to migration. However, the material found at Monkodonja, particularly the ceramics, shows no influences from this region. Isolated finds so far only testify to contacts in the direction of Gargano, Apulia [13].

Pollen analyses suggest that at the same time the peninsula began to be settled with the Castellieri, humans began to massively intervene in the landscape, leading to lasting changes. Forests were cleared to obtain building and heating materials as well as for metallurgy and pottery production, transforming the landscape from dense forests to open mixed forests with shrubs and herbs, creating a completely different environment compared to the Neolithic and Copper Age. Human intervention in the environment not only altered the landscape but most likely also affected the habitat of many animal species. Pollen diagrams for Istria in the 7th–2nd millennia BC support that significant human-induced changes in the landscape correlate with the Early Bronze Age in this region.

Around the middle of the 2nd millennium BC, or in the 15th century BC, the fortified settlement of Monkodonja was destroyed, presumably in an armed attack, and was never resettled. During a similar period, other large fortified settlements on the west coast of the Adriatic in Apulia were also destroyed, but they were later rebuilt and resettled. It is not yet possible to determine exactly how close the destruction of the individual settlements occurred in time. However, it seems tempting to see them as the result of far-reaching events and upheavals around the middle of the 2nd millennium BC, which occurred over a relatively short period and affected other parts of the Adriatic region. That settlement activities in Istria, particularly in the area around Monkodonja, did not completely cease after the destruction of the central settlement is indicated by the continued use of the necropolis of the settlement on the neighboring hill of Mušego. The study of surface-collected ceramics and materials from various sondages and smaller excavations in the area south of the Limski kanal and between Rovinj and Bale seems to indicate that most of the surrounding settlements were occupied in a similar time period to Monkodonja and not later, as the spectrum of ceramic forms is identical. Stratified settlement finds that can be clearly assigned to the Late Middle Bronze Age and the Late Bronze Age are not yet available. On the one hand, this results in a gap in knowledge about the ceramic forms after the 15th century BC; on the other hand, no larger finds of settlement ceramics have yet come to light that could potentially fill this gap. Overall, the impression is of a settlement decline after the 15th century BC, but this is still difficult to assess based on the current state of research.

The emergence of the Castellieri around 2000 BC could, in one interpretation, be associated with the abrupt climate changes of the 4.2-kiloyear event and/or the Avellino eruption of Mount Vesuvius in Campania, both of which are believed to have caused population shifts or migrations. However, given the limitations of the current data on population increases and the presumed foreign origin of possible new arrivals in Istria at the beginning of the 2nd millennium BC, it is not yet possible to draw definitive conclusions regarding causality. Similarly, for our second question—what might have triggered a presumed 'crisis' in the Adriatic region around the middle of the 2nd millennium BC, potentially leading to socioeconomic stress, population shifts, and associated conflicts—only speculative assumptions are available at this point, which require further investigation by future research. The Minoan eruption in the southern Aegean at the end of the 17th century BC, while a significant event, appears too early to be directly linked to a presumed crisis in the Adriatic around the middle of the 2nd millennium BC or in the 15th century BC. However, it is possible that both short-term and long-term effects of the eruption influenced regional developments over time, though these potential effects would have interacted with other factors.

While the questions raised in this study about the emergence of the Castellieri settlements and a possible "migration wave" at the beginning of the 2nd millennium BC remain unanswered due to limited available data, the climate changes, natural disasters, and their links to human activity observed during this period could provide a foundation for future research. Future studies should aim to refine these ideas with more precise data to establish clearer causal relationships.

Funding: This paper was supported by the Slovenian Research Agency (ARIS) as part of the research group "Raziskave kulturnih formacij/Research of Cultural Formations (P6-0278 (A), 2019–2027)" at Alma Mater Europaea University, Maribor.

Data Availability Statement: Data are contained within the article.

Conflicts of Interest: The author declares no conflict of interest.

References

1. Hänsel, B.; Mihovilić, K.; Teržan, B. *Monkodonja. Istraživanje Protourbanog Naselja Brončanog Doba Istre. Knjiga 1—Iskopavanje i nalazi Građevina/Forschungen zu Einer Protourbanen Siedlung der Bronzezeit Istriens. Teil 1—Die Grabung und der Baubefund*; Monografije i katalozi Arheološki muzej Istre 25; Arheološki muzej Istre u Puli: Pula, Croatia, 2015.
2. Mihovilić, K. *Histri u Istri/Gli Istri in Istria/The Histri in Istria*; Monografije i katalozi Arheološki muzej Istre 23; Arheološki muzej Istre u Puli: Pula, Croatia, 2013.
3. Križman, M. *Antička Svjedočanstva o Istri*; Povijest Istre I; Čakavski sabor: Pula, Rijeka, Croatia, 1979.
4. Kozličić, M. Jadranske plovidbene rute od staroga do prvih stolećja novoga vijeka. *Histria Antiq.* **2012**, *21*, 13–20.
5. Buršić-Matijašić, K. Gradina Monkodonja. Tipološko-statistička obrada keramičkih calaza srednjobrončanodobne istarske gradine Monkodonja kod Rovinja/The Monkodonja Hillfort. In *A Typological and Statistical Analysis of Pottery Finds from the Middle Bronze Age Hillfort of Monkodonja near Rovinj*; Monografije i katalozi Arheološki muzej Istre 9; Arheološki muzej Istre u Puli: Pula, Croatia, 1998.
6. Sakara Sučević, M. Prazgodovinska Keramika med Miljskim Zalivom in Porečjem Mirne. Ph.D. Thesis, Univerza na Primorskem, Fakulteta za Humanistične Študije, Koper, Slovenia, 2012.
7. Buršić-Matijašić, K. *Gradine Istre. Povijest Prije Povijesti*; ZN "Žakan Juri": Pula, Croatia, 2007.
8. Teržan, B.; Mihovilić, K.; Hänsel, B. Eine protourbane Siedlung der älteren Bronzezeit im istrischen Karst. *PZ* **1999**, *74*, 154–193. [CrossRef]
9. Flego, S.; Rupel, L. *Prazgodovinska Gradišča Tržaške Pokrajine*; Založništvo tržaškega tiska: Trieste, Italy, 1993.
10. Karoušková-Soper, V. *The Castellieri of Venezia Giulia, North-Eastern Italy (2nd–1st Millennium B.C.)*; BAR International Series, British Archaeological Reports 192; BAR Publishing: Oxford, UK, 1983.
11. Lonza, B. *Appunti sui Castellieri Dell'istria e della Provincia di Trieste. Società per la Preistoria e Protostoria Della Regione Friuli-Venezia Giulia*; Quaderno 2; Edizioni Italo Svevo: Trieste, Italy, 1977.
12. Marchesetti, C. *I Castellieri Preistorici di Trieste e della Regione Giulia*; Museo Civico di Storia Naturale: Trieste, Italy, 1903.
13. Hellmuth Kramberger, A. The role of Istria in the Bronze and Iron Age communications network of the Adriatic considered by the examples of the Gradinas Monkodonja and Monbrodo. *Origini* **2021**, *XLV*, 195–212.
14. Burton, R.F. *Note Sopra i Castellieri o Rovine Preistoriche Della Penisola Istriana*; Appolonia: Capodistria, Italy, 1877.
15. Komšo, D. Kargadur—Eine Siedlung aus dem frühen- und mittleren Neolithikum Istriens. *Mitt. Berliner Ges. Anthr. Ethn. Urgesch.* **2006**, *27*, 111–118.
16. Legnani, F.; Stradi, F. Gli scavi nella caverna dei Ciclami nel Carso triestino. Atti della VII Riunione Scientifica dell' IIPP. *Atti Delle Riun. Sci. Dell'istituto Ital. Preist. Protostoria* **1963**, *7*, 31–38.
17. Gilli, E.; Montagnari Kokelj, E. La Grotta dei Ciclami nel Carso Triestino (Materiali degli scavi 1959–1961). *Atti Della Soc. Per Preist. Protostoria Della Reg. Friuli-Venezia Giulia* **1992**, *7*, 65–162.
18. Komšo, D. *Pećina Laganiši—Mjesto Života i Smrti/Laganiši Cave—A Place of Life and a Place of Death*; Arheološki Muzej Istre u Puli: Pula, Croatia, 2008.
19. Forenbaher, S.; Kaiser, T.; Miracle, P.T. Pupićina cave pottery and the neolithic sequence in northeastern Adriatic. *Atti Della Soc. Per Preist. Protostoria Della Reg. Friuli-Venezia Giulia* **2003**, *14*, 61–102.
20. Blečić Kavur, M.; Komšo, D. Tajni grad—Grad tajni Oporovina i sustav špilja u kanjonu Lovranske Drage/A secret city—A city of secrets. Oporovina and the cave system in the canyon at Lovranska draga. *HistriaA* **2015**, *46*, 75–97.
21. Čuka, M. Prapovijesna keramika iz pećine Garbinovice kod Poreča: Probno arheološko istraživanje iz 2013. Godine/Prehistoric Pottery from Garbinovica Cave near Poreč: A Trial Archaeological Exploration Performed in 2013. *HistriaA* **2014**, *45*, 5–62.
22. Čuka, M. Tipološka obrada odabranih ulomaka prapovijesne keramike iz Pećine ispod sela Srbani/A Typological Analysis of Selected Prehistoric Pottery Fragments from Pećina Ispod Sela Srbani. *HistriaA* **2009**, *40*, 13–44.
23. Forenbaher, S.; Rajić Šikanjić, P.; Miracle, P.T. Lončarija iz Vele Peći kod Vranje (Istra). *HistrArch* **2006**, *37*, 5–46.
24. Hellmuth Kramberger, A. Consideration of cultural transfer in case of Bronze Age tripods from Istria and their use. In *Kulturna Formacija in Kulturni Spomin—Jubilejna Publikacija in Rezultat Raziskovalne Skupine "Raziskave Kulturnih Formacij"/9. Znanstvena Konferenca za Mednardono Udeležbo. Za Človeka gre: Digitalna Transformacija v Znanosti, Izobraževanju In umetnosti. Zbornik Recenziranih Prispevkov za Področje Raziskav Kulturnih Formacij*; Hellmuth Kramberger, A., Perko, V., Eds.; AMEU-ECM, Alma Mater Press: Maribor, Slovenia, 2022; pp. 14–30.
25. Hellmuth Kramberger, A. "Sonnenbarken" und "Mondscheiben" im bronzezeitlichen Istrien? Zu zwei besonderen Schalen mit verziertem Boden aus der Gradina von Monkodonja nahe Rovinj, Kroatien. *Archaeol. Austriaca* **2020**, *104*, 153–168. [CrossRef]
26. Hellmuth Kramberger, A. *Monkodonja. Istraživanje Protourbanog Naselja Brončanog doba Istre Knjiga 2—Keramika s Brončanodobne Gradine Monkodonja/Forschungen zu Einer Protourbanen Siedlung der Bronzezeit Istriens Teil 2—Die Keramik aus der Bronzezeitlichen Gradina Monkodonja*; Monografije i katalozi Arheološki muzej Istre 28; Arheološki Muzej Istre u Puli: Pula, Croatia, 2017.
27. Gradišča. *Gradišča v Zahodni in Osrednji Sloveniji: Zbornik s Posvetovanja o Gradiščih, Pivka, 24. maja 2017*; Zavod Krasen Kras: Gorjansko, Slovenia, 2018.

28. Mušić, B.; Medarić, I. Geofizička istraživanja na Monkodonjii / Geophysikalische Untersuchungen. In *Monkodonja. Istraživanje Protourbanog Naselja Brončanog Doba Istre. Knjiga 1—Iskopavanje i Nalazi Građevina/Forschungen zu Einer Protourbanen Siedlung der Bronzezeit Istriens. Teil 1—Die Grabung und der Baubefund*; Hänsel, B., Mihovilić, K., Teržan, B., Eds.; Monografije i katalozi Arheološki muzej Istre 25; Arheološki Muzej Istre u Puli: Pula, Croatia, 2015; pp. 391–421.
29. Mihovilić, K. Povijest istraživanja brončanodobnis grobnih tumula na području Iste/Zur Forschungsgeschite bronzezeitlicher Grabhügel Istriens. In *Monkodonja. Istraživanje Protourbanog Naselja Brončanog Doba Istre Knjiga 4—Mušego/Mon Sego, Grobni Tumuli—Nekropola Gradine Monkodonja i Brončanodobni Tumuli Istre /Forschungen zu Einer Protourbanen Siedlung der Bronzezeit Istriens Teil 4—Mušego/Mon Sego, Grabhügel—Eine Nekropole von der Gradina Monkodonja und die Bronzezeitlichen Grabhügel Istriens*; Mihovilić, K., Teržan, B., Eds.; Monografije i katalozi Arheološki muzej Istre 37; Arheološki Muzej Istre u Puli: Pula, Croatia, 2022; pp. 15–30.
30. Cummer, W.W.; Schofield, E. Keos III. Ayia Irini: House A. In *Results of Excavations Conducted by the University of Cincinnati Under the Auspices of the American School of Classical Studies at Athen*; Zabern: Mainz am Rhein, Germany, 1984.
31. Mihovilić, K.; Teržan, B.; Hänsel, B.; Matošević, D.; Becker, C. *Rovinj Prije Rima/Rovino Prima dei Romani/Rovinj vor den Römern*; Oetker Vogel Verlag: Kiel, Germany, 2001.
32. Hänsel, B.; Mihovilić, K.; Teržan, B. Fortification concepts of the bronze age hillforts in Istria. In *Bronze Age Fortresses in Europe: Proceedings of the Second International LOEWE Conference, 9–13 October 2017 in Alba Julia*; Hansen, S., Krause, R., Eds.; Universitätsforschungen zur prähistorischen Archäologie 335/Prähistorische Konfliktforschung 3; Habelt: Bonn, Germany, 2019; pp. 99–115.
33. Balbo, A.L.; Andrič, M.; Rubinić, J.; Moscariello, A.; Miracle, P.T. Palaeoenvironmental and Archaeological Implications of a Sediment Core from Polje Čepić, Istria, Croatia. *Geol. Croat.* **2006**, *59*, 109–124. [CrossRef]
34. Andrič, M. Paleookolje v Sloveniji in severnemu delu hrvaške Istre v pozni prazgodovini / The vegetation of Slovenia and northern Istria in late prehistory. *Arheol. Vestn.* **2004**, *55*, 509–522.
35. Hänsel, B.; Mihovilić, K.; Teržan, B. *Monkodonja. Istraživanje Protourbanog Naselja Brončanog Doba Istre. Knjiga 3 Nalazi od Metala, Gline, Kosti i Kamena kao i Ljudskih i Životinjskih Kostiju/Forschungen zu Einer Protourbanen Siedlung der Bronzezeit Istriens. Teil 3 Die Funde aus Metall, Ton, Knochen und Stein Sowie die menschlichen und Tierischen Knochen*; Monografije i katalozi Arheološki muzej Istre 34; Arheološki Muzej Istre u Puli: Pula, Croatia, 2020.
36. Hänsel, B.; Matošević, D.; Mihovilić, K.; Teržan, B. Zur Sozialarchäologie der befestigten Siedlung von Monkodonja (Istrien) und ihrer Gräber am Tor. *PZ* **2009**, *84*, 151–180. [CrossRef]
37. Cazzella, A.; Moscoloni, M.; Recchia, G. *Coppa Nevigata e Lárea Umida alla foce del Candelaro Durante l'eta del Bronzo*; Edizioni del Parco: Foggia, Italy, 2012.
38. Cassano, S.M.; Cazzella, A.; Manfredini, A.; Moscoloni, M. *Coppa Nevigata e il suo Territorio. Testimonianze Archeologiche dal VII al II Millennio a.C.*; Edizioni Quasar: Rome, Italy, 1987.
39. Scarano, T.; Roca, I. *Le fortificazioni Della Media età del Bronzo. Strutture, Contesti, Materiali*; Claudio Grenzi Editore: Foggia, Italy, 2012.
40. Walter, H.; Felten, F. *Alt-Ägina III. Die Vorgeschichtliche Stadt: Befestigungen, Häuser, Funde*; Verlag Philipp von Zabern: Mainz am Rhein, Germany, 1981.
41. Marinatos, S. *Kreta, Thera und das Mykenische Hellas*; Hirmer Verlag München: München, Germany, 1973.
42. Toškan, B.; Achino, K.F.; Becker, C. Životinjski ostaci / Animal remains from Monkodonja. In *Monkodonja. Istraživanje Protourbanog Naselja Brončanog Doba Istre. Knjiga 3 Nalazi od Metala, Gline, Kosti i Kamena kao i Ljudskih i Životinjskih Kostiju/Forschungen zu Einer Protourbanen Siedlung der Bronzezeit Istriens. Teil 3 Die Funde aus Metall, Ton, Knochen und Stein Sowie die Menschlichen und Tierischen Knochen*; Hänsel, B., Mihovilić, K., Teržan, B., Eds.; Monografije i katalozi Arheološki muzej Istre 34; Arheološki Muzej Istre u Puli: Pula, Croatia, 2020; pp. 457–526.
43. Hellmuth Kramberger, A. Resource management in the hillfort settlement of Monkodonja near Rovinj, Istria in the Early and Middle Bronze Age. In *Preistoria e Protostoria del Caput Adriae*; Borgna, E.E., Càssola Guida, P., Corazza, S., Eds.; Studi di Preistoria e Protostoria 5; Istituto Italiano di Preistoria e Protostoria: Florence, Italy, 2018; pp. 371–380.
44. Jung, R. *ΧΡΟΝΟΛΟΓΙΑ COMPARATA. Vergleichende Chronologie von Südgriechenland und Süditalien von ca. 1700/1600 bis 1000 v. u. Z.*; Veröffentlichungen der Mykenischen Kommission 26; Verlag der Österreichischen Akademie der Wissenschaften: Vienna, Austria, 2006.
45. Reimer, P.; Baillie, M.G.L.; Bard, E.; Bayliss, A.; Beck, J.W.; Bertrand, C.J.H.; Blackwell, P.G.; Buck, C.E.; Burr, G.S.; Cutler, C.B.; et al. 1 IntCal04 terrestrial radiocarbon age calibration, 0–26 cal kyr BP. *Radiocarbon* **2004**, *46*, 10129–11058.
46. Reimer, P.; Baillie, M.G.L.; Bard, E.; Bayliss, A.; Beck, J.W.; Blackwell, P.G.; Bronk Ramsey, C.B.; Buck, C.E.; Burr, G.S.; Edwards, R.L.; et al. 26 IntCal09 and Marine09 Radiocarbon Age Calibration Curves, 0–50,000 Years cal BP. *Radiocarbon* **2009**, *51*, 1111–1150. [CrossRef]
47. Teržan, B.; Teßmann, B.; Goslar, T. Radiokarbonsko datiranje tri grobna Tumula na Mušegu / Radiokarbondatierungen der drei Tumuli von Mušego. In *Monkodonja. Istraživanje Protourbanog Naselja Brončanog Doba Istre Knjiga 4—Mušego/Mon Sego, Grobni Tumuli—Nekropola Gradine Monkodonja i Brončanodobni Tumuli Istre /Forschungen zu Einer Protourbanen Siedlung der Bronzezeit Istriens Teil 4—Mušego/Mon Sego, Grabhügel—Eine Nekropole von der Gradina Monkodonja und die bronzezeitlichen Grabhügel Istriens*; Mihovilić, K., Teržan, B., Eds.; Monografije i katalozi Arheološki muzej Istre 37; Arheološki Muzej Istre u Puli: Pula, Croatia, 2022; pp. 173–179.
48. Hänsel, B.; Teržan, B.; Mihovilić, K. Radiokarbondaten zur älteren und mittleren Bronzezeit Istriens. *PZ* **2007**, *82*, 23–50. [CrossRef]
49. Teßmann, B. Antropološka istraživanja ljudskih ostataka iz rano- do srednjobrončanodobnog naselja Monkodonja / Die anthropologischen Untersuchungen der menschlichen Überreste aus der früh- bis mittelbronzezeitlichen Siedlung von Monkodonja. In

Monkodonja. Istraživanje Protourbanog Naselja Brončanog Doba Istre. Knjiga 3 Nalazi od Metala, Gline, Kosti i Kamena kao i Ljudskih i Životinjskih Kostiju/Forschungen zu Einer Protourbanen Siedlung der Bronzezeit Istriens. Teil 3 Die Funde aus Metall, Ton, Knochen und Stein Sowie Die menschlichen und tierischen Knochen; Hänsel, B., Mihovilić, K., Teržan, B., Eds.; Monografije i katalozi Arheološki muzej Istre 34; Arheološki Muzej Istre u Puli: Pula, Croatia, 2020; pp. 529–567.

50. Fuller, B.T.; Riehl, S.; Linseele, V.; Marinova, E.; De Cupere, B.; Bretschneider, J.; Richards, M.P.; Van Neer, W. Agropastoral and dietary practices of the northern Levant facing Late Holocene climate and environmental change: Isotopic analysis of plants, animals and humans from Bronze to Iron Age Tell Tweini. *PLoS ONE* **2024**, *19*, e0301775. [CrossRef]
51. Molloy, B. Was There a 3.2 ka Crisis in Europe? A Critical Comparison of Climatic, Environmental, and Archaeological Evidence for Radical Change during the Bronze Age–Iron Age Transition. *J. Archaeol. Res.* **2023**, *31*, 331–394. [CrossRef]
52. Hazell, C.J.; Pound, M.J.; Hocking, E.P. High-resolution Bronze Age palaeoenvironmental change in the Eastern Mediterranean: Exploring the links between climate and societies. *Palynology* **2022**, *46*, 2067259. [CrossRef]
53. Demény, A.; Kern, Z.; Czuppon, G.; Németh, A.; Schöll-Barna, G.; Siklósy, Z.; Leél-Őssy, S.; Cook, G.; Serlegi, G.; Bajnóczi, B.; et al. Middle Bronze Age humidity and temperature variations, and societal changes in East-Central Europe. *Quat. Int.* **2019**, *504*, 80–95. [CrossRef]
54. Weiberg, E.; Finné, M. Resilience and persistence of ancient societies in the face of climate change: A case study from Late Bronze Age Peloponnese. *World Archaeol.* **2018**, *50*, 584–602. [CrossRef]
55. Meller, H.; Arz, H.W.; Jung, R.; Risch, R. (Eds.) *2200 BC—Ein Klimasturz als Ursache für den Zerfall der Alten Welt?/2200 BC—A Climatic Breakdown as a Cause for the Collapse of the Old World? 7. Mitteldeutscher Archäologentag vom 23. bis 26. Oktober 2014 in Halle (Saale)/7th Archaeological Conference of Central Germany October 23–26, 2014 in Halle (Saale). Tagungen des Landesmuseums für Vorgeschichte Halle 12/I*; Landesamt für Denkmalpflege und Archäologie Sachsen-Anhalt: Halle, Germany, 2015.
56. Weninger, B.; Clare, L.; Rohling, E.J.; Bar-Yosef, O.; Böhner, U.; Budja, M.; Bundschuh, M.; Feurdean, A.; Gebel, H.-G.; Jöris, O.; et al. The Impact of Rapid Climate Change on prehistoric societies during the Holocene in the Eastern Mediterranean. *Doc. Praehist.* **2009**, *36*, 7–59. [CrossRef]
57. Mayewski, P.A.; Rohling, E.E.; Stager, J.C.; Karlén, W.; Maasch, K.A.; Meeker, L.D.; Meyerson, E.A.; Gasse, F.; van Kreveld, S.; Holmgren, K.; et al. Holocene climate variability. *Quat. Res.* **2004**, *62*, 243–255. [CrossRef]
58. Ran, M.; Chen, L. The 4.2 ka BP climatic event and its cultural responses. *Quat. Int.* **2019**, *521*, 158–167. [CrossRef]
59. Bini, M.; Zanchetta, G.; Perşoiu, A.; Cartier, R.; Català, A.; Cacho, I.; Dean, J.R.; Di Rita, F.; Drysdale, R.N.; Finnè, M.; et al. The 4.2 ka BP Event in the Mediterranean region: An overview. *Clim. Past* **2019**, *15*, 555–577. [CrossRef]
60. Carolin, S.A.; Walker, R.T.; Day, C.C.; Ersek, V.; Sloan, A.; Deed, M.W.; Talebian, M.; Henderson, G.M. Precise timing of abrupt increase in dust activity in the Middle East coincident with 4.2 ka social change. *Proc. Natl. Acad. Sci. USA* **2019**, *116*, 67–72. [CrossRef]
61. Kaniewski, D.; Marriner, N.; Cheddadi, R.; Guiot, J.; Van Campo, E. The 4.2 ka BP event in the Levant. *Clim. Past* **2018**, *14*, 1529–1542. [CrossRef]
62. Kuzucuoğlu, C.; Marro, C. (Eds.) *Sociétés Humaines et Changement Climatique à la fin du Troisième Millénaire: Une Crise a-t-elle eu lieu en Haute Mésopotamie? Actes du Colloque de Lyon, 5–8 Décembre 2005*; Varia Anatolica 19; Anatoliennes: Paris, France, 2007.
63. Migowski, C.; Stein, M.; Prasad, S.; Negendank, J.F.W.; Agnon, A. Holocene climate variability and cultural evolution in the Near East from the Dead Sea sedimentary record. *Quat. Res.* **2006**, *66*, 421–431. [CrossRef]
64. Cullen, H.M.; de Menocal, P.B.; Hemming, S.; Hemming, G.; Brown, F.H.; Guilderson, T.; Sirocko, F. Climate change and the collapse of the Akkadian empire: Evidence from the deep sea. *Geology* **2000**, *28*, 379–382. [CrossRef]
65. Weiss, H.; Courty, M.-A.; Wetterstrom, W.; Guichard, F.; Senior, L.; Meadow, R.; Curnow, A. The genesis and collapse of third millennium North Mesopotamian Civilization. *Science* **1993**, *261*, 995–1004. [CrossRef] [PubMed]
66. Salzmann, E. *Silver, Copper and Bronze in Early Dynastic Ur, Mesopotamia*; Der Anschnitt Beiheft 44; Leidorf: Bochum, Germany, 2019.
67. Maran, J. *Kulturwandel auf dem Griechischen Festland und den Kykladen im Späten 3. Jahrtausend v. Chr.*; Universitätsforschungen zur prähistorischen Archäologie 53; Habelt: Bonn, Germany, 1998.
68. Claussen, M.; Kubatzki, C.; Brovkin, V.; Ganopolski, A.; Hoelzmann, P.; Pachur, H.J. Simulation of an abrupt change in Saharan vegetation in the mid-Holocene. *Geophys. Res. Lett.* **1999**, *26*, 2037–2040. [CrossRef]
69. Bar-Matthews, M.; Ayalon, A.; Kaufman, A. Middle to Late Holocene (6500 Yr. Period) Paleoclimate in the Eastern Mediterranean Region from Stable Isotopic Composition of Speleothems from Soreq Cave, Israel. In *Water, Environment and Society in Times of Climatic Change: Contributions from an International Workshop Within the framework of International Hydrological Program (IHP) UNESCO, Held at Ben-Gurion University, Sede Boker, Israel from 7–12 July 1996*; Issar, A.S., Brown, N., Eds.; Springer: Dordrecht, The Netherlands, 1998; pp. 203–214.
70. Lemcke, G.; Sturm, M. δ18o and Trace Element Measurements as Proxy for the Reconstruction of Climate Changes at Lake Van (Turkey): Preliminary Results. In *Third millennium B.C. Climate Change and Old World Collapse*; Nüzhet Dalfes, H., Kukla, G., Weiss, H., Eds.; Nato ASI Subseries I/ ASII 49; Springer: Berlin, Germany; pp. 178–196.
71. Bar-Matthews, M.; Ayalon, A.; Kaufman, A. Late quaternary paleoclimate in the eastern Mediterranean region from stable isotope analysis of speleothems at Soreq Cave, Israel. *Quat. Res.* **1997**, *47*, 155–168. [CrossRef]
72. Frumkin, A. The Holocene climatic record of the salt caves of Mount Sedom, Israel. *Holocene* **1991**, *1*, 191–200. [CrossRef]

73. Kaniewski, D.; Paulissen, E.; Van Campo, E.; Weiss, H.; Otto, T.; Bretschneider, J.; Van Lerberghe, K. Late second–early first millennium BC abrupt climate changes in coastal Syria and their possible significance for the history of the Eastern Mediterranean. *Quat. Res.* **2010**, *74*, 207–215. [CrossRef]
74. Jung, R. Push and Pull Factors of the Sea Peoples between Italy and the Levant. In *An Archaeology of Forced Migration. Crisis-induced Mobility and the Collapse of the 13th c. BCE Eastern Mediterranean*; Driessen, J., Ed.; Aegis 15; Presses Universitaires de Louvain: Louvain-La-Neuve, Belgium, 2018; pp. 279–312.
75. Knapp, A.B.; Manning, S.W. Crisis in Context: The End of the Late Bronze Age in the Eastern Mediterranean. *Am. J. Archaeol.* **2016**, *120*, 99–149. [CrossRef]
76. Kaniewski, D.; Van Campo, E.; Van Lerberghe, K.; Boiy, T.; Vansteenhuyse, K.; Jans, G.; Nys, K.; Weiss, H.; Morhange, C.; Otto, T.; et al. The Sea Peoples, from Cuneiform Tablets to Carbon Dating. *PLoS ONE* **2011**, *6*, e20232. [CrossRef]
77. Finné, M.; Holmgren, K.; Shen, C.-C.; Hu, H.-M.; Boyd, M.; Stocker, S. Late Bronze Age climate change and the destruction of the Mycenaean Palace of Nestor at Pylos. *PLoS ONE* **2017**, *12*, e0189447. [CrossRef]
78. Di Vito, M.A.; Talamo, P.; de Vita, S.; Rucco, I.; Zanchetta, G.; Cesarano, M. Dynamics and effects of the Vesuvius Pomici di Avellino Plinian eruption and related phenomena on the Bronze Age landscape of Campania region (Southern Italy). *Quat. Int.* **2018**, *499*, 231–244. [CrossRef]
79. Sevink, J.; van Bergen, M.J.; van der Plicht, J.; Feiken, H.; Anastasia, C.; Huizinga, A. Robust date for the Bronze Age Avellino eruption (Somma-Vesuvius): 3945 ± 10 calBP (1995 ± 10 calBC). *Quat. Sci. Rev.* **2011**, *30*, 1035–1046. [CrossRef]
80. Passariello, I.; Livadie, C.A.; Talamo, P.; Lubritto, C.; D'Onofrio, A.; Terrasi, F. 14C Chronology of Avellino Pumices Eruption and Timing of Human Reoccupation of the Devastated Region. *Radiocarbon* **2009**, *51*, 803–816. [CrossRef]
81. Manning, S.W. *A Test of Time and A Test of Time Revisited: The Volcano of Thera and the Chronology and History of the Aegean and East Mediterranean in the Mid-Second Millennium BC*; Oxbow Books: Oxford, UK, 2014.
82. Friedrich, W.L. The Minoan Eruption of Santorini around 1613 B.C. and its consequences. In *1600—Kultureller Umbruch im Schatten des Thera-Ausbruchs?/1600—Cultural change in the shadow of the Thera-Eruption? 4. Mitteldeutscher Archäologentag vom 14. bis 16. Oktober 2011 in Halle (Saale)*; Meller, H., Bertemes, F., Bork, H.-R., Risch, R., Eds.; Tagungen des Landesmuseums für Vorgeschichte Halle 9; Landesamt für Denkmalpflege und Archäologie Sachsen-Anhalt: Halle, Germany; pp. 37–48.
83. Oppenheimer, C. Nature and impacts of the Minoan Eruption of Santorini. In *1600—Kultureller Umbruch im Schatten des Thera-Ausbruchs?/1600—Cultural change in the shadow of the Thera-Eruption? 4. Mitteldeutscher Archäologentag vom 14. bis 16. Oktober 2011 in Halle (Saale)*; Meller, H., Bertemes, F., Bork, H.-R., Risch, R., Eds.; Tagungen des Landesmuseums für Vorgeschichte Halle 9; Landesamt für Denkmalpflege und Archäologie Sachsen-Anhalt: Halle, Germany; pp. 49–58.
84. Kutschera, W. Dating of the Thera/Santorini volcanic eruption. In *1600—Kultureller Umbruch im Schatten des Thera-Ausbruchs?/1600—Cultural Change in the Shadow of the Thera-Eruption? 4. Mitteldeutscher Archäologentag vom 14. bis 16. Oktober 2011 in Halle (Saale)*; Meller, H., Bertemes, F., Bork, H.-R., Risch, R., Eds.; Tagungen des Landesmuseums für Vorgeschichte Halle 9; Landesamt für Denkmalpflege und Archäologie Sachsen-Anhalt: Halle, Germany, 2013; pp. 59–64.
85. Jung, R. The time around 1600 B.C in Southern Italy: New powers, new contacts and new conflicts. In *1600—Kultureller Umbruch im Schatten des Thera-Ausbruchs?/1600—Cultural Change in the Shadow of the Thera-Eruption? 4. Mitteldeutscher Archäologentag vom 14. bis 16. Oktober 2011 in Halle (Saale)*; Meller, H., Bertemes, F., Bork, H.-R., Risch, R., Eds.; Tagungen des Landesmuseums für Vorgeschichte Halle 9; Landesamt für Denkmalpflege und Archäologie Sachsen-Anhalt: Halle, Germany, 2013; pp. 235–251.
86. Orsi, G.; Cioni, R.; Di Renzo, V. The Campanian Plain during the Bronze Age: Development of volcanism and impact of the Vesuvius Avellino eruption in a densely populated area. In *1600—Kultureller Umbruch im Schatten des Thera-Ausbruchs?/1600—Cultural Change in the Shadow of the Thera-Eruption? 4. Mitteldeutscher Archäologentag vom 14. bis 16. Oktober 2011 in Halle (Saale)*; Meller, H., Bertemes, F., Bork, H.-R., Risch, R., Eds.; Tagungen des Landesmuseums für Vorgeschichte Halle 9; Landesamt für Denkmalpflege und Archäologie Sachsen-Anhalt: Halle, Germany; pp. 117–134.
87. Manning, S.W.; Höflmayer, F.; Moeller, N.; Dee, M.W.; Bronk Ramsey, C.; Fleitmann, D.; Higham, T.; Kutschera, W.; Wild, E.M. Dating the Thera (Santorini) eruption: Archaeological and scientific evidence supporting a high chronology. *Antiquity* **2014**, *88*, 1164–1179. [CrossRef]
88. Bekić, L. Sustav gradina na Rovinjskom području. *Histria Archaeol.* **1996**, *27*, 19–92.
89. Hellmuth Kramberger, A.; Müller, S.; Čuka, M. Bronze Age hillforts around Rovinj on the western coast of the Istrian peninsula and their pottery. *IpoTESI Preist.* **2022**, *15*, 121–138.
90. Hellmuth Kramberger, A.; Müller, S.; Čuka, M. Monbrodo—Prapovijesna Gradina blizu plaže Cisterna, južno od Grada Rovinja, u svijetlu novih istraživanja/Monbrodo—A prehistoric hillfort near Cisterna Beach south of the city of Rovinj in the light of new researches. *Histria Archaeol.* **2018**, *49*, 13–37.
91. Müller, S. *The Latest Research on Bronze Age Hillforts in Istria*; Busan University of Foreign Studies Mediterranean Regional Studies Center: Busan, Republic of Korea, 2018; pp. 151–154.
92. Müller, S.; Čuka, M.; Hellmuth Kramberger, A. Monbrodo—Nova istraživanja gradine južno od Rovinja u blizini uvale cisterna/Monbrodo—New research on the hillfort south of the Cisterna bay near Rovinj. *Histria Archaeol.* **2016**, *47*, 21–55.
93. Gerling, C.; Price, T.D. Lokalni stanovici vs. doseljenici: Izotopske analize iz četiri ukopa ljudskih ostataka/Ortskonstanz versus Einwanderung: Isotopenanalysen an vier menschlichen Bestattungen. In *Monkodonja. Istraživanje Protourbanog Naselja Brončanog Doba Istre. Knjiga 1—Iskopavanje i Nalazi Građevina/Forschungen zu Einer Protourbanen Siedlung der Bronzezeit Istriens. Teil 1—Die Grabung und der Baubefund*; Monografije i katalozi Arheološki muzej Istre 25; Arheološki Muzej Istre u Puli: Pula, Croatia, 2015; pp. 231–237.

94. Budd, P.; Montgomery, J.; Barreiro, B.; Thomas, R.G. Differential diagenesis of strontium in archaeological human dental tissues. *Appl. Geochem.* **2000**, *15*, 687–694. [CrossRef]
95. Sharp, Z. *Principles of Stable Isotope Geochemistry*; Pearson Prentice Hall: Hoboken, NJ, USA, 2007.
96. Chenery, C.A.; Pashley, V.; Lamb, A.L.; Sloane, H.J.; Evans, J.A. The oxygen isotope relationship between the phosphate and structural carbonate fractions in human bioapatite. *Rapid Commun. Mass Spectrom.* **2012**, *26*, 309–319. [CrossRef]
97. Villalba-Mouco, V.; Haak, W. Kratki izvještaj o tijeku analize drevnog DNK-a iz ljudskih ostataka s Monkodonje i drugih brončanodobnih nalazišta Istre/Brief progress report on ancient DNA analyses of human remains from Monkodonja and other bronze age sites in Istria, Croatia. In *Monkodonja. Istraživanje Protourbanog Naselja Brončanog Doba Istre Knjiga 4—Mušego/Mon Sego, Grobni Tumuli—Nekropola Gradine Monkodonja i Brončanodobni Tumuli Istre/Forschungen zu einer Protourbanen Siedlung der Bronzezeit Istriens Teil 4—Mušego/Mon Sego, Grabhügel—Eine Nekropole von der Gradina Monkodonja und die Bronzezeitlichen Grabhügel Istriens*; Mihovilić, K., Teržan, B., Eds.; Monografije i katalozi Arheološki muzej Istre 37; Arheološki Muzej Istre u Puli: Pula, Croatia, 2022; pp. 165–171.
98. Primorac, D.; Šarac, J.; Havaš Auguštin, D.; Novokmet, N.; Bego, T.; Pinhasi, R.; Šlaus, M.; Novak, M.; Marjanović, D.Y. Chromosome Story—Ancient Genetic Data as a Supplementary Tool for the Analysis of Modern Croatian Genetic Pool. *Genes* **2024**, *15*, 748. [CrossRef]
99. Urem, D. *Limska Gradina—Keramika s Područja Nekropole/The Limska Gradina Hillfort—Pottery from the Cemetery Area*; Monografije i katalozi Arheološki muzej Istre 21; Arheološki Muzej Istre u Puli: Pula, Croatia, 2012.
100. Hansen, S.; Krause, R. (Eds.) Materialisierung von Konflikten/Materialisation of Conflicts. In *Beiträge der Dritten Internationalen LOEWE-Konferenz vom 24. bis 27. September 2018 in Fulda/Proceedings of the Third International LOEWE Conference, 24–27 September 2018 in Fulda*; Universitätsforschungen zur prähistorischen Archäologie 346/Prähistorische Konfliktforschung 4; Habelt: Bonn, Germany, 2020.
101. Link, T.; Peter-Röcher, H. (Eds.) *Gewalt und Gesellschaft. Dimensionen der Gewalt in ur- und frühgeschichtlicher Zeit/Violence and Society. Dimensions of Violence in Pre- and Protohistoric Times. Internationale Tagung an der Julius-Maximilians-Universität Würzburg 14.–16. März 2013*; Universitätsforschungen zur prähistorischen Archäologie 259; Habelt: Bonn, Germany, 2014.
102. Peter-Röcher, H. Gewalt und Krieg im prähistorischen Europa. In *Beiträge zur Konfliktforschung auf der Grundlage Archäologischer, Anthropologischer und Ethnologischer Quellen*; Universitätsforschungen zur prähistorischen Archäologie 143; Habelt: Bonn, Germany, 2007.
103. al-Maqdissi, M.; Badawy, M.; Bretschneider, J.; Hameeuw, H.; Jans, G.; Vansteenhuyse, K.; Voet, G.; Van Lerberghe, K. The Occupation Levels of Tell Tweini and their Historical Implications. In Proceedings of the 51st Rencontre Assyriologique Internationale, Held at the Oriental Institute of the University of Chicago, Chicago, IL, USA, 18–22 July 2005; Biggs, R.D., Myers, J., Roth, M.T., Eds.; Studies in Ancient Oriental Civilization. The Oriental Institute of the University of Chicago: Chicago, IL, USA, 2008; pp. 341–350.
104. Jung, R. Chronological problems of the Middle Bronze Age in Southern Italy. In *Le Bronze Moyen et l'origine du Bronze Final en Europe Occidentale (XVIIe–XIIIe Siècle av. J.C.)*; Lachenal, T., Mordant, C., Nicholas, T., Véber, C., Eds.; Mémoires d'Archéologie du Grand-Est; Association pour la Valorisation de l'Archéologie du Grand-Est: Strasbourg, France, 2017; pp. 623–642.
105. Belardelli, C. *Coppa Nevigata. Materiali da Scavi e Rinvenimenti 1903–1909*; Grandi contesti e problem della Protostoria Italiana 8; All 'Insegna del Giglio: Florence, Italy, 2004.
106. Cazzella, A.; Moscoloni, M.; Recchia, G. Coppa Nevigata (Manfredonia, Foggia. In *L 'età del bronzo recente in Italia. Atti del Congresso Nazionale di Lido di Camaiore, 26–29 Ottobre 2000*; Cocchi Genick, D., Ed.; M. Baroni: Viareggio, Italy; pp. 151–158.

quaternary

MDPI

Technical Note

^{87}Sr/^{86}Sr Isotope Ratio as a Tool in Archaeological Investigation: Limits and Risks

Mattia Rossi, Paola Iacumin * and Gianpiero Venturelli

Department of Chemical, Life and Environmental Sustainability Sciences, Earth Sciences Section, University of Parma, 43100 Parma, Italy; mattia.rossi@unipr.it (M.R.); giampiero.venturelli@gmail.com (G.V.)
* Correspondence: paola.iacumin@unipr.it

Abstract: During the last forty years, the use of strontium isotopes in archaeology and biogeochemical research has spread widely. These isotopes, alone or in combination with others, can contribute to trace past and present environmental conditions. However, the interpretation of the isotopic values of strontium is not always simple and requires good knowledge of geochemistry and geology. This short paper on the use of strontium isotopes is aimed at those who use this tool (archaeologists, but not only) but who do not have a thorough knowledge of mineralogy, geology, and geochemistry necessary for a good understanding of natural processes involving these isotopes. We report basic knowledge and suggestions for the correct use of these isotopes. The isotopic characteristics of bio-assimilable strontium depend not so much on the isotopic characteristics of the bulk rock as, rather, on those of its more soluble minerals. Before studying human, animal and plant remains, the state of conservation and any conditions of isotopic pollution should be carefully checked. Samples should be collected according to random sampling rules. The data should be treated by a statistical approach. To make comparisons between different areas, it should be borne in mind that the study of current soils can be misleading since the mineralogical modification of soil over time can be very rapid.

Keywords: ^{87}Sr/^{86}Sr; archaeology; environment; mineral dissolution; bioavailable Sr

Citation: Rossi, M.; Iacumin, P.; Venturelli, G. ^{87}Sr/^{86}Sr Isotope Ratio as a Tool in Archaeological Investigation: Limits and Risks. *Quaternary* **2024**, *7*, 6. https://doi.org/10.3390/quat7010006

Academic Editor: Matthew Peros

Received: 26 August 2023
Revised: 22 December 2023
Accepted: 5 January 2024
Published: 11 January 2024

1. Introduction

The use of strontium isotopes in archaeological and, more generally, in environmental research, dates back to the 1980s and 1990s of the last century (e.g., [1–3]). Bentley [4] made a very good introduction to and evaluation of the use these isotopes in archaeology. The reader is addressed to this paper for historical information. More recently, several authors critically discussed the application of strontium isotopes to archaelogy and the determination and mapping of bioavailable strontium isotopes (e.g., [5–9]), reporting observations which are, in part, summarised in this paper. Although the use of strontium isotopes is widespread, sometimes in connection with other isotopes, not all archaeologists and biologists can be expected to have a physical-chemical, mineralogical, and geological background to correctly manage strontium isotope data. From experience, we can state that good management of isotopic and chemical data in archaeology and environmental investigation needs basic knowledge not so much on the analytical technique, but rather on the physical meaning of the isotopes of the elements used. This is the reason why basic equations regarding radioactive decay, isotope mixing, and mineral dissolution are introduced in the text. Comprehension of these equations needs mathematical knowledge which constitutes a normal background of any university student.

This paper may be considered as a technical report (in the original significance of the greek τέχνη) where we express our point of view as geochemists and geologists; although, we know that, as stated by Pollard ([10], p. 634), "helpful and constructive critiques … outside the fraternity were not always welcomed". In a very simple and schematic way, we try to summarize theoretical and practical suggestions, in part separately presented in several papers (e.g., [5–10]), for scientists who have no experience in strontium isotope geochemistry but use it for practical purposes. In particular, hereafter, we want to summarise (i) the basic geochemical concepts necessary for the correct application of strontium isotopes, and (ii) the limits and the risks of using these isotopes.

In order not to weigh down the text and bore the reader, bibliographic citations in the text are few, but essential. In fact, we believe that in science, a redundancy of bibliographic citations frequently distracts the reader from the topic addressed. Chemical definitions are reported in Appendix A, whereas accessory calculations, which are necessary for a good understanding of the deep significance of the mathematical relations of the text, are reported in the Supplementary Material. Consequently, the text could be read at two different levels.

2. Strontium and Rubidium Isotopic Abundance, and Decay of ^{87}Rb

The term "isotopic abundance" is frequently used ambiguously in several papers. Thus, we remember that the isotopic abundance of the isotope ^{A}E of the element E is defined as the ratio $X_{^{A}E} = n_{^{A}E} / n_E$, where n is the number of nuclides ^{A}E and of the total atoms of the element E present in the system of interest (see Appendix A). According to the literature, the approximate isotopic abundances of strontium [11] and rubidium [12] in most terrestrial materials are the following: $X_{^{84}Sr} \approx 0.55–0.58\%$, $X_{^{86}Sr} \approx 9.75–9.99\%$, $X_{^{87}Sr} \approx 6.94–7.14\%$, $X_{^{88}Sr} \approx 82.29–82.75\%$; $X_{^{85}Rb} \approx 72.17\%$, $X_{^{87}Rb} \approx 27.83\%$ ($X_{^{85}Rb}/X_{^{87}Rb} = 2.596$).

The strontium geochemical cycle is very different in respect, for instance, to hydrogen, oxygen, carbon, and nitrogen which scientists frequently use for their research. The distribution of hydrogen, oxygen, and carbon isotopes depends mostly on the cycle of these elements in the hydrosphere and atmosphere, whereas the distribution of strontium isotopes depends mostly on the solid earth and its complex surface rock distribution. Therefore, the correct application of the isotopes of strontium needs basic knowledge of mineralogy, geology, and geochemistry.

At the low temperatures of the Earth's surface environment, isotopes of hydrogen, oxygen, carbon, and nitrogen undergo fractionation. On the contrary, at any temperature, the potential fractionation of strontium isotopes, which have high numbers of atomic mass, is very low, if present, and not detected by analytical present-day technology. In other words, whereas the isotope ratios $n_{^{2}H}/n_{^{1}H}$, $n_{^{18}O}/n_{^{16}O}$, $n_{^{13}C}/n_{^{12}C}$, and $n_{^{15}H}/n_{^{14}N}$ in different substances, coexisting at a given temperature, are generally different, the isotope ratio $n_{^{87}Sr}/n_{^{86}Sr}$ is the same.

Natural change in the $n_{^{87}Sr}/n_{^{86}Sr}$ isotope ratio is due to the unstable atomic nucleus of ^{87}Rb, which decays to ^{87}Sr with emission of a β^- particle:

$$^{87}Rb \rightarrow {}^{87}Sr + \beta^- + \overline{\nu} + Q \quad (1)$$

where $\overline{\nu}$ is an anti-neutrino and Q is the decay energy (for physical-chemical definition see IUPAC—International Union of Pure and Applied Chemistry. Radioactive decay is a spontaneous nuclear transformation. The speed of this transformation at a generic time t is proportional to the amount of ^{87}Rb which is present in the system at that time. In the mathematical language, for a system which does not exchange elements with the environment, the instantaneous "speed of decay", $v_{decay(t)}$, at the time t is usually defined by the following relation,

$$v_{decay(t)} = -\frac{dn_{^{87}Rb}}{dt} = \frac{dn_{^{87}Sr}}{dt} = \lambda\, n_{^{87}Rb(t)} \quad (2)$$

$$\lambda = (1.3972 \pm 0.0045) \times 10^{-11} \times a^{-1}$$

where $dn_{^{87}Rb}$ and $dn_{^{87}Sr}$ indicate very small variations of $n_{^{87}Rb}$ and $n_{^{87}Sr}$ in a very small time interval dt (approximate mathematical definition), λ is a constant called "decay constant" (the probability that an atom of ^{87}Rb has to decay in one year), and the letter *a* indicates the years. In agreement with relation (1), the minus sign in front of $dn_{^{87}Rb}$ is present because, during decay, $n_{^{87}Rb}$ decreases with time, whereas $n_{^{87}Sr}$ increases.

From Equation (2), after simple mathematical manipulation (see Supplementary Material S1), the following general relations are obtained,

$$n_{^{87}Rb(t)} = n_{^{87}Rb(t_o)} e^{-\lambda(t-t_o)} \tag{3}$$

and

$$n_{^{87}Rb(t_o)} = n_{^{87}Rb(t)}\, e^{\lambda(t-t_o)} \tag{4}$$

where t_o and t are the time of the beginning of decay computation and the present time, respectively, and e is the Neper's number (e = 2.71828182 ...). Subtracting $n_{^{87}Rb(t)}$ from (4) and remembering that the loss of ^{87}Rb is equal to the gain of ^{87}Sr , we also write:

$$n_{^{87}Sr(t)} - n_{^{87}Sr(t_o)} = n_{^{87}Rb(t_o)} - n_{^{87}Rb(t)} = n_{^{87}Rb(t)}\left(e^{\lambda(t-t_o)} - 1\right) \tag{5}$$

where the difference $n_{^{87}Sr(t)} - n_{^{87}Sr(t_o)}$ is the number of ^{87}Sr nuclides generated by decay in the time interval $t - t_o$.

Consider now that in a system containing rubidium, only the numbers of ^{87}Sr and ^{87}Rb atoms change during the time. Thus, since for a closed system (no matter is changed with the environment) the number of the ^{86}Sr atoms is a constant value, independent on the time, dividing (5) by the constant $n_{^{86}Sr}$ value, we obtain:

$$\frac{n_{^{87}Sr(t)}}{n_{^{86}Sr}} - \frac{n_{^{87}Sr(t_o)}}{n_{^{86}Sr}} = \frac{n_{^{87}Rb(t)}}{n_{^{86}Sr}}\left(e^{\lambda(t-t_o)} - 1\right) \tag{6}$$

Relation (6) is usually written in terms of isotopic abundances as follows:

$$\left(\frac{X_{^{87}Sr}}{X_{^{86}Sr}}\right)_t - \left(\frac{X_{^{87}Sr}}{X_{^{86}Sr}}\right)_{t_o} = \left(\frac{X_{^{87}Rb}}{X_{^{86}Sr}}\right)_t \left(e^{\lambda(t-t_o)} - 1\right)$$

or, more simply,

$$\left(\frac{^{87}Sr}{^{86}Sr}\right)_t - \left(\frac{^{87}Sr}{^{86}Sr}\right)_{t_o} = \left(\frac{^{87}Rb}{^{86}Sr}\right)_t \left(e^{\lambda(t-t_o)} - 1\right)$$

3. The Strontium and Rubidium in Minerals

3.1. General

In the common minerals, strontium and rubidium are mostly enveloped ("coordinated") by oxygen, and their bond with oxygen is prevalently ionic (electrostatic forces). Thus, strontium and rubidium may be regarded as Sr^{2+} and Rb^+ ions that are linked to oxygen, O^{2-}. The oxygen atoms are distributed around Sr^{2+} and Rb^+ and placed at the apex of regular or distorted polyhedrons which define, in this way, the shape of Sr^{2+} and Rb^+ lattice sites. Sr^{2+} replaces calcium, Ca^{2+}, and, in minor amount, Na^+ and K^+ in several minerals. For instance, this substitution occurs in common calcium sulphates (gypsum, $CaSO_4 \cdot 2H_2O$, and anhydrite, $CaSO_4$), carbonates (e.g., calcite and aragonite, $CaCO_3$, and dolomite, $CaMg(CO_3)_2$), plagioclase ($Na_{1-x}Ca_xAl_{1+x}Si_{3-x}O_8$), K-feldspar ($KAlSi_3O_8$), and apatite (simplified formula: $Ca_3(PO_4)_3(F, Cl, OH)$), a mineral that in the form of carbonate hydroxyl apatite, is the inorganic component of bones. It is noteworthy that the ionic radii of Sr^{2+} may change a little bit in relation to the number of the enveloping oxygen atoms. For instance, in aragonite, Ca^{2+} is present in ninefold coordination, allowing aragonite to accept easily larger cations such as Sr^{2+}; in this coordination, Ca^{2+} has an ionic radium of about 1.18 (1 =1 ångström = 10^{-10} m) [13]. On the contrary, in calcite, Ca^{2+} and Sr^{2+} exist in sixfold coordination with an ionic radium of about 1.00 and 1.18 , respectively.

Rubidium Rb^+ mostly replaces K^+ in trioctahedral micas, including illite. The number of effective coordination ranges from eleven to seven [14], the ionic radii for K^+ and Rb^+ being about 1.56 to 1.70 and 1.46 to 1.58 , respectively. In K-feldspar, K^+ and Rb^+ are sevenfold coordinated [15] with ionic radii of about 1.46 and 1.56 .

To conclude, the crystal lattice features reported above are relevant for strontium and rubidium distribution in different minerals. In particular, the ionic radium is very important for element substitution; elements with similar ionic radii may substitute each other in the crystal lattice.

The order of magnitude of strontium and rubidium concentration in some common rock-forming minerals is reported in Table 1.

Table 1. Order of magnitude of Sr and Rb concentration in some main rock-forming minerals.

Minerals	Chemical Formula	Sr (ppm wt)	Rb (ppm wt)
Gypsum	$CaSO_4 \cdot 2\,H_2O$	>1000	≈0
Anhidrite	$CaSO_4$	>1000	≈0
Plagioclase	$Na_{1-x}Ca_xAl_{1+x}Si_{3-x}O_8$	200–1000 (1)	5–40 (1)
K-felspar	$KAlSi_3O_8$	50–800 (1)	200–800 (1)
Calcite	$CaCO_3$	100–700	≈0
Aragonite	$CaCO_3$	100–1000	≈0
Dolomite	$CaMg(CO_3)_2$	100–500	≈0
Phyllosilicates (*)	Largely variable	<100 (1)	100–2000 (1)

ppm wt = part per million in weight; our evaluation from the abundant literature; (1) [16]; (*) clay minerals (in particular, montmorillonite) may adsorb Sr^{2+} on the surface of their crystals; thus, the concentration of Sr^{2+} may be higher than about 100 ppm wt.

3.2. Strontium Isotopes in Minerals and Whole Rock

In this paper, the term rock is also used for soil which, from a mineralogical and petrographic point of view, may be mostly regarded as unconsolidated rock commonly containing organic material. Several papers, which use strontium isotopes for archaeological reconstruction, sometimes address their attention to the isotopic composition of the whole rock or soil. However, it is noteworthy that bulk rock/soil is not important for the acquisition of isotopes by plants and animals. Rather, the different minerals and their solubility in aqueous solutions are relevant. Hereafter, we approach this topic.

3.3. Minerals as Constituents of the Rock

We may observe a rock from two different points of view: (i) mineralogical composition and (ii) physical state (consolidated and unconsolidated).

3.3.1. Monomineralic Consolidated/Unconsolidated Rock

The rock is formed only by a mineral (e.g., limestone and dolomitic rock consisting of the minerals calcite or aragonite, and the mineral dolomite, respectively) generated at a defined time in an isotopically homogeneous environment (for instance, limestone precipitated from marine water at a defined geological time). In case after deposition the rocks behave as a closed system, the rock is expected to maintain the initial macroscopic isotope homogeneity during the time. To summarise, the isotopic ratio may change over time, but at a given time, it will be the same in all the portions of the rock.

3.3.2. Polymineralic Consolidated/Unconsolidated Rock

(i) The rock is mineralogically heterogeneous containing several minerals, which, at the time of their generation, had the same isotope ratio. This approximately occurs, for instance, during crystallization of a magmatic silicate magma on the surface or in the interior of the Earth (note that this is strictly true only in case the different minerals become closed systems at the same temperature; see [17]). In this case, the variation in the isotope ratio of the single mineral from the time of its formation to the time of the measurements depends largely on the original ^{87}Rb content of the single mineral. Summarising: (a) at the time of the rock generation, all the minerals had the same isotope ratio which, however, (b) during the time, change their value with different speed because of their different ^{87}Rb original content. The rock initially was mineralogically heterogeneous and homogeneous for the isotope ratio, whereas at the time of interest, it was both mineralogically and isotopically heterogeneous.

(ii) At the time of its formation, the rock contains several minerals which already have different isotope ratios. This is typical for clastic rocks, such as soil, moraines and terrigenous sediments, or clastic consolidated rocks (e.g., silt, sandstones, conglomerates) which derive by disruption of older rocks. In this case, the variation in the strontium isotope ratio in the different minerals will be due both (a) to the original isotope values of the different rock-forming minerals and, obviously, (b) to the time which elapsed from the rock generation to the present. Summarising: the rocks, are mineralogically and isotopically heterogeneous both at the time of their formation and at the time of interest.

3.4. *Variation in Strontium Isotopes in Different Minerals (i) and Strontium Isotopes in the Total Rock (Tot)*

3.4.1. Different Decay Speed of ^{87}Rb in the Different Minerals

As reported above, usually, the different rock-forming minerals have different initial ^{87}Rb content, and thus, according to relations (2) and (5), the speed of ^{87}Sr generation is different for different minerals. Two minerals, for instance, calcite (simplified formula: $CaCO_3$) and muscovite (simplified formula: $KAl_2[AlSi_3O_{10}]\ (OH)_2$) both have the same initial $n_{^{87}Sr(t_o)}/n_{^{86}Sr}$ ratio (= 0.7060) but different rubidium content, as reported in Table 2.

Table 2. Data for an *exemplum fictum* of calculation of strontium isotope ratio for calcite (Cc) and muscovite (Mu).

	W_i	C^i_{Sr}	$n_{^{87}Rb(t)}/n_{^{86}Sr}$	$n_{^{87}Sr(t_o)}/n_{^{86}Sr}$	$n_{^{87}Sr(t)}/n_{^{86}Sr}$
Calcite (Cc)	0.30	0.0800	0.0952	0.7060	*0.7061*
Muscovite (Mu)	0.70	0.0070	39.11	0.7060	*0.7333*

$W_i = Q_i / (Q_{Cc} + Q_{Mu}) = Q_i/Q_{tot}$, weight fraction of the rock-forming mineral Cc or Mu generically indicated as i, where Q_i is the mass of the single mineral and Q_{tot} the mass of the total rock, consisting of the minerals Cc and Mu. $C^i_{Sr} = Q^i_{Sr}/Q_i$, weight concentration of strontium in Cc or Mu, where Q^i_{Sr} is the mass of strontium in the single mineral and Q_i is the mass of the single mineral Cc or Mu. (t), refers to data at the time t of interest; (t_o), refers to data at the initial time. In *italics*, the isotope ratios obtained after $t - t_o$ = 50 Ma (million years). Concentration of strontium in the total rock $C^{tot}_{Sr} = W_{Cc} \times C^{Cc}_{Sr} + W_{Mu} \times C^{Mu}_{Sr} = 0.30 \times 0.0800 + 0.70 \times 0.0070 = 0.0289$ = 289 ppm wt (ppm = parts per million).

Using relation (6), after $t - t_o$ = 50 million years from their generation, calcite will assume $n_{^{87}Sr(t)}/n_{^{86}Sr}$ equal to 0.7061 and muscovite equal to 0.7333. The variation for calcite, with a small rubidium amount, is so low because the decay speed of ^{87}Rb is very low for this mineral (see relation (3)), whereas for muscovite, the amount of rubidium is high, and thus, the speed of decay is high.

3.4.2. Relation of the Isotope Data for Minerals and for Whole Rock

The minerals 1, 2, ..., φ indicated generically by the letter i, form the total rock indicated as tot. If the isotopic features of the different minerals are known, for the different minerals and total rock, we may write the following approximate isotope balance relation (see Supplementary Material S2):

$$\frac{n^{tot}_{^{87}Sr}}{n^{tot}_{^{86}Sr}} \cong \sum_{i=1}^{\varphi} \left(\frac{C^{i}_{Sr}}{C^{tot}_{Sr}} W_i \frac{n^{i}_{^{87}Sr}}{n^{i}_{^{86}Sr}} \right) \quad (7)$$

Relation (4) is a particular expression of a mixing equation which gives a "weighted mean" of the isotope ratio $n_{^{87}Sr}/n_{^{86}Sr}$ for the total rock. The $\frac{C^{i}_{Sr}}{C^{tot}_{Sr}} W_i$ values are mathematical weights of the different addends, i.e., values which estimate the contribution of the $n_{^{87}Sr}/n_{^{86}Sr}$ ratio of the different minerals to the isotopic values of the total rock. C^{i}_{Sr} is the weight concentration of strontium in the generic mineral, i, C^{tot}_{Sr} in the total rock, tot, and W_i the weight fraction of the mineral i in the rock (for symbols, see Table 2). As an example, consider a rock formed by calcite and muscovite with the isotopic features reported in Table 2. We write:

$$\frac{n^{tot}_{^{87}Sr}}{n^{tot}_{^{86}Sr}} \cong \frac{C^{Cc}_{Sr}}{C^{tot}_{Sr}} W_{Cc} \frac{n^{Cc}_{^{87}Sr}}{n^{Cc}_{^{86}Sr}} + \frac{C^{Mu}_{Sr}}{C^{tot}_{Sr}} W_{Mu} \frac{n^{Mu}_{^{87}Sr}}{n^{Mu}_{^{86}Sr}}$$
$$= \frac{0.08}{0.0289} \times 0.30 \times 0.7061 + \frac{0.007}{0.0289} \times 0.70 \times 0.7333 = 0.7107$$

This value is far both from those of calcite and muscovite.

3.5. *Selective Mineral Dissolution and Its Important Role on the Strontium Isotopes Values in the Water Solution*

Frequently, people retain that the bulk rock isotopic feature is relevant for the isotopic characters of the bioavailable strontium. This, however, is not correct. Hereafter, we show that the isotopic features of the single minerals are important.

It is noteworthy that the speed of mineral dissolution (kinetics of dissolution) is different for the different minerals and depends on several factors (e.g., temperature and pressure, size and morphology of the grains, pH of the aqueous solution, saturation condition of the solution in the mineral phase considered, etc.). For instance, we consider the kinetics of calcite and muscovite dissolution in pure water in conditions far from the solution saturation in these minerals. According to the literature (e.g., [18,19], and references therein), at about 25 °C and for pH in the range 5 to 9.5, the dissolution rate of calcite is in the order of magnitude of 10^{-6} mole m^{-2} s^{-1}, i.e., one meter square of the mineral releases 10^{-6} mole of calcite in a second. In turn, for muscovite at 25 °C, the value is about 10^{-11} to 10^{-12} mole m^{-2} s^{-1} ([20], and reference therein). Thus, the rate of dissolution of calcite is at least one hundred thousand times the rate of muscovite! Even if muscovite has very high $n_{^{87}Sr}/n_{^{86}Sr}$ values in respect to calcite, the isotopic value of the bioavailable strontium (strontium dissolved in water of the soil) will be very close to the value for calcite. An example of this calculation is reported in Supplementary Material S3, where we demonstrated that in a system consisting of calcite and of muscovite with a high $n_{^{87}Sr}/n_{^{87}Sr}$ ratio, the isotope ratio of a water solution which dissolved calcite and muscovite will have an isotope ratio very close to that of calcite, not to that of the whole rock.

Calcite, aragonite, and Ca-sulphates are the most common Sr-rich rock-forming minerals which exhibit the highest solubility. Thus, these minerals, when present, are expected to mostly affect the isotope ratio of the circulating waters, as demonstrated above for calcite. This is well documented in several places. For instance, in the "Prosecco wine" area in Northern Italy [21], the main Sr-bearing minerals present in the soil are Ca-carbonate, dolomite, phyllosilicates, and plagioclase, where some phyllosilicates may assume strontium via surface sorption. We considered four places investigated by Aviani [21] and Petrini et al. [22]: Nardin-Lison, Sant'Anna, Lonigo, Pittarello. For all these localities, strontium for the isotopic determinations was extracted using both CH_3 $COONH_4$ (ammonium acetate) 1 M (pH ≈ 7) and HCl 2.5 N (very acidic solution). The reason for using solutions with different pH values is to consider the different conditions that may be present in the soil in natural conditions: from neutral to acidic (the second is the case when there is a lot of organic substance that decomposes). In practice, in the laboratory, two extreme and opposite conditions of dissolution were considered. For instance, at pH $\approx$ 7, Ca-carbonates and Ca-sulphates are dissolved a moderate to low amount, whereas at a very low pH, carbonates are strongly dissolved together with Al- and Fe-hydroxides. On the contrary, phyllosilicates are always slightly dissolved. In the "Prosecco wine" area, statistically, it cannot be excluded that the average values obtained for solutions at pH $\approx$ 7 and at very low pH values for the same soil sample are the same ($p_{\text{same mean}} = 0.20$, $p_{\text{same median}} = 0.23$). Since in the area there is no evidence of Ca-sulphate occurrence, and the phyllosilicates are, in both cases, poorly dissolved, the data obtained suggest the dominant role of carbonates. It is noteworthy, however, that different extraction methods in general do not give the same values of isotope ratio (see Section 4.2).

Moreover, the analysis of the whole soil after total dissolution gives largely higher strontium isotope values (this is not a realistic condition since, in nature, the dissolution of all the minerals that make up a rock does not occur). This demonstrates that the phyllosilicates present in the soil are only slightly dissolved when treated with neutral (acetate) or acid (HCl) solutions and represent residuals of old rocks with a high strontium isotope ratio. Thus, the total rock isotope data cannot be used for the evaluation of strontium isotopic features of available strontium.

To conclude, we recall again what we stated above: the isotopic features of the whole-rock are not significant in determining the isotope ratio in plants and animals; rather, the single phases present in the rock are relevant.

3.6. Sorption/Desorption and Minerals

Clay minerals, such as smectite, illite, and kaolinite, as well as organic matter may sorb strontium from circulating water solutions because of the cation exchange capacity of these substances. This process is well known in the scientific literature (see, for instance, [23–25]). Generally, strontium sorption increases as strontium concentration and total salinity of the solution decrease. The $n^{\text{sorbed}}_{^{87}\text{Sr}} / n^{\text{sorbed}}_{^{86}\text{Sr}}$ value of the sorbed strontium is determined by (i) strontium present in the most soluble minerals of the rock and/or (ii) strontium carried by water coming from the environment. The last origin of strontium may be dominant when the minerals present in the rocks are only slightly soluble, thus transferring into the solution only a very small amount of strontium. This, for instance, may sometimes occur when the rocks consist of common rock-forming silicates.

4. Strontium Isoscapes and Their Use in Archaeology

4.1. General

Geochemical prospecting includes any method of mineral exploration based on systematic measurement of one or more chemical properties of a naturally occurring material ([26], and reference therein). It is a very old method to recognize possible geochemical anomalies in a territory and, thus, to identify areas with mineral deposits. Conceptually, isoscape, a term recently used by West et al. [27], is no more than a graphical representation, obtained with the aid of a geographic information system (GIS), of the results of isotope geochemical prospecting.

The use of strontium isoscapes in archaeological and biogeochemical sciences needs great caution for the following reason:

(i) Isoscapes generally refer to large-scale grid sampling with cells in the order of tens or hundreds of km^2, and in nature, variation in the strontium isotope ratio does not necessarily merge continuously from one value to another, but it may be sharp, even between neighboring sites when they are located on the boundary between different geological formations. A good example of this condition is reported by Montgomery et al. [5]. Moreover, frequently, sampling for isoscapes is not randomly distributed in the area of interest. For instance, in Italy, covering an area of

302.073 km^2, Lugli et al. [28] used 1920 data of the $^{87}Sr/^{86}Sr$ ratio: on average, one sample for 157 km^2; in France, Willmes et al. [29] used data referring to 840 sites for an area of 551.626 km^2: one sampling grid for 657 km^2! The criteria for sampling, including the sampled materials, is another important point. In the isoscape of Italy, Lugli et al. [28] include data related to 'plant', 'water', 'biomineral' (i.e., bones, teeth, and bio-calcareous shells), 'food', 'soil' (including both exchangeable soil fractions and bulk soils) and 'rock' (mainly evaporites, metamorphic and magmatic rocks, and a few sedimentary bulk rocks). Thus, at best, the most common isoscapes can only give generic indications for wide areas.

(ii) As a general rule, the sampling grid used to perform isoscapes and local investigations should be the same. This, of course, is practically impossible to obtain because isoscapes are usually made at large scale, as stated above. Thus, for local investigation, scientists should use isoscapes with great caution and integrate them with values obtained from more detailed random sampling.

4.2. *Archaeological Investigation and Present-Day Environmental Condition*

For a detailed discussion about problem of material selection for the strontium isotope measurements of archaeological and present-day environmental samples, we address the reader to Holt et al. [7]. In addition to the observations of Holt et al. [7], we remember that the determination of bioavailable strontium in soil in some amount depends on the method used for strontium extraction. For instance, for the same soil sample, we obtained a value of 0.707743 using extraction with water at pH $\cong$ 5.7 (pure water initially in equilibrium with CO_2 of the atmosphere, $P_{CO_2} \cong 10^{-3.5}$ bar) and a value of 0.708139 using extraction with CH_3COONH_4 (ammonium acetate) 1 M. This discrepancy may generate a bias in the data distribution.

In archaeological research, the first important question is: does soil always maintain the isotopic features it had in the past in an area of interest? We need a correct answer to this question when we want to use a present-day isoscape as an indicator of past conditions. In addition to climate modification, geomorphological, and mineralogical changes due to erosion of the soil and/or deposition of allogenic material by water flows, wind, anthropic land management, and chemical processes may generate important geochemical variations. For instance, soil rich in organic material may produce a large increase in CO_2 partial pressure in water solutions migrating in the soil, which increases the solubility of calcite. If the leaching proceeds for a sufficiently long time, calcite can disappear completely; in this case, this mineral no longer contributes to the bioavailable strontium. Moreover, in soil free of soluble minerals, the most important role may be played by strontium-bearing water solutions coming from the environment (precipitation or other external sources). Examples of strontium isotope variation due to environmental water and/or to mineralogical variations in the soil over time, are reported by Åberg [3].

5. Investigating and Planning Strontium Isotope Research

We remind the reader that any investigation must be performed keeping in mind what we want to know from strontium isotopes: in particular, archaeologists need data referring to the past, not to the present; biologists possibly need data referring to the past and/or to the present. This implies different type of sampling and strontium archives.

(i) Sampling should be random to avoid bias of the data obtained on the statistical population sample. Homogenized sampling, where several samples are collected in a defined small area and then reduced to only one homogenized sample for analysis [7], in our opinion, is not a good method, because in this way, the variance in the data population for the area of interest is reduced. This could make definition of allogenous samples and comparison with other areas impossible.

(ii) What does "same strontium isotope ratio" mean? Modern technology furnishes strontium isotope data with analytical uncertainty on the fifth or even on the sixth decimal digit, whereas, also for small areas (up to one km^2 as an order of magnitude), isotope data, at best, may exhibit variation in the fourth decimal digit (see, for instance, [24]). If that is so, two samples could be considered as having approximately the same value if they do not differ on the fourth digit. Thus, in principle, the identification of exotic samples not belonging to the population of data related to materials coming from a defined area should consider the variability in isotope data for the area of interest.

(iii) Are the sampled biological remains from the same site? The answer depends on (a) the spatial definition of "same site" and (b) how we define the belonging of biological remains to the area of interest.

  (a) Spatial definition depends, of course, on the aim of the investigation. In other words, we go back to the investigation scale. For instance, Cavazzuti et al. [30], in studying human settlements located in the Po plain (Northern Italy), assume that the settlements, although far, occur in a very "homogeneous" (it is not clear what they mean: isotopically homogeneous, mineralogically homogeneous, or both?) flat area without geographical barriers. They define three different areas around each settlement with a radius of 5 km ("site catchment area"), from 5 to 20 km ("immediate hinterland"), and from 20 to 50 km ("broader hinterland") and they compare the strontium isotope values to the background of these three areal categories.

  (b) The belonging of human or, in general, animal remains to the area of interest depends on how we operationally define this belonging. Operatively, the minimum time of belonging may be evaluated trough the mean residence time of calcium or strontium in bones. The mean residence time, however, depends on the bone type and on the age of the individuals. For instance, the turnover for femur is about 25–30 years, whereas for ribs is about 5–10 years. Therefore, using the ribs of an individual from another area, the individual will be found to belong to the area of interest approximately 5–10 years after its arrival. Instead, using the femur, it will appear to belong to that area after 25–30 years. Thus, in a defined area, using contemporaneously data from femur and rib, the variability in the strontium data may increase. This is not sufficiently considered in the scientific papers.

(iv) Before sampling, the geology and mineralogy of the area should be carefully considered to give an idea of the dominant mineral sources.

  (a) An accurate geological analysis suggests that, reasonably, the soil of the area of interest did not change its mineralogical and geochemical characteristics from the time of the settlement to the present. Under this condition, the isotopic prospecting of the available strontium of the present-day soil and plants is the most elementary way for determining the isotope reference background which the biological remains of the area of interest may be compared to.

  (b) There is evidence or suspect that the mineralogy and geochemistry of the area are not preserved; obviously, present day material cannot be used to define the geochemical background of the area. In this case, different biological remains (teeth, bones, shells, seeds, et cetera) may be used. In the event of an area of interest having no archaeological evidence (different types of burial, funerary objects, behaviour of the different animal species, etc.), suggesting a different provenance of human or animal remains, all the last ones must be considered as *potentially* belonging to the settlement of interest. It is evident that in this way, the variability in the isotope data could be significantly expanded, and some external individuals could be attributed to the settlement.

(v) If possible, the variability in the data should be defined using a high number of analyses obtained from different individuals (indicatively, more than 15; the use of too few samples may be misleading). For each area, in case the data have normal distribution, data far from the prevalent distribution values could be identified with some statistical method. For instance:

$$\text{background} = \text{average of the data, } \frac{\sum x}{n} = \overline{x}$$

where $\sum x$ = sum of the data x obtained on random samples from the area of interest, and n = number of data.

$$s = \text{experimental standard deviation of the data, } \sqrt{\frac{(x - \overline{x})^2}{n - 1}},$$

here, assumed very similar to the theoretical standard deviation, σ, is the entire population of data.

Threshold = background + k × standard deviation, $\overline{x}$ + k s, where k is the "coverage factor". Frequently, for practical problems, k = 3 is assumed. Using k = 3, there is only a very small probability (less than about 1%) that a sample will be mistakenly excluded from the group to which it belongs. Theoretically, in place of k, an appropriate value of the "Student's t" should be used. A qualitative indication of the probability that the single value is outside the investigated population, is given by the so-called "contrast":

$$\text{Contrast} = \frac{x}{\text{threshold}}$$

In case the distribution of data is not normal, non-parametric statistics may be used, for instance, the "kernel density estimation" (KDE) (see [31], statistical software PAST 4.15). Supplementary Material S4 report an exemplum fictum of calculation.

(vi) In general, we can only establish if the analyzed individuals *may* belong to the same group, not that they do belong to the same group. In fact, samples settled on different areas with similar geological formations exhibit the same isotopic values, even if the areas are far from one another. This happens, for example, if the individuals come from areas located on carbonate formations of a very similar geological age and with a similar genetic and diagenetic history. This is an important limit for the use of strontium isotopes alone. For example, in the Illasi valley, Lessini mountains, NE of Verona (Italy), plants grown on hydrothermalised carbonate formations from the Late Carnian to Liassic ages have very similar isotopic values (about 0.7083 ± 0.0003, our unpublished data), even if they are located many kilometers away from one another. On the contrary, samples coming from the same hypothetical locality straddling Cretaceous and Late Carnian–Liassic formations exhibit significant isotopic differences already on the fourth digit (Cretaceous carbonate, 0.7077 against Late Carnian–Liassic carbonates, 0.7083) (see also [5]).

6. Summary

We tried to explain how the use of strontium isotopes may be sometimes problematic in archaeological research. These are the main points to be used as a vade mecum when approaching strontium isotope study in archaeology:

(i) We cannot assume that the current geochemical, mineralogical, and geological conditions of the investigated area are the same as in the past because variation in the surface conditions is frequent also during a short time.

(ii) The use of large-scale isoscapes is risky because local investigation is usually performed on a smaller scale.

(iii) Before studying human, animal, and plant remains, an accurate control of their diagenetic condition is essential because pollution of the samples by environmental strontium-bearing material with different isotope ratios is very easy (for instance, diagenesis with dissolution/deposition of carbonate).

(iv) The samples (soil, human/animal remains, plants, etc.) should be selected randomly. Usually, this is not considered in the literature.

(v) To reach a reliable scientific conclusion, the investigation of a large number of remains and related measurements is necessary. Without a large number of data, comparison between different areas is risky (statistically insignificant).

(vi) If samples fall outside the prevalent distribution interval, we can state that they do not belong to the same group. The individuals falling in the prevailing distribution interval do not necessarily belong to the same group; we can only state that it is not excluded they belong to the same group.

Supplementary Materials: The following supporting information can be downloaded at: https://www.mdpi.com/article/10.3390/quat7010006/s1.

Author Contributions: Conceptualization, all authors contributed equally; investigation, M.R.; data curation, all authors contributed equally; writing—original draft preparation, M.R.; writing—review and editing, P.I. and G.V. All authors have read and agreed to the published version of the manuscript.

Funding: This research received no external funding.

Data Availability Statement: The data presented in this study are available in article and Supplementary Material here.

Acknowledgments: We are grateful to the referees for their comments and suggestions.

Conflicts of Interest: The authors declare no conflicts of interest.

Appendix A

Nuclide: generic nuclear species with number of protons Z and neutrons N.

Element, ($_{Z}\mathbf{E}$): atomic species defined by the number of protons; for example, $_{38}\mathbf{Sr}$, generically indicated as **Sr**.

Isotope ($^{A}_{Z}\mathbf{E}_{N}$): atom of the same element E with defined N value (A = Z + N, Atomic Mass Number); for example, $^{86}_{38}\mathbf{Sr}_{48}$ for simplicity indicated as $^{86}\mathbf{Sr}$.

Isotopic abundance: $\mathbf{X}_{\mathbf{A}_{\mathbf{E}}} = \mathbf{n}_{\mathbf{A}_{\mathbf{E}}}/\mathbf{n}_{\mathbf{E}}$, where $\mathbf{n}_{\mathbf{A}_{\mathbf{E}}}$ and $\mathbf{n}_{\mathbf{E}}$ are the number of atoms of the isotope $^{\mathbf{A}}\mathbf{E}$ of the element E and the total atoms of the element E in the material of interest, respectively; for example, $\mathbf{n}_{^{86}\mathbf{Sr}}$ and $\mathbf{n}_{\mathbf{Sr}}$.

References

1. Ericson, J.E. Strontium isotope characterization in the study of prehistoric human ecology. *J. Hum. Evol.* **1985**, *14*, 503–514. [CrossRef]
2. Price, T.D.; Grupe, G.; Schröter, P. Reconstruction of migration patterns in the Bell Beaker period by stable strontium isotope analysis. *Appl. Geochem.* **1994**, *9*, 413–417. [CrossRef]
3. Åberg, G. The use of natural strontium isotopes as tracers in environmental studies. *Water Air Soil Pollut.* **1995**, *79*, 309–322. [CrossRef]
4. Bentley, R.A. Strontium isotopes from the Earth to the archaeological skeleton: A review. *J. Archaeol. Methods Theory* **2006**, *13*, 135–187. [CrossRef]
5. Montgomery, J. Passports from the past: Investigating human dispersals using strontium isotope analysis of tooth enamel. *Ann. Hum. Biol.* **2010**, *37*, 325–346. [CrossRef] [PubMed]
6. Slovak, N.M.; Paytan, A. Applications of Sr isotopes in archaeology. In *Handbook of Environmental Isotope Geochemistry*; Advances in Isotope Geochemistry; Baskaran, M., Ed.; Springer: Berlin/Heidelberg, Germany, 2011; pp. 743–768.
7. Holt, H.; Evans, J.A.; Madgwick, R. Strontium (^{87}Sr/^{86}Sr) mapping: A critical review of methods. *Earth-Sci. Rev.* **2021**, *216*, 103593. [CrossRef]
8. Lengfelder, F.; Grupe, G.; Stallauer, A.; Huth, R.; Söllner, F. Modelling strontium isotopes in past biospheres—Assessment of bioavailable ^{87}Sr/^{86}Sr ratios in local archaeological vertebrates based on environmental signatures. *Sci. Total Environ.* **2019**, *648*, 236–252. [CrossRef]
9. Nafplioti, A. Moving forward: Strontium isotope mobility research in the Aegean. *Medit. Archaeol. Archeom.* **2021**, *21*, 165–179.
10. Pollard, A. Isotopes and impact: A cautionary tale. *Antiquity* **2011**, *85*, 631–638. [CrossRef]
11. Meija, J.; Coplen, T.B.; Berglund, M.; Brand, W.A.; De Bièvre, P.; Gröning, M.; Holden, N.E.; Irrgeher, J.; Loss, R.D.; Walczyk, T.; et al. Isotopic compositions of the elements 2013 (IUPAC Technical Report). *Pure Appl. Chem.* **2016**, *88*, 293–306. [CrossRef]
12. Catanzaro, E.J.; Murphy, T.J.; Garner, E.L.; Shields, W.R. Absolute isotopic abundance ratio and atomic weight of terrestrial rubidium. *J. Res. Natl. Bur. Stand. A Phys. Chem.* **1969**, *73A*, 511–516. [CrossRef] [PubMed]

13. Shannon, R.D. Revised effective ionic radii and systematic studies of interatomic distances in halides and chalcogenides. *Acta Cryst.* **1976**, *A32*, 751–767. [CrossRef]
14. Cibin, G.; Mottana, A.; Marcelli, A.; Brigatti, M.F. Potassium coordination in trioctahedral micas investigated by K-edge XANES spectroscopy. *Mineral. Petrol.* **2005**, *85*, 67–87. [CrossRef]
15. Downs, R.T.; Andalman, A.; Hudacsko, M. The coordination numbers of Na and K atoms in low albite and microcline as determined from a procrystal electron-density distribution. *Am. Min.* **1996**, *81*, 1344–1349. [CrossRef]
16. Wedephohl, K.H. *Handbook of Geochemistry*; Springer: Berlin/Heidelberg, Germany, 1974; Volume 2.
17. Braun, J.; van der Beek, P.; Batt, G. *Quantitative Thermochronology: Numerical Methods for the Interpretation of Thermochronological Data*; Cambridge University Press: Cambridge, UK, 2006; ISBN 978-0-521-83057-7.
18. Cabillas, P.; Köhler, S.; Prieto, M.; Chaïrat, C.; Oelkers, E.H. Experimental determination of the dissolution rates of calcite, aragonite, and bivalves. *Chem. Geol.* **2005**, *216*, 59–77. [CrossRef]
19. Fan, X.; Batchelor-McAuley, C.; Yang, M.; Compton, R.G. Single calcite particle dissolution kinetics: Revealing the influence of mass transport. *ACS Meas. Sci. Au* **2022**, *2*, 422–429. [CrossRef]
20. Lammers, K.; Smith, M.M.; Susan, A.; Carroll, S.A.B. Muscovite dissolution kinetics as a function of pH at elevated temperature. *Chem. Geol.* **2007**, *466*, 149–158. [CrossRef]
21. Aviani, U. Applicazione della Sistematica Isotopica dello Sr alla Tracciabilità e alla Qualificazione di Prodotti Vitivinicoli: Studio sul Prosecco Veneto. Ph.D. Thesis, University of Trieste, Trieste, Italy, 2013.
22. Petrini, P.; Sansone, L.; Slejko, F.F.; Buccianti, A.; Marcuzzo, P.; Tomasi, D. The $^{87}Sr/^{86}Sr$ strontium isotopic systematics applied to Glera vineyards: A tracer for the geographical origin of the Prosecco. *Food Chem.* **2015**, *170*, 138–144. [CrossRef]
23. Wahlberg, J.S.; Baker, J.H.; Vernon, R.V.; Dewar, R.S. *Exchange Adsorption of Strontium on Clay Minerals*; US Geological Survey Bulletin 1140-C; U.S. Government Publishing Office: Washington, DC, USA, 1965.
24. Wallace, S.H.; Shaw, S.; Morris, K.; Small, J.S.; Fuller, A.J.; Burke, I.T. Effect of groundwater pH and ionic strength on strontium sorption in aquifer sediments: Implication for ^{90}Sr mobility at contaminated nuclear sites. *Appl. Geochem.* **2012**, *27*, 1482–1491. [CrossRef]
25. Ning, Z.; Ishiguro, M.; Koopal, L.K.; Sato, T.; Kashlwagi, J. Strontium adsorption and penetration in kaolinite at low Sr^{2+} concentration. *Soil Sci. Plant Nutr.* **2017**, *63*, 14–17. [CrossRef]
26. Hawkes, H.E. *Principles of Geochemical Prospecting*; U.S. Geological Survey Bulletin 1000-F; U.S. Government Publishing Office: Washington, DC, USA, 1957; pp. 223–355.
27. West, J.B.; Sobek, A.; Ehleringer, J.R. A Simplified GIS Approach to modeling global leaf water isoscapes. *PLoS ONE* **2008**, *3*, e2447. [CrossRef] [PubMed]
28. Lugli, F.; Cipriani, A.; Bruno, L.; Ronchetti, F.; Cavazzuti, C.; Benazzi, S. A strontium isoscape of Italy for provenance studies. *Chem. Geol.* **2022**, *587*, 120624. [CrossRef]
29. Willmes, M.; McMorrow, L.; Kinsley, L.; Armstrong, R.; Aubert, M.; Eggins, S.; Falguères, C.; Maureille, B.; Moffat, I.; Grün, R. The IRHUM (Isotopic Reconstruction of Human Migration) database—Bioavailable strontium isotope ratios for geochemical fingerprinting in France. *Earth Syst. Sci. Data* **2014**, *6*, 117–122. [CrossRef]
30. Cavazzuti, C.; Skeates, R.; Millard, A.R.; Nowell, G.; Peterkin, J.; Bernabò Brea, M.; Cardarelli, A.; Salzani, L. Flow of people in villages and large centers in Bronze Age Italy through strontium and oxygen isotopes. *PLoS ONE* **2019**, *14*, e0209693. [CrossRef]
31. Hammer, Ø.; Harper, D.A.T.; Ryan, P.D. *PAST: Paleontological Package for Education and Data Analysis*; University of Oslo: Oslo, Norway, 2001.

quaternary

Article

Prehistoric Astronomical Observatories and Paleoclimatic Records in Bulgaria Estimate Astroclimate during 4000–4500 BCE: A Critical Assessment

Alexey Stoev [1], Yavor Shopov [2,3,4], Penka Maglova [1,*], Ognyan Ognyanov [1] and Lyubomira Raykova [1]

1 Stara Zagora Department, Space Research and Technology Institute, Bulgarian Academy of Sciences, 6000 Stara Zagora, Bulgaria
2 University Centre for Space Research and Technologies, Faculty of Physics, Sofia University "St. Kliment Ohridski", 1504 Sofia, Bulgaria
3 Medical Physics and Biophysics Department, Medical University of Sofia, 1431 Sofia, Bulgaria
4 Institute for Systems Science, Durban University of Technology, P.O. Box 1334, Durban 4000, South Africa
* Correspondence: seac2021stz@abv.bg; Tel.: +395-876-25-34-13

Abstract: Prehistoric astronomical observatories include a specific type of rock-cut monuments from the Mountainous Thrace in Bulgaria, with a specific shape and orientation in space, which are part of the characteristic representatives of the archeoastronomical sites on the Balkan Peninsula from the period of 4000–4500 BCE. Earlier societies focused on the triad "astronomical instrument"—celestial objects—trained observers. When choosing sites for the construction of oriented stone complexes for astronomical observations, prehistoric people were interested in the number of clear days and nights within the tropical year, which is connected with the paleoclimate of the region and to the astroclimate, which determines the possibility of observing heavenly bodies. Here we examine 13 prehistoric astronomical observatories using the methods of archaeoastronomy in order to determine the period of their operation. Since the existence of a large number of such objects is indirect evidence of a good astroclimate, we make an assessment of the paleoclimate in the relevant era in the Bulgarian lands in order to find out if it was suitable for astronomical observations. The estimations are made according to the geological data and solar insolation luminescence proxy records of the evolution of cave speleothems from Duhlata cave in the village of Bosnek, Pernik municipality, which is still the only available experimental record of past solar insolation in Europe covering the last 20,000 years. The number of clear days and nights are estimated, and a critical assessment of the possibility of successful observations of the Sun during equinoxes and solstices is made using the methods of "horizon" astronomy and meridional culminations. It is also shown that the climate at the end of the Ice Age was cooler than today. About 11,700 years ago (11,700 radiocarbon years before 1950 CE or 11,700 BP), the climate began to warm, and forest vegetation developed on the territory liberated from the glaciers. During the Upper Atlantic (6–8 thousand years BCE), the average annual temperature on the Balkan Peninsula and in particular in Bulgaria was about 2–2.5 °C higher than it is today. This climate allows some very good astroclimatic conditions for observations of the Sun near the horizon and increases the accuracy of the observational data in determining the time of occurrence in its extreme positions on the horizon. We show that changes in climate (and astroclimate accordingly) influence the type of prehistoric astronomical observatories.

Keywords: prehistoric observatories; archaeoastronomy; astroclimate; paleoclimatic records; cave speleothems

Citation: Stoev, A.; Shopov, Y.; Maglova, P.; Ognyanov, O.; Raykova, L. Prehistoric Astronomical Observatories and Paleoclimatic Records in Bulgaria Estimate Astroclimate during 4000–4500 BCE: A Critical Assessment. *Quaternary* **2023**, *6*, 6. https://doi.org/10.3390/quat6010006

Academic Editors: Ioannis Liritzis and James B. Innes

Received: 10 August 2022
Revised: 5 December 2022
Accepted: 20 December 2022
Published: 5 January 2023

1. Introduction

The beauty of the sky has always attracted our eyes. Undoubtedly, man has observed the starry sky and celestial bodies since ancient times. Astronomy is part of culture and society; it has a clear social character. Archaeoastronomy studies the ways in which people in the past have understood the phenomena in the sky, how they used them and what role the sky played in their cultures [1].

With the advent of man, the Earth's climate has changed under the influence of a complex of factors, including human activity. The replacement of natural vegetation with agricultural crops, deforestation and/or the burning of forests and the emergence and growth of settlements change the natural albedo (reflection of the solar radiation) of the territory. This causes reflectivity changes, which are the reason for changes in the Earth's surface heat balance. An attempt has been made to answer the question, "When did these processes begin?" One of the reasons is agriculture, which is associated with a sedentary lifestyle, and which first appeared in the Neolithic (10–11 thousand years BCE) [2]. On the territory of Bulgaria, Neolithic settlements appeared about 8.1–8.6 thousand years ago (radiocarbon years before 1950 CE). At the beginning of the Neolithic and later in the Eneolithic (Chalcolithic, Copper Age), from 4000–4500 BCE, human influence on the environment was limited, so climate change during this period was little affected by anthropogenic activity.

Every major region in the world has its own individual destiny for development in prehistory. Almost all of them go through certain stages in the formation of societies, religions and technologies. All previous studies show that the basic notions of time and timing are developed during these periods, and that the foundations of these culture share common features and characteristics of the hearths of civilization. The architecture of the special facilities for the observation of celestial objects studied here shows the skills of the people of that time in planning, creating construction technology and making long-term observations.

Rock monuments of this type, despite their initial antiquity and diversity, can arise only in the presence of a sufficiently developed triad of "astronomical instrument—celestial objects—trained observers", as well as in the presence of certain socio-cultural, economic and spatial-climatic conditions. As a specific cultural phenomenon, rock-cut monuments for astronomical observations span many different periods of time in different regions of the world. From the creation of the first rock sanctuaries by various communities with places for the contemplation of heaven and prayers, to the emergence of much later, extremely complex rock structures for targeted observations, long periods of historical time pass. It can be tentatively believed that the first astronomical rock structures appeared on Earth at the beginning of the 6th millennium BCE [3,4].

The prehistoric astronomical observatories consisting of rock-cut monuments from the Mountainous Thrace in Bulgaria are characteristic representatives of the archaeoastronomical sites on the Balkan Peninsula from the period 4000–4500 BCE. When choosing sites for the construction of these oriented stone complexes for astronomical observations, prehistoric people were interested in the number of clear days and nights within the tropical year, which is connected with the paleoclimate in the region and with its astroclimate, which determines the possibility of observing heavenly bodies.

The aim of our research is to examine 13 prehistoric astronomical observatories using the methods of archaeoastronomy and determine the period of their functioning, which is indirect evidence of a good astroclimate. In order to find out if the astroclimate was suitable for astronomical observations, we make an assessment of the paleoclimate in the relevant era in the Bulgarian lands. The estimations are made according to geological data and solar insolation luminescence proxy records of the evolution of cave speleothems from Duhlata cave in the village of Bosnek, Pernik municipality, which is still the only available experimental record of past solar insolation in Europe covering the last 20,000 years.

2. Geographical, Climatic and Socio-Cultural Preconditions for the Appearance of Astronomical Observation Facilities in Prehistory

Today, we know that, for the emergence of long-term settlements and economic structures in prehistoric times, a mild climate and the availability of fresh water and natural shelters are required, as well as fertile soils and materials for the production of tools. These conditions mostly correspond to the strip of the globe between 43° and 23° North latitude. The fertile valleys of the great rivers Tigris and Euphrates, the Yellow River and the Yangtze, the Indus, the Nile, the Mediterranean coast and the mountainous foothills of present-day Syria and Iran find themselves in this vast and favorable territory. In this strip, one can find the largest number of the ancient settlements, religious sites and production centers that serve as the basis for the emergence of all major ancient civilizations on the Eurasian continent, beginning in the 9th millennium BCE. It was at that time that the most significant migrations of ancient peoples took place. Traces of the movement of pastoralists and farmers, merchants and builders of cult ensembles spread throughout Europe. The area in which those tribes and peoples existed depended directly on the size of the favorable territory.

Climate in the region of ancient settlements is one of the basic conditions for determining their size and population. The migration of tribes of farmers and pastoralists is often associated with the search for new habitable lands. The socio-cultural sphere of this period is characterized by the completion of the formation of pagan cults with a hierarchy of deities. Most often, the pantheon of gods is governed by a supreme god (the Sun). Specific ornaments of a solar or astral nature play an important role, carrying significant and diverse information, which often includes a magical component. It has become the structural backbone of almost all prehistoric arts. An agricultural calendar is emerging in many of these areas. Cult components sooner or later begin to be associated with the agricultural calendar, gradually becoming the behavioral basis of prehistoric society. During the same period, cults are built and actively spread, especially the cult of the Sun and natural forces, as well as the cult of ancestors. Cult structures and complexes are realized variously: from small areas in the structure of houses or settlements to large cult ensembles, which become the ritual center of the surrounding villages.

The history of the prehistoric tribes that inhabited the Bulgarian lands and the Balkan Peninsula during the Neolithic and Eneolithic Ages is an integral part of the history of Southeast Europe. Unlike in the past, when technical possibilities did not allow such a detailed study of prehistory, today it is possible, and new horizons and data are being discovered. Of particular importance are the achievements of the radiocarbon method, dendrochronology and the study of cave speleothems, which give prehistory real chronological boundaries [5–7].

Figure 1 depicts several of the so-called "cultural-territorial zones", which are determined primarily by the geographical factor. During the Neolithic and Eneolithic epochs, three such zones emerged: Zone I—North, Zone II—Forest and Zone III—South. Each of them has its own way of life, and the settlement in it is engaged in various activities of life. Between the individual zones, there are others that carry the legacies of these three zones interwoven within them. During the Neolithic era, several closely related ethnocultural areas can be traced, having formed at the end of the 9th millennium BCE. They are the result of the Neolithic consolidation of the Neolithic agricultural and livestock economy [8,9].

In the south, the cultures with Early Neolithic ceramics stand out (the groups in Thessaly and the Eastern Mediterranean; the groups from Karanovo—Kremikovtsi—Starchevo, etc.). The characteristic features of these areas are painted ceramics, ground dwellings, well-organized settlement structures and trade relations. Exceptions include the groups to the north, where ground dwellings and dugouts meet in one place. This is obviously due to the close coexistence of the two cultures [8].

The second area covers the region between the Balkans and the Carpathians. It includes the cultures Ovcharovo, Krish and others. Their characteristic features are pottery with impurities of chaff, an almost complete absence of painted pottery and a settlement built only on the plateaus, and their main residential forms are the dugout and the ground dwelling. At the end of the Neolithic era (end of the 8th millennium BCE), Thrace probably experienced its first demographic explosion. The population began a gradual movement to the north, where the Neolithic population was much less compact. That is why here, at the level of the cultures of Polyanitsa, Sava, etc., in the Balkan Northeast, ground buildings appear, close to each other, and settlement mounds appear, which until then were typical only for the southern regions [8,9].

Figure 1. Scope of cultural-territorial zones on the Balkan Peninsula: Zone I—North, Zone II—Forest and Zone III—South.

Studies on the structure and planning of a Neolithic settlement can be given only if a large part of the settlement space has been discovered. In recent years, this has proved to be very real and possible, and many hitherto unknown facts about the structure of settlements, the manner of the construction of housing and its orientation and many other facts have been established. The plans of entire Neolithic sites, such as Karanovo, Golyamo Delchevo I, Podgorica and many others, have been revealed. Scattered throughout Bulgaria, they give a realistic idea of the peculiarities of the housing construction, settlement traditions, culture and technologies of their inhabitants [8,9].

3. Ancient Astronomical Observatories in Bulgaria (Horizon and Meridional Astronomy)

In ancient times, when choosing sites for the construction of oriented stone complexes (rock-cut monuments) for astronomical observations, prehistoric people were interested in the number of clear days and nights within the tropical year. The set of atmospheric conditions that affect the quality and quantity of astronomical observations of a particular object determine the clear days and nights. The most important of them are the transparency of the air, the degree of its homogeneity and the ability to obtain a long series of observations. The "horizon" and "meridian" astronomical practices during the Eneolithic are less strongly influenced by the astroclimate, mainly due to the lack of optical instruments. The observations were made with the naked eye, using sighting and projection devices made mainly of stone. However, the need for a good astroclimate also applies to them, mainly due to the need to accurately record the equinoxes and solstices throughout the year. This was a particularly important procedure for creating and maintaining the prehistoric calendar.

A lot of prehistoric observatories are found in the mountainous regions of Bulgaria (Figure 2). They can be defined as spatial structures of different reliefs, rock shapes and cuts connected in a certain functional dependence with the horizon, the sky and the heavenly objects and phenomena. Archaeoastronomical research has been made taking into account the structural elements and orientation of the rock-cut monuments as well as archaeological artifacts found on their territory.

Figure 2. Location of the studied rock-cut monuments with astronomical meaning on the territory of today's Bulgaria: 1—Magura cave, 2—Bailovo cave, 3—Buzovgrad, 4—Zaychi vrah, 5—Lilyach, 6—Markov Kamak and Tsarevi Porti, 7—Kozi Kamak, 8—Belintash, Parmakla Kaya cave and Angel Voyvoda, 9—Kovil, 10—Harman Kaya, Tangardak Kaya cave and Tatul.

Prehistoric observatories show the skills of their makers to build such facilities, to conduct astronomical observations and to apply them. Generally, we can find them in high places. There is usually a spring around, and they are near populated areas. Prehistoric observatories take the form of megalithic structures such as sanctuaries and tombs. Very often, images of astronomical objects and phenomena (the Sun, the Moon, bright stars, comets, eclipses) can be found in these places. Solar, lunar and stellar images and solar calendars have been found in the cave complexes near the villages of Baylovo and Lipnitsa, Sofia district and the village of Tsarevets, Mezdra municipality; in the Topchika cave, Asenovgrad municipality; and in the Magura cave near the village of Rabisha, Belogradchic municipality [10–14].

Exploring the prehistoric observatories, we get an idea of the astronomical knowledge and observations, of the customs and beliefs, and of the art, the worldview and the values of the people and societies that lived in the ancient Bulgarian lands [15–27].

3.1. Methodology of Archaeoastronomical Investigations

The methodology of archaeoastronomical research examines the ways, methods and aims of investigation in the field of archaeoastronomical reconstructions [16].

3.1.1. Methodology for the Detection and Study of Structural Orientations in Rock-Cut Monuments (Method of "Horizon" Astronomy)

The basis of the methodology for finding the coincidences of terrain markers (structural orientations) with the points on the horizon at which the Sun rises (or sets) for an observer from a selected place is the assumption that prehistoric people empirically recorded these positions within the tropical year (the time between two identical positions of the sun at sunrise or sunset), which makes it possible to make a calendar. To restore the mechanism of ancient observations, we need to measure the azimuth of the point of sunrise (sunset) from the prehistoric monument and the specific place of observation needed for determining the declination of the sun by the formula

$$\sin \delta = \sin \varphi \sin h + \cos \varphi \cdot \cos h \cdot \cos A \quad (1)$$

where A and h are the azimuth and the height of the Sun above the true horizon at the moment of sunrise (sunset). The latitude is denoted by φ and the declination of the Sun—by δ. The height of the visible horizon is measured with a Theo 010 theodolite, with a centi-centigrade accuracy. When the horizon line is more than 5–8 km away, the curvature of the earth's surface is taken into account, as well as the refraction and parallax of the observed object. The sum of all these factors gives the following formula:

$$h = h_{hor} - r_{terra} + \pi - R \quad (2)$$

where h is the real height, h_{hor}—the apparent height of the horizon, r_{terra}—correction for the curvature of the earth's surface, π—parallax and R—the refraction of the luminary [17].

Obtaining the values of h allows us to find the azimuth of that point on the horizon, where the center of the Sun's disk will fall, with the horizon line dividing it into two equal parts. For the moment of the first touch of the disk to the line of the visible horizon, we must add 0.25°. Since the apparent diameters of the Sun and the Moon are the same, the corrections used in determining sunrises (sunsets) are the same for both bodies.

3.1.2. Methodology for the Detection and Study of Light and Shadow Effects in Rock-Cut Monuments (Method of Meridional Culminations)

The methodology for studying light calendar effects is related to the detection and precise documentation of the main lines (horizontal and vertical, orientations) forming the projection hole. Observing and measuring a "light" spot on a dark surface is easier than directly observing the sun's disk on the horizon line [18]. This procedure is facilitated when the spot has the correct geometric shape and is designed on appropriate "reception markers"—elements of the rock relief or the framework of the monument. By definition, "light projection" in a calendar sense means the coincidence of a selected edge of a light projection with a deliberately constructed "reception" marker [19]. Only two celestial bodies, the Sun and the Moon, have great enough intensities of incident light to produce such effects. The "effects" are studied empirically at characteristic points. The following constants for the facility need to be measured and calculated in advance:

1. Astronomical azimuth of the object (A—most often this is the main axis of the facility (the cave) or other auxiliary directions);
2. Latitude (φ—interpolated by map or accurate GPS);
3. Height of the Sun (h—from the characteristic "receiving markers"; dimensions and orientation of the projection opening are also needed.

With the resulting A, φ and h (the latter is adjusted by +16′ to correspond to the lowest point of the Sun relative to the horizon), according to formula (1), we calculate the declination δ of the Sun and find in the astronomical calendar the corresponding day of the year on which this light projection is realized.

According to the formula:

$$\cos A = (\sin \delta - \sin \varphi \cdot \sin h)/\cos \varphi \cdot \cos h \quad (3)$$

we calculate the azimuth A of the sunrise on the eastern horizon during the equinoxes and solstices.

In the process of research, it is quite possible also to detect a characteristic lunar declination (minimum = $[-(\varepsilon - 5°9')]$, or maximum = $[+(\varepsilon + 5°9')]$), where ε is the inclination of the ecliptic and 5°9′ is the inclination of the moon's orbit relative to the ecliptic.

The hypothesis of the presence of light calendar effects in a monument is always related to the position of the Sun (or Moon) relative to the axis of the facility, between the centers of two entrances, or exactly in the axis of a corridor (gallery). This axis is marked from the middle of the lower outer edge of the lintel (the upper threshold of the hole to be projected) at the desired lower point—the middle of the lower threshold of the step or rock carving constituting the artificial marker. By selecting appropriate "reception markers" that must meet certain objective criteria, it is quite possible to detect calendar relationships with the annual movement of the Sun. Moreover, in certain cases, it is possible to astronomically date the chronological boundaries of the object's existence along the height of the Sun in culmination during the winter solstice.

Due to the lack of archaeoastronomical research on a larger number of monuments of this type, it is not known whether one, two or more calendar dates were used. If we assume that the date "remembered" by a marker was only one, the most significant—for example, the solstice—we can assume that the other dates are determined by movable screens in the contour of the opening. From the previous archaeoastronomical studies of rock-cut monuments and cult caves [3,4,10–27], we can conclude that the height of the Sun, designated as a shadow or light spot, is much more important than the horizontal orientation of the near and far benchmarks to sunrise or sunset.

Facilities for observing the Sun have been found at the following rock-cut monuments: *Belintash*, near the village of Mostovo, Plovdiv district, *Zaychi vruh* (Figure 3), near the village of Kabile, Yambol district; *Tatul*, near the village of Tatul, Momchilgrad municipality; *Tangarduk Kaya* (Figure 4), Kurdjali district; *Harman Kaya* (Figure 5), near the village of Dolna Chobanka, Momchilgrad municipality; *Buzovgrad* (Figure 6), Kazanluk municipality; *Tzarevi porti*, the village of Kovachevitsa, Garmen municipality; *Markov kamak*, Blagoevgrad municipality, *Lilyach*, Kyustendil municipality; *Kovil*, Krumovgrad municipality, *Baylovo*, Gorna Malina municipality; *Magura*, Belogradchik municipality and *Parmakla Kaya*, near the village of Nochevo, Asenovgrad municipality [14–27].

Figure 3. Zaychi vruh (Cabyle). Artificially hewn out of the rock trenches, oriented East–West and North–South, used for observations of the sun during equinoxes and culminations of bright luminaries. An additionally leveled rock, located in the northeast, allows determination of the summer solstice. This device could be used for measuring time intervals longer or shorter than a day, after [25].

Figure 4. Tangarduk Kaya cave sanctuary. The additionally processed gallery of the natural cave is oriented North–South, and it is convenient for observation of Sun culminations. The light projection of the entrance could be used for defining the summer and winter solstices or for determining the longest and the shortest day of the year, after [19].

Figure 5. Harman Kaya rock-cut monument. Observations of the Sun from the northeastern, artificially leveled site of the rock complex. The eastern peripheral part offers convenient reliefs which, after a little further processing, were used as sighting devices. This is a typical system for observations of sunrises and sunsets (during solstice or equinox), which coincide with characteristic points of the local horizon(1—summer solstice, 2—spring and autumn equinox, 3—winter solstice), after [26].

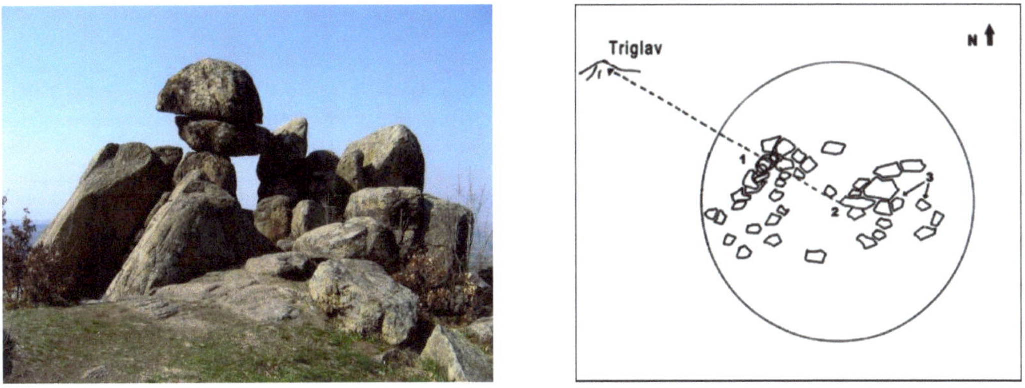

Figure 6. Buzovgrad rock sanctuary. Scheme of the basic structural elements of the sanctuary: 1—trilith, 2—"throne", 3—"sacrificial altars". The Triglav peak seen through the megalith's aperture. It is supposed that observations of the solar disc during sunset on the day of the summer solstice were made when solar disc touched the line of the visible horizon, after [20].

Figures 3–6 give several examples of facilities for observation of the sun (rises and sets) in its extreme positions on the visible horizon (solstices and equinoxes) and meridian culminations.

According to Ruggles [28], the rising and setting positions of the sun, moon and planets are not affected by the precession of equinoxes (or only the precession) but do change over time, by a smaller amount, owing to the fact that the obliquity of the ecliptic (ε) changes slowly with time. Over the past few millennia, it has been slowly decreasing, from about 24.15° in 5000 BCE to 23.45° now (27, [28] Table 31.3), but over a longer timescale (of about 41,000 years), it oscillates between limits of about 24.4° and 22.2°. A maximum was reached in about 6000 BCE, and a minimum will be reached in about 14,000 CE. Compared with the shifts in the stellar rising and setting positions due to precession, the differences in the sun's rising and setting position are small: for example, in temperate zones around 2000–6000 BCE, the sun rose and set further north at the June solstice, and the same amount further south at the December solstice, than now. The corresponding azimuth difference is about 1°.

The archaeoastronomical dating of the studied rock-cut monuments associated with long-term astronomical observations is made using the rising and setting positions of the sun in its extreme point (solstice and equinox). The maximal height of the sun in culmination during solstices (where the solar declination δ is at its maximum, and equals the obliquity of the ecliptic ε) is calculated by the formula:

$$hmax = 90 - \varphi + \varepsilon \quad (4)$$

Using the Equations (1)–(4), we calculate the obliquity of the ecliptic ε for every studied object. The period of creation and operation of the rock-cut monument is determined by the comparison of the derived obliquity of the ecliptic ε with the values presented in Table 1.

Table 1. The obliquity of the ecliptic, after [28].

Years	5000 BCE	4000 BCE	3000 BCE	2000 BCE	1000 BCE	CE 1/1 BCE	1000 CE	2000 CE
ε	24.15°	24.1°	24.0°	23.9°	23.8°	23.7°	23.55°	23.45°

Table 2 presents the rock-cut monuments associated with long-term astronomical observations and astronomical practices (prehistoric astronomical observatories), their orientations towards astronomically significant points on the visible horizon, as well as the observed phenomena—the rise, set or noon culmination of the observed object—and archaeoastronomical dating.

Table 2. Prehistoric astronomical observatories, orientation of the astronomical facilities there, supposed astronomical phenomena observed on the visible horizon—the rise, set or noon culmination of the observed object—and archaeoastronomical dating.

No	Rock-Cut Monument Type of Observations	Orientation of the Facility	Observed Phenomenon	Archaeoastronomical Dating
1.	Belintash [15] horizon observations	Summer solstice	Sunrise	2000 BCE
2.	Zaychi vruh (Cabyle) [25] horizon observations meridional culminations	Summer solstice and equinoxes	Sunrise, sunset, stellar and solar culminations	2200 BCE
3.	Tatul [27] horizon observations	Winter solstice and equinoxes	Sunrise	2200 BCE
4.	Tangarduk kaya cave [19] meridional culminations	Winter solstice	Solar culminations	3000 BCE
5.	Harman kaya [26] horizon observations	Summer and winter solstices and equinoxes	Sunrise, sunset	2500 BCE
6.	Buzovgrad [20] horizon observations	Summer solstice	Sunset	1800 BCE
7.	Tsarevi porti [22] horizon observations	Summer solstice	Sunrise	2000 BCE
8.	Markov kamak [21] horizon observations	Summer solstice	Sunrise	2200 BCE
9.	Lilyach [24] horizon observations meridional culminations	Summer and winter solstices and equinoxes	Sunrise, sunset, stellar and solar culminations	2200 BCE
10.	Kovil [23] horizon observations	Summer solstice and equinoxes	Sunrise	2800 BCE
11.	Baylovo cave [3] horizon observations	Summer and winter solstices and equinoxes	Sunrise	3000 BCE
12.	Magura cave [14] horizon observations	Summer and winter solstices and equinoxes	Sunrise	3000 BCE
13.	Parmakla kaya cave [19] meridional culminations	Winter solstice	Solar culmination	3000 BCE

Two groups of rock-cut monuments used for astronomical observations have been identified. We see that, in the first group of prehistoric observatories (Belintash, Zaychi vrah (Cabyle), Tatul, Harman Kaya, Buzovgrad, Tsarevi Porti, Markov Kamak, Lilyach, Kovil, Bailovo cave and Magura cave), ancient observers used the **method of "horizon" astronomy**. In the second group (Zaychi vrah (Cabyle), Tangarduk kaya cave, Lilyach and Parmakla kaya cave), they used the **method of meridional culminations** of the same celestial bodies. From Table 2, it is evident that these are the earliest prehistoric observatories (dated to about 3000 BCE).

4. Paleoclimate

After the dating of the studied prehistoric astronomical observatories, in order to find out if the astroclimate was suitable for astronomical observations, we make an assessment of the paleoclimate in the relevant era in the Bulgarian lands.

Paleoclimate is the climate of the Earth in an individual region, at a particular geological or prehistoric time. Paleoclimatology is the science of studying the climate in the past. It is a very interesting and relevant, multidisciplinary field combining history, anthropology, archaeology, chemistry, physics, geology and atmospheric and oceanic sciences. Paleoclimate studies use the geological and biological evidence preserved in sediments, rocks, tree rings, corals, ice sheets and other climate archives to reconstruct past climates in terrestrial and aquatic environments around the world [29]. Clues about past climate conditions are obtained from *proxy indicators*, which are indirect forms of evidence that can be used to infer the climate [30,31]. These include:

- Isotopic Geochemical Studies: the study of isotopic ratios in rocks, bubbles from ice cores, deep-sea sediments, etc. [32,33];
- Dendrochronology: the study of tree ring growth;
- Pollen Distribution: the study of plant types during the relevant climatic era and the distribution of pollen found in sediments, ice, rocks, caves, etc.;
- Study of lake sediments (Lake Varves): (like dendrochronology, but with lake sediments-a varve is an annual layer of mud in the sediment);
- Coral Bed Rings;
- Fossils in different geological and historical layers: studies of geological settings, etc. [34];
- Historical documents and artifacts that testify to the emergence and development of civilizations, etc.

Radiocarbon dating is one of the most useful absolute dating methods used in archaeology, geology, sedimentology and many other sciences. The ability to date minute samples via accelerator mass spectrometry (AMS) means that paleobotanists and paleoclimatologists can use radiocarbon dating directly on pollen purified from sediment sequences, or on small quantities of plant material or charcoal. The dating of organic material recovered from strata of interest can be used to correlate strata in different locations that appear to be similar on geological grounds. Dating material from one location gives date information about the other location, and the dates are also used to place strata in the overall geological timeline [35]. Radiocarbon dating measurements produce ages in "radiocarbon years", which should be converted to calendar ages by a process of **calibration**.

As an example, we consider the calibration of the beginning of Holocene, the current geological epoch, which is determined to have begun about 11,700 years ago [36].

A sample from the Two Creeks Fossil Forest, Wisconsin was used in an interlaboratory test (the work of over 70 laboratories). These tests produced a median age of 11,788 $\pm$ 8 BP (2σ confidence) which, when calibrated, gives a date range of 13,730 to 13,550 cal BP or 11,730 to 11,550 BCE [37]. The Two Creeks radiocarbon dates are now regarded as a key result in developing the modern understanding of North American glaciation at the end of the Pleistocene [38].

Because the information needed to convert radiocarbon ages to calendar ages is constantly being improved, it was decided that radiocarbon ages and not calendar ages would become the standard method of recording results. This has the advantage that the thousands of dates published in articles prior to any given update do not have to be re-calculated.

The astroclimate is a set of meteorological and climatic conditions that determine the possibility of the observation of cosmic bodies. In fact, this set of atmospheric conditions mainly affects the quality of astronomical observations. The most important of them are the transparency of the air, the degree of its homogeneity (influencing the "sharpness" of the image of objects), the amount of background glow of the atmosphere, the daily temperature drops and the strength of the wind. In modern astronomy, the astroclimate is a combination of factors that distort the shape of the wave front emitted by celestial objects passing through the Earth's atmosphere.

4.1. Data from Cave Speleothems

It is well established that variations in the total amount of solar radiation at the Earth's surface (insolation) produce global changes in the climate. It has been proven in many ways that global climate change is due to variations in the total solar radiation reaching the earth's surface (earth's surface insolation or ***solar insolation***—the amount of electromagnetic energy (solar radiation) falling on the earth's surface) [39,40].

Past climate conditions can be estimated by studying the evolution of secondary Karst formations as speleothems, secondary mineral deposits formed in caves that serve as natural records of the solar insolation over very long time spans [39,41]. The obtained time series have durations of hundreds of thousands of years and should be calibrated using instrumental records. Thus, a large number of global change parameters can be reconstructed [39,42]. Speleothems provide high quality, well preserved and undisturbed records, but they are typically interrupted by numerous hiatuses caused by the disappearance of the infiltrating water solutions that produce speleothem growth due to droughts, etc. Speleothems that grow continuously for tens of thousands of years are exceptionally rare, while those that grow continuously for hundreds of thousands of years are unique. Such a speleothem is the one from the Duhlata Cave in the village of Bosnek region of Bulgaria [43], which produced the record on Figure 7.

Temperature, past precipitation, the nature of the soil and the vegetation cover, pollution, air composition, glaciation, fluvial erosion and deposition and groundwater flows can be usually read from the luminescence of cave speleothems and deposits.

Luminescence is the property of cave minerals most sensitive to depositional conditions [41]. Many speleothems exhibit luminescence when exposed to ultraviolet (UV) light sources or other high-energy beams. Depending on the excitation source, there are specific kinds of luminescence: "Photoluminescence" (excited by UV and other light sources), "X-ray luminescence" (by X-rays), "Cathodoluminescence" (by electron beam), "Thermoluminescence" (by heat), "Candoluminescence" (by flames) and "Triboluminescence" (by crushing). Different types of excitation may excite different luminescent centers—electron defects of the crystal lattice; admixture ions substituting ions in the crystal lattice or incorporated in cavities of that lattice; inclusions of other minerals; or fluid inclusions, molecules, ions or radicals adsorbed inside of the lattice [36]. Some or all of them may exist in a single speleothem. If the emission proceeds only during the excitation, then it is called "fluorescence", if it proceeds after the termination of the excitation then it is called "phosphorescence". Some luminescent centers produce only fluorescence, but others produce both fluorescence and phosphorescence.

Calcite speleothems frequently display luminescence, which is produced by the calcium salts of the humic and fulvic acids derived from the soils above the cave.

The method of Laser Luminescent MicroZonal Analysis [43] gives information about changes in the mineral-forming conditions and gives high resolution records of the annual rainfalls and annual temperatures in the past, allowing for the reconstruction of climate and solar activity variation.

The intensity of speleothem luminescence shows a direct connection with ***solar insolation***: The luminescence of calcite speleothems precipitated in caves depends exponentially upon the soil temperatures, which are determined primarily by solar infrared radiation in the case when **the cave is covered only by grass** or by air temperatures where there is **forest or bush cover**. In the first case, the microzonality of the luminescence detected in speleothems can be used as an indirect solar insolation index, and, in the second, as a paleotemperature proxy. Thus, in terms of the dependence on cave site conditions, we may speak about "solar sensitive" and "temperature sensitive" paleoluminescence in speleothem records, as in tree ring records.

Figure 7 shows part of the ***solar insolation luminescence proxy record*** from the Duhlata Cave, located in the region of the village of Bosnek, Pernik municipality, Bulgaria [43], which is still the only available experimental record of past solar insolation in Europe covering the last 20,000 years. The cave is covered only by grass, and the soil temperature there reaches 55 °C during the summer time, which means that this speleothem luminescence record represents only the solar insolation. It demonstrates a dramatic minimum of insolation after the Holocene maximum, from 7000 ± 2000 years ago (7000 ± 2000 BP or radiocarbon years before 1950 CE) [44]. From 4200 BCE to 600 CE, the insolation levels were below the levels during the Younger Dryas event (Figure 7), which is considered to be the end of the Last Glacial Period (LGP), a return to glacial conditions in the period 12,900–11,700 years BP [45], after which Pleistocene ended and Holocene, the current geological epoch, began with a sharp climatic warming [37]. However, it did not lead to corresponding minima of the temperatures in the region as recorded in the paleotemperature reconstructions for Bulgaria for the same period [46,47], showing that, from 5000 BCE to 100 BCE, the mean annual temperatures in Bulgaria were higher than today (Figure 8).

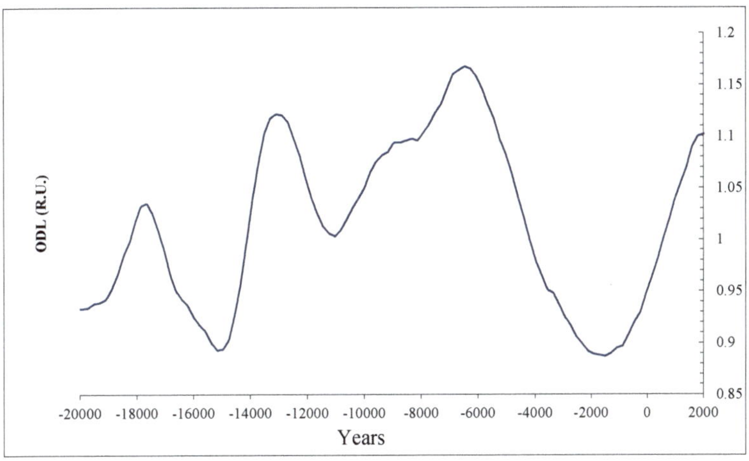

Figure 7. Luminescence proxy record of the Solar Insolation (Duhlata Cave, Bulgaria) (Optical Density of Luminescence (ODL) in decimal logarithm relative units (R.U.) depending on time, after [43].

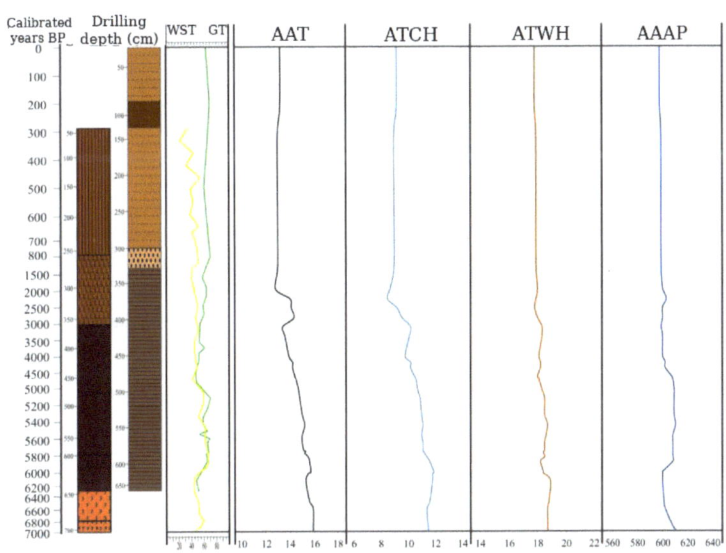

Figure 8. Reconstruction of paleoclimatic variables: WST-wood and shrub taxa (%), GT—grass taxa (%); AAT—average annual temperature (°C); ATCH—average temperature of the cold half of the year (°C); ATWH—average temperature of the warm half of the year (°C); AAAP—average annual amount of precipitation (mm/m^2), after [46].

4.2. Paleogeological and Pollen Data

Glaciers appeared in Northern Eurasia during the Pliocene, 4–5 million years ago, but the climate of the Eastern Mediterranean remained warm. A permanent drought was also recorded in Southeastern Europe at the end of the Pliocene. According to floristic data, it has been proven [48] that the average July temperature in the last 4 million years in Eastern Europe has decreased by approximately 7 °C. As a result, at the end of the Pliocene, the height of the snow line in the region decreased by 750 m. Although the climate as a whole began to change towards the cooling and strengthening of its continental character, at the beginning of the Pleistocene, the territory of today's steppe zone in southern Europe still had a Pliocene-dominated forest-steppe landscape. The territory of today's Bulgaria fell into the wide transitional zone between the periglacial areas and the pluvial belt, due to which the influences of both periglacial and warmer and wetter climatic conditions alternated rhythmically during the Quaternary period on its territory.

The Holocene epoch started about 9700 BCE, when the glacial cycle of the Pleistocene ended. According to the data from a number of geographical locations, and in particular from drilling cores in the Greenland ice sheet [49], the Pleistocene-Holocene boundary reflected the first significant warming after the end of the Younger Dryas. During the Holocene, in the so-called Holocene maximum, the average annual temperatures in North-East Bulgaria from 5000–3800 years BCE were 2–3 °C higher in comparison to the modern ones [46].

According to pollen data, the Holocene climate in Bulgaria was warmest around 5000–4500 BCE. This warm weather covered the transition from the Neolithic to the Eneolithic in Bulgaria. Subsequently, the climate became colder and wetter, and the Black Sea level dropped by 4–5 m at 3500–2000 BCE [50]. During the Holocene, in the period 4000–3000 BCE, forest cover of the lands predominated over the steppes, followed by an advance of grasslands in N-E Bulgaria. During the last 2000 years, all climatic parameters have remained without significant variations in the region [49]. However, due to the Mediterranean influence in the Rhodope, Pirin, Osogovo and Sredna Gora mountains, the amount frosty weather is 20–30% less than in the Balkan Mountains. The average duration of frost-free time (without negative temperatures) in the plains and hills of Bulgaria was 180–225 days per year.

5. Astroclimate

In prehistory, in the absence of optical instruments, the astroclimate is mainly influenced by the number of cloudless days and nights, the dust in the atmosphere due to volcanic eruptions, forest fires, stormy winds and, of course, the levels of solar activity and cosmic rays.

By definition, the astroclimate is a set of environmental conditions that determine the quality of astronomical observations. Basic and mutually independent conditions for determining the astroclimate are:

- The purity and transparency of the atmosphere;
- The light pollution of the natural environment;
- The seismicity of the region [51,52].

The first two of these conditions form a background glow in the sky, depending on the composition and amount of pollution in the atmosphere (dust, aerosols, moisture) and the amount of light pollution in it. On the other hand, the factors that affect the astroclimate are both natural and artificial (anthropogenic).

Natural factors include microclimate and meteorological conditions, the presence of clouds, fog, lightning, atmospheric pressure, temperature and humidity, the system of local and prevailing winds, temperature anomalies in the air and on the land surface, terrain, natural seismic activity and natural sources of air pollution. The natural factors determining the astroclimate exist as a given and, in some cases, can be weakened or strengthened by human activity.

Anthropogenic factors include artificial light sources, heat, artificial inhomogeneities of the area (areas with different albedo and heat capacity), artificial sources of air pollution, electromagnetic interference and seismic noise and artificial obstacles to air flow (due to land reclamation, land afforestation, etc.). Anthropogenic factors are usually negative, degrading the quality of the astroclimate and reducing the quality of astronomical observations.

Finally, we can make the conclusion that, around 8000 BCE, the climate in our lands was very favorable. During this period, the average annual temperature on the Balkan Peninsula and in particular in Bulgaria was about 2–2.5 °C higher than today [53]. This climate provides very good astroclimatic conditions for observations of the Sun, Moon, planets and bright stars, increases the effectiveness of observational data in determining the time of sunrises and sunsets (extreme positions of celestial bodies on the horizon) and noon culminations.

6. Conclusions

Here we examine 13 prehistoric astronomical observatories using the methods of archaeoastronomy in order to determine the period of their operation. In choosing sites for the construction of oriented stone complexes for astronomical observations, prehistoric people were interested in the number of clear days and nights within the tropical year. The aim of our investigation is to make an assessment of the paleoclimate in the relevant era in the Bulgarian lands in order to find out if it was suitable for astronomical observations. The estimations are made according to the geological data and solar insolation luminescence proxy records of the evolution of cave speleothems from the Duhlata cave in the village of Bosnek, Pernik municipality.

Two groups of rock-cut monuments used for astronomical observations have been identified. In the first group of prehistoric astronomical observatories (Belintash, Zaychi vrah (Cabyle), Tatul, Harman Kaya, Buzovgrad, Tsarevi Porti, Markov Kamak, Lilyach, Kovil, Bailovo cave and Magura cave), ancient observers used the **method of "horizon" astronomy** because of the climatic optimum (stable average annual temperature and low humidity), with good visibility of the Sun, the Moon and bright stars on the horizon line at sunrise and sunset. In the second group (Zaychi vrah (Cabyle), Tangarduk kaya cave, Lilyach and Parmakla kaya cave), observers used the **method of meridional culminations** of the same celestial bodies, as it was difficult to observe sunrises and sunsets on the horizon due to the high humidity, precipitation and the predominance of cloudy days. From Table 2, it is evident that earliest prehistoric observatories (dated about 3000 BCE) were designed for the observation of meridional culminations. This can be explained by the results from the reconstructions of the average annual temperatures and the average annual amounts of precipitation over the last 7000 years [46]. During this period of the emergence of prehistoric observatories in Bulgarian lands, the temperatures and precipitation levels were comparatively higher, as was the air humidity, and observations were only possible at noon and midnight.

The dynamics of climatic variables (temperature, humidity, sunshine) for several climatic periods has been investigated using paleogeological data [48], pollen analysis [46,47] and the analysis of cave speleothems from the territory of Bulgaria [43]. The solar insolation luminescence proxy record from the Duhlata Cave, placed in the region of the village of Bosnek, Bulgaria, is derived using the method of Laser Luminescent MicroZonal Analysis [43], and it is still the only available experimental record of past solar insolation in Europe covering the last 20,000 years. Climatic parameters for the last 8000 years are reconstructed, and shorter time intervals absolutely attached to the chronological boundaries of the historical periods established by archaeological excavations are defined from them. From about 5000–3800 years BCE, the average annual temperatures in Bulgaria were 2–3 °C higher in comparison to modern ones. Such a climate gives very good astroclimatic conditions for observations of the Sun near the horizon and allows increased accuracy in determining the time of occurrence of its extreme positions on the horizon.

Author Contributions: Conceptualization, A.S., P.M. and Y.S.; methodology, A.S., Y.S. and P.M.; software, O.O.; investigation, A.S., P.M., Y.S., O.O. and L.R.; data curation, A.S. and Y.S.; writing—original draft preparation, A.S., P.M., Y.S., O.O. and L.R.; visualization, A.S., P.M. and Y.S.; supervision, A.S. and P.M.; project administration, Y.S.; funding acquisition, Y.S. All authors have read and agreed to the published version of the manuscript.

Funding: This research was funded by the Ministry of Education under the National Program on "Protection of the environment and reducing the risk of unfavorable events and natural disasters", WP I.6.10 "Geological records of Quaternary climate change".

Data Availability Statement: Not applicable.

Acknowledgments: We thank everyone who gave us information about the rock-cut monuments and all the participants in the archaeoastronomical expeditions to study the sites.

Conflicts of Interest: The authors declare no conflict of interest.

References

1. Liritzis, I.; Vlachos, A. Skyscape impact to cultural astronomy. *Sci. Cult.* **2022**, *8*, 131–155.
2. Bellwood, P. *First Farmers: The Origins of Agricultural Societies*; Blackwell Publishing: Oxford, UK, 2005; ISBN 0-631-20565-9.
3. Stoev, A.D.; Maglova, P.V. Astronomy in the Bulgarian Neolithic. In *Handbook of Archaeoastronomy and Ethnoastronomy*; Ruggles, C.L.N., Ed.; Springer Reference; Springer: New York, NY, USA, 2015; ISBN 978-1-4614-6140-1. [CrossRef]
4. Maglova, P.V.; Stoev, A.D. Thracian Sanctuaries. In *Handbook of Archaeoastronomy and Ethnoastronomy*; Ruggles, C.L.N., Ed.; Springer Reference; Springer: New York, NY, USA, 2015; ISBN 978-1-4614-6140-1. [CrossRef]
5. Hajdas, I.; Ascough, P.; Garnett, M.H.; Fallon, S.J.; Pearson, C.L.; Quarta, G.; Spalding, K.L.; Yamaguchi, H.; Yoneda, M. Radiocarbon dating. *Nat. Rev. Methods Primers* **2021**, *1*, 62. [CrossRef]
6. Smith, D.; Lewis, D. Dendrochronology. In *Encyclopedia of Quaternary Science*; Elsevier: Amsterdam, The Netherlands, 2007; pp. 459–465. [CrossRef]
7. Harmon, R.S.; Schwarcz, H.P.; Gascoyne, M.; Hess, J.W.; Ford, D.C. Paleoclimate Information From Speleothems: The Present As A Guide To The Past. In *Studies of Cave Sediments*; Sasowsky, I.D., Mylroie, J., Eds.; Springer: Dordrecht, The Netherlands, 2007; pp. 199–226. [CrossRef]
8. Todorova, H.; Vaysov, I. *New Stone Age in Bulgaria*; Nauka i Izkustvo: Sofia, Bulgaria, 1993; 287p. (In Bulgarian)
9. Todorova, H. *The Stone-Copper Age in Bulgaria (Fifth Millennium BC)*; Nauka i Izkustvo: Sofia, Bulgaria, 1986. (In Bulgarian)
10. Stoev, A.; Stoytchev, T. Lunar Observatories on the Bulgarian Lands. *Interdiscip. Stud.* **1991**, *18*, 137–144. (In Bulgarian)
11. Gerassimova–Tomova, V.; Stoytchev, T.; Stoev, A. Astronomical Symbolism at Rock-Cave Art of Painting. *Interdiscip. Stud.* **1991**, *18*, 203–213. (In Bulgarian)
12. Stoytchev, T.; Stoev, A. Astronomical Interpretation of Some Graffiti from the Bulgarian Caves. *Interdiscip. Stud.* **1991**, *18*, 194–202. (In Bulgarian)
13. Muglova, P.; Stoev, A.; Stoeva, M. Research on Drawings Representing Celestial Phenomena and Cosmological Elements from Cave Sanctuary from the Neolithic. In Proceedings of the 12th International Congress of Speleology, La Chaux de Fonds, Switzerland, 10–17 August 1997; Symp. 2: Archaeology and Paleontology in Caves. Volume 3, pp. 95–96.
14. Maglova, P.; Stoev, A.; Spasova, M. Eneolithic solar calendar in the Magura cave near the village of Rabisha, Belogradchik municipality, Bulgaria. In *Harmony and Symmetry. Celestial Regularities Shaping Human Culture, Proceedings of the SEAC 2018 Conference in Graz*; Draxler, S., Lippitsch, M.E., Wolfschmidt, G., Eds.; Tredition: Hamburg, Germany, 2020; pp. 45–53. ISBN -978-3-347-14632-7.
15. Muglova, P.; Stoev, A.; Radoslavova, T. Thracian rock sanctuary "Belintash" near Mostovo village, Plovdiv district—A model of an ancient solar observatory. *Interdiscip. Stud.* **1990**, *17*, 156–167.
16. Potemkina, T.M.; Maglova, P.V.; Stoev, A.D. On the methodology of archaeoastronomical research, Worldview of ancient and traditional societies of Eurasia. In *Memory of V.N. Chernetsov, Collection of Articles, Russian Academy of Sciences, Institute of Archeology*; TAUS Publishing House: Moscow, Russia, 2006; pp. 189–197. ISBN 5-903011-04-7.
17. Kulikov, K.A. *A Course in Spherical Astronomy*; Nauka: Moscow, Russia, 1974; 232p. (In Russian)
18. Ruggles, C.L.N. Pushing Back the Frontiers or Still Running Around in the Same Circles? 'InterpretativeArchaeoastronomy' Thirty Years On. *Proc. Int. Astron. Union* **2011**, *7*, 1–18. [CrossRef]
19. Stoeva, P.; Stoev, A.; Spasova, M. Late Chalcolithic solar-chthonic rock-cut structures for time measuring in the Eastern Rhodopes, Bulgaria, In Mediterranean Archaeology and Archaeometry, special issue dedicated to the proceedings of SEAC 2015 conference "Astronomy in past and present cultures" Rome, 9–13 November 2015. *Mediterr. Archaeol. Archaeom.* **2016**, *16*, 401–406. [CrossRef]
20. Stoev, A.; Muglova, P.; Spasva, M. Sun in the cult practices on the territory of the megalithic sanctuary "Buzovgrad". In Proceedings of the First International Symposium "Ancient cultures in South-East Europe and the Eastern Mediterranean. Megalithic Monuments and Cult Practices", Blagoevgrad, Bulgaria, 11–14 October 2012; pp. 207–216.
21. Stoev, A.; Maglova, P. Astronomical markers and orientations of the rock-cut monument "Markov Kamak" in the Tsarev Peak region, Rila Mountain. In *"Boundaries of Cultural Studies"-Collection of Works on the Occasion of 15 Years Since the Foundation of the*

Specialty "Cultural Studies" at South-West University "Neofit Rilski"; Neofit Rilski University Press: Blagoevgrad, Bulgaria, 2010; pp. 72–92. ISBN 978-954-680-693-2.
22. Stoev, A.; Maglova, P.; Markov, V.; Spasova, M. Prehistoric rock sanctuary with arch near the village of Kovachevitsa, Bulgaria: Spatial orientation and solar projections. *Mediterr. Archaeol. Archaeom.* **2016**, *16*, 413–419. [CrossRef]
23. Maglova, P.; Stoev, A.; Spasova, M.; Benev, B. Eneolithic Projection System for Astronomical Observations on the Territory of Megalithic Sanctuary near the Village of Kovil (Municipality of Krumovgrad, Kurdzhali District). In Proceedings of the Second International Symposium "Megalithic Monuments and Cult Practices", Blagoevgrad, Bulgaria, 12–15 October 2016; pp. 55–64.
24. Stoev, A.; Maglova, P.; Markov, V.; Spasova, D.; Genov, A. Structure of the sacred space, astronomical orientation and functional evolution of the rock-cut monument near the village of Lilych, Kyustendil region, Bulgaria. In *Harmony and Symmetry. Celestial Regularities Shaping Human Culture, Proceedings of the SEAC 2018 Conference in Graz*; Draxler, S., Lippitsch, M.E., Wolfschmidt, G., Eds.; Tredition: Hamburg, Germany, 2020; pp. 39–44. ISBN -978-3-347-14632-7.
25. Stoev, A.D.; Varbanova, Y. Positional systems for solar and lunar observations in the archaic cultures in Bulgaria. In *Astronomical Traditions in Past Cultures, Proceedings of the First Annual General Meeting of the European Society for Astronomy in Culture (SEAC), Smolyan, Bulgaria, 31 August–2 September 1993*; IA-BAS: Sofia, Bulgaria, 1996; pp. 93–100.
26. Stoev, A.; Maglova, P.; Stoeva, M.; Tashev, Y.; Videnov, B.; Velkov, R.; Velkov, V. An archaeoastronomical investigation of theHarman kaya rock-cu tmegacomplex in the region of Dolna Chobanka village, Momchilgrad municipality. In *Thracia 15 in Honour of Alexander Fol's 70th Anniversary*; Institute of Thracology, BAS, Tangra TanNakRa Publhing House: Sofia, Bulgaria, 2003; pp. 333–339.
27. Stoeva, P. Ancient Solar Observations from Rock Sanctuaries at East Rhodopes, Bulgaria. In Proceedings of the International Conference under the Patronage of UNESCO "Astronomy and World Heritage: Across Time and Continents", Kazan, Russia, 19–23 August 2009; Marov, M., Ed.; Publishing House of Kazan University: Kazan, Russia, 2016; pp. 145–152, ISBN 978-5-00019-618-2.
28. Ruggles, C.L.N. Long-Term Changes in the Appearanceof the Sky. In *Handbook of Archaeoastronomy and Ethnoastronomy*; Ruggles, C.L.N., Ed.; Springer Science+Business Media: New York, NY, USA, 2015. [CrossRef]
29. Bradley, R.S. *Palaeoclimatology*, 2nd ed.; Academic Press: San Diego, CA, USA, 1999.
30. Available online: https://www.globalchange.umich.edu/globalchange1/current/lectures/kling/paleoclimate/index.html (accessed on 10 October 2022).
31. Available online: https://www.usgs.gov/programs/climate-research-and-development-program/science/paleoclimate-research (accessed on 12 October 2022).
32. Bowen, R. *Isotopes in the Earth Sciences*; Elsevier: London, UK, 1988; 662p, ISBN 1-85166-145-X.
33. Rollinson, H. *Using Geochemical Data, Evaluation, Presentation, Interpretation*; Routledge: London, UK, 1993; eBook Published 1 October 2013. [CrossRef]
34. Garrison, E. *Techniques in Archaeological Geology*, 2nd ed.; Springer: Berlin, Germany, 2016.
35. Godwin, H. The Croonian Lecture: Radiocarbon dating and Quaternary history in Britan. *Proc. R. Soc. Lond. B Biol. Sci.* **1961**, *153*, 287–320. [CrossRef]
36. Heaton, T.J.; Blaauw, M.; Blackwell, P.G.; Ramsey, C.B.; Reimer, P.J.; Scott, E.M. The IntCal20 approach to radiocarbon calibration curve construction: A new methodology using Bayesian splines and errors-in-variables. *Radiocarbon* **2020**, *62*, 821–863. [CrossRef]
37. Taylor, R.E.; Bar-Yosef, O. *Radiocarbon Dating*, 2nd ed.; Left Coast Press: Walnut Creek, CA, USA, 2014; ISBN 978-1-59874-590-0.
38. Macdougall, D. *Nature's Clocks: How Scientists Measure the Age of Almost Everything*; University of California Press: Berkeley, CA, USA, 2008; ISBN 978-0-520-24975-2.
39. Shopov, Y.Y.; Stoykova, D.; Tsankov, L.; Sanabria, M.; Lundberg, J.; Georgiev, L.; Forti, P.; Georgieva, D. Verification of the Causes of Glaciations and Sea Level Changes Using the Records of Calcite Speleothems. *Int. J. Speleol.* **2000**, *29*, 71–75. [CrossRef]
40. Liritzis, I.; Galloway, R.B. Solar-Climatic effects on Lake/Marine sediment radioactivity variations. In: Holocene Cycles. Climate, Sea Levels, and Sedimentation. *J. Coast. Res.* **1995**, *10*, 63–71.
41. Tarashtan, A.N. *Luminescence of Minerals*; Naukova Dumka: Kiev, Ukraine, 1978. (In Russian)
42. Shopov, Y.Y.; Stoykova, D.; Tsankov, L.; Sanabria, M.; Georgieva, D.; Ford, D.; Georgiev, L. Influence of Solar Luminosity over Geomagnetic and Climatic Cycles as Derived from Speleothems. *Int. J. Speleol.* **2004**, *33*, 19–24. [CrossRef]
43. Stoykova, D.; Shopov, Y.Y.; Gabreva, D.; Tsankov, L.T.; Yonge, C.J. Origin of the Climatic Cycles from Orbital to Sub-Annual Scales. *J. Atmos. Sol. Terr. Phys.* **2008**, *70*, 293–302. [CrossRef]
44. Marcott, S.A.; Shakun, J.D.; Clark, P.U.; Mix, A.C. A reconstruction of regional and global temperature for the past 11,300 years. *Science* **2013**, *339*, 1198–1201. [CrossRef]
45. Rasmussen, S.O.; Andersen, K.K.; Svensson, A.M.; Steffensen, J.P.; Vinther, B.M.; Clausen, H.B.; Siggaard-Andersen, M.-L.; Johnsen, S.J.; Larsen, L.B.; Dahl-Jensen, D.; et al. A new Greenland Ice core chronology for the last glacial termination. *J. Geophys. Res.* **2006**, *111*, D06102. [CrossRef]
46. Vergiev, S.; Filipova-Marinova, M. Pollen-based paleoclimate reconstructions of North-Eastern Bulgaria during the last 7000 years using modern analog technique (MAT). *Rev. Bulg. Geol. Soc.* **2020**, *81*, 155–157. [CrossRef]
47. Vergiev, S. Pollen-based paleoclimate reconstructions of Beloslav Lake (Northeastern Bulgaria) during the last 6000 years using modern analog technique (MAT). *Rev. Bulg. Geol. Soc.* **2021**, *82*, 132–134. [CrossRef]
48. Evlogiev, J. The boundary Neogene-Quaternary and stratigraphy of the Early Pleistocene in the Tethys, the CentralParathetys and Ocean. *Rev. Bulg. Geol. Soc.* **1995**, *56*, 87–104. (In Bulgarian)

49. Walker, M.; Johnsen, S.; Rasmussen, S.O.; Steffensen, J.P.; Popp, T.; Gibbard, P.; Hoek, W.; Lowe, J.; BjoËrck, S.; Cwynar, L.; et al. The Global StratotypeSection and Point (GSSP) for the base of the Holocene Series/Epoch (Quaternary System/Period) in the NGRIP ice core. *Episodes* **2008**, *31*, 264–267. [CrossRef]
50. Filipova-Marinova, M. Archaeological and paleontological evidence of climate dynamics, sea-level change, and coastlinemigration in the Bulgarian sector of the Circum-Pontic Region. In *The Black Sea Flood Question: Changes in Coastline, Climate, and Human Settlement*; Yanko-Hombach, V., Gilbert, A.S., Panin, N., Dolukhanov, P.M., Eds.; Springer: Dordrecht, The Netherlands, 2007; pp. 453–482.
51. Kucherov, N.I. Astroclimate, Znanie ed.: Moscow, Russia, 1962; pp. 10–24.
52. Rozenberg, E. *Twilight: A Study in Atmospheric Optics*; translated from Russian; Plenum: New York, NY, USA, 1966; 358p.
53. Baltakov, G.; Kenderova, R. *Quaternary Paleogeography*; Maleo: Varna, Bulgaria, 2003; 324p. (In Bulgarian)

quaternary

Article

Tracing Raw Material Sources of Prehistoric Stone Artefacts by Non-Invasive Techniques: The Case of the Early Bronze Age (3rd Mill. BCE) Site of Vathy, Astypalaia, Greece

Maria Kokkaliari [1,*], Eugenia Adam [2], Andreas Vlachopoulos [3] and Ioannis Iliopoulos [1]

1 Department of Geology, University of Patras, 26504 Patras, Greece
2 Ephorate of Antiquities of Ioannina, Ministry of Culture and Sports, 45221 Ioannina, Greece
3 Department of History and Archaeology, School of Philosophy, University of Ioannina, 45110 Ioannina, Greece
* Correspondence: kokkaliari_m@upnet.gr

Abstract: Recent findings of archaeological research in the Vathy gulf area, Astypalaia Island, indicate its continuous habitation since prehistoric times, most importantly in the transitional period from the Final Neolithic to the Early Bronze Age (late 4th/early 3rd millennium BC). The evaluation of the prehistoric stone artefacts from Vathy using non-invasive analytical methods (Near Infrared Spectroscopy—NIR), in combination with the mineral-petrographic characterization of the main lithological formations of the island, is expected to provide important information about raw material procurement and possible exchange networks. The geological study of the island combined with the analytical methods applied to the archaeological artefacts and the geological samples led to the identification of both local and allogenic materials. The possible locations of raw material sources were established and the origin of allogenic materials was estimated. The stone artefacts made of local geo-materials consist mainly of calcitic sandstone, shale, marl, and limestone/marble, comprising the largest part of the lithological formations of the island, as well as pumice and volcanic rocks of varying chemical composition. By means of a portable microscope and NIR spectroscopy, we were further able to identify allogenic geo-materials including chalcedony, mica schist, bauxite and meta-bauxite, steatite, and paragonite. Based on the mineralogical and petrographic characterization of the stone artefacts, a first attempt is made to evaluate the possible raw material sources and to identify potential intra-island modes of stone exploitation.

Keywords: geomaterials; near infrared spectroscopy—NIRS; lithic artefacts; non-invasive methods; raw materials; archaeometry; Vathy; Astypalaia Island

Citation: Kokkaliari, M.; Adam, E.; Vlachopoulos, A.; Iliopoulos, I. Tracing Raw Material Sources of Prehistoric Stone Artefacts by Non-Invasive Techniques: The Case of the Early Bronze Age (3rd Mill. BCE) Site of Vathy, Astypalaia, Greece. *Quaternary* **2022**, *5*, 42. https://doi.org/10.3390/quat5040042

Academic Editor: Leszek Marks

Received: 29 July 2022
Accepted: 4 October 2022
Published: 9 October 2022

1. Introduction

Archaeometry focuses on the application of analytical techniques in archaeological finds to provide information about dating, prospecting, analysis of technology and provenance, as well as reconstruction of a coeval environment [1], thus providing another scope for the interpretation of cultural and archaeological growth and evolution [2–5]. Many researchers [6–12] have used analytical techniques in a variety of archaeological findings to acquire knowledge about the source rock material (including volcanic rocks, chert, clayey raw materials, sandstones, etc.), their provenance, or even their utility.

For this reason, the use of non-invasive techniques is a matter of great importance, contributing significantly to the archaeological desiderata arising, while keeping the findings intact. The study of archaeomaterials include fundamental techniques widely used for their chemical identification (i.e., X-ray Fluorescence analysis—XRF, RAMAN Spectroscopy) or their textural analysis (i.e., Scanning Electron Microscope—SEM).

According to Hunt [13], reflectance spectroscopy can be defined as the technique that uses the energy in certain wavelength regions of the electromagnetic spectrum to analyze minerals. The operating principle of this methodology is based on the process of energy absorption in the near infrared region (350–2500 nm) to describe changes in molecular level, measuring the energy of the bond vibrations between the cations and the molecules. Molecular vibrations are in fact internal movements of atoms related to changes of the length and the angle of their bonds.

The bonds vibrate differently, at discrete wavelengths, as a function of their length, representing consequently the presence of certain NIR-active chemical compounds, leading to specific mineralogical components. Vibration modes are generally distinguished in stretching and bending vibrations because the relative positions of the atoms are not absolutely fixed and are, on the contrary, constantly fluctuating. The stretching vibrations are caused by changes in the distance between atoms along their bond axis. The bending vibrations are related to the change in the angle between two bonds, and characterized by scissoring, rocking, wagging and twisting motions. When a molecule absorbs infrared radiation, the molecular vibrations trigger a net change in the dipole moment, depending on the value of the charge difference and the distance between the two charged centers. Even though the study of molecular vibrations was originally based on Hook's Law, describing a simple harmonic oscillator in quantum physics can be considered only approximately simple harmonic due to the inelastic nature of the molecular bonds. Normal, or fundamental absorptions occur when a molecule is excited from ground to the lowest-energy excited state (v = 1). Overtones occur when fundamental mode is excited with two or more quanta of energy (2 v, 3 v etc.), resulting in a higher energy state. Finally, combination bands result from two or more fundamental vibrations that are excited simultaneously.

This methodology is based on diffuse reflectance spectroscopy and is considered a non-invasive method as it can be applied in-situ, directly to any material and is appropriate when no sample preparation is needed, retrieving characteristic spectra from the visible to the near infrared region, thus providing information about its chemical composition. When infrared light penetrates the sample, it can be reflected, or scattered, following a chain path until finally exiting the sample. This whole process will cause the initial energy to be lost, since part of it will be absorbed by the material. The detector will collect the new infrared radiation, now containing some new information. Diffuse reflection occurs since the light penetrates the surface layer of the particles, excites the vibrational regions of the target molecule, and then scatters in all directions. Thus, the reflection spectrum produced, depends on the composition of the sample.

In the field of archaeometry, only few researchers have applied NIR spectroscopic analysis to a variety of materials, including ceramics [14], ancient wood [15], paper [16], or even with the use of imaging spectroscopy in paintings [17] etc. Tanyaş et al. [18], studied the spectral properties of chlorite stone bowls to explore the regional locations of the source rocks, incorporating ASTER satellite data, and also emphasizing in petrographic and geochemical analysis, that critically supplement the spectroscopic analysis.

Although NIR spectroscopy has numerous advantages, being economical and a direct method without any necessary sample preparation, its non-invasive nature, as in most methodologies, necessitates the contribution of other methodologies as well, to provide a reliable evaluation of the material. Spectral signatures can be affected by many factors, causing the problematic interpretation of spectral signatures. These factors mainly deal with the presence of multiple mineral phases resulting in the presence of overlapping spectral absorptions, as well as the textural characteristics of the material (i.e., the presence of porosity, the grain size of the minerals) that may affect the spectra. In this study, we attempt to thoroughly examine a variety of archaeological artefacts, regarding the use of non-invasive techniques, such as NIR spectroscopy, in the identification of the source materials to draw conclusions about their provenance. The study of the archaeological artefacts was also based on microscopic observation, providing further information about the microstructure of the stone tools and their mineralogical composition in order to eliminate the above concerns.

Astypalaia is one of the Dodecanese islands in Southern Aegean (Figure 1). Although important finds from historical times have been located on the island, its prehistory was until recently overlooked. Chance finds at the Pyrgos peninsula overlooking the Vathy gulf ("Vathy") in the Northeastern part of the island in 2008 and extensive surveys and excavations since 2011 [19,20] led to the location of archaeological remains from prehistoric to modern times, indicating almost unceasing habitation. The prehistoric occupation is as early as the Final Neolithic/Early Bronze Age (4th–3rd millennia BC); the location and subsequent excavation of infant jar burials and the discovery of a prehistoric fortified acropolis with important petroglyphs [19–21] constitute Vathy a diachronic palimpsest and place Astypalaia within the broader frame of insular prehistory as known from the other Dodecanese islands and mostly the Cyclades.

Figure 1. Oblique satellite view (Google Earth Pro) of Greece and Astypalaia Island in the Aegean Sea.

During the surveys, a total of 117 ground-stone implements [22,23] were recovered (Table 1); the collection includes mostly utilitarian items (suggesting fishing and processing activities, e.g., grinding and pounding of edibles or other materials), along with two stone beads. A retouched geometric artefact, originally thought to be of flint is also included in the analyzed sample (a final total of 118 artefacts).

Table 1. Typology of the assemblage recovered during the surveys.

Type	No of
discs	2
pull weights	1
pumice	2
adze	1
grinding slabs	2
pounders/polishers	5
utilized pebbles	14
stone beads	2
pebbles	54
bauxite	2
serpentine	1
flint nodule	1
copper slags	3
unidentified	27
geometric tool	1
Total	118

Contacts with other Aegean islands are to be found in the chipped stone assemblage made from obsidian of mostly Melian origin (macroscopic identification [24]) as well as in pottery imported from the Cyclades [25]. With that in mind, it was of interest to investigate the provenance of the raw materials employed in the manufacture of the ground tools and implements that were also recovered at the site [26].

The evaluation of the stone artefacts using non-invasive analytical methods (Near Infrared Spectroscopy—NIRS), combined with the mineral-petrographic characterization of the main lithological formations of the island, is expected to provide important information about the commercial transactions of that time and the expertise of the islanders on stone exploitation. Information about the geologic structure of the island in combination with the evaluation of the archaeological and geological materials led to the identification of local and non-local materials.

2. Materials and Methods

2.1. Archaeological Artefacts

The collection of artefacts retrieved during archaeological fieldwork at Vathy comprises lids, grindstones, pounders, rubbers, polishing pebbles, floaters, two beads, a chisel, and a single bifacial geometric tool (Figure 2). Numerous obsidian tools [24] and marble [25] are also included in the lithic finds, constituting other subjects of research. During our research, 118 stone artefacts were studied through the in-situ application of non-invasive, non-destructive methods since rapid technological development provide important information without interfering in the archaeological artefacts.

Figure 2. Representative photographs of the stone artefacts analyzed in the present study.

The artefacts were studied macroscopically with the aid of a portable microscope (Wifi ProScope mobile, Bodelin Technologies, Oregon City, OR, USA) and the capture of micro-photographs (Figure 3a). The mineralogical identification was determined through the application of Near Infrared Spectroscopy (NIR) using an SM-3500 specTERRA (Spectral Evolution Inc., Lawrence, MA, USA) portable apparatus, with a field of view (FOV)

2 × 1.5 cm, in contact with the surface of the samples, and a total number of 1200 of spectra were evaluated (Figure 3b). The data acquired ranges from 0.35 to 2.5 μm with a spectral resolution according to the software, between 3.5 and 8 nm. The spectra were finally produced by the instrument at 1 nm intervals, using the DARWin SP Data Acquisition software (Spectral Evolution Inc., Lawrence, MA, USA), which is accompanied by the United States Geological Survey (USGS spectral library) mineral standards database. Sample spectra were first calibrated relative to a (5 × 5 cm in size standard panel (Spectral Evolution, Inc. Lawrence, MA, USA).

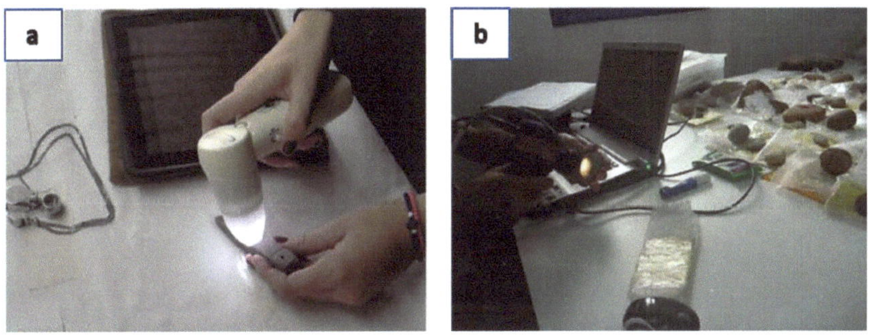

Figure 3. Non-destructive analytical techniques employed for the evaluation of the stone artefacts studied: (**a**) use of the WiFi ProScope mobile microscope paired with iPad; (**b**) use of Near Infrared (NIR) portable spectrometer SM-3500 specTERRA.

The spectra are characterized by absorption features, due to energy-matter interactions at this region caused by molecular vibrations at specific frequencies. Consequently, the absorption features allow us to determine which functional molecular group participates in the structure of the mineral. The major spectral features in the NIR region are a result mostly of the water content and the vibrations of the hydroxyl groups, such as Al-OH, Fe-OH, Mg-OH [27]. Therefore, it is easy to identify hydrous minerals, such as amphiboles, mica, clay minerals, etc. Minerals, such as quartz and feldspars, do not include hydroxyl or water in their structure, thus they cannot be identified through NIRS. The main characteristic absorption features imprinted through NIR spectroscopy are listed in Table 2.

Table 2. Major absorption features [27] of the minerals identified in the present study.

Wavelength (nm)	Mechanism	Mineral Group
~1400	OH- and water	clay minerals, hydroxyls, zeolites
~1900	water	smectites
~2200	Al-OH	clay minerals, amphiboles, mica
~2250	Fe-OH	epidote, biotite, tourmaline, chlorite, phlogopite
~2300–2400	Mg-OH	amphiboles, epidote, chlorite
~2340	$CO_3{}^{2-}$	carbonates

2.2. *Raw Materials*

Destructive methods were used in the lithologic specimens, including petrographic study and mineralogical identification. Through the petrographic observation, the lithologic samples of Astypalaia Island were studied in detail, providing further information about the texture of the samples as well as the grain size of the minerals and the mineral evaluation was verified through the XRPD.

Petrographic thin sections were prepared and studied by means of transmitted light microscopy (Zeiss AxioScope A.1, Carl Zeiss Microscopy Deutschland GmbH, Oberkochen, Deutschland) in order to determine the rock type and mineral mode. The mineralogy of the samples was further verified in the Mineral and Rocks Research Laboratory, Department of

Geology, University of Patras, Greece, by means of X-ray Powder Diffraction (BRUKER D8 Advance, Bruker AXS, Karlsruhe, Germany) using Ni filtered Cu-Kα radiation, operating at 40 kV and 40 mA and employing a Bruker Lynx Eye fast detector. Samples were step-scanned from 2° to 70° 2θ and a time/step of 0.2 s. Samples were powdered (<10 μm) in a vibration disc mill using an agate grinding set and randomly mounted in a sample holder. The crystalline phases were identified using the DIFFRACplus EVA software (Bruker-AXS, Bruker-AXS, Madison, WI, USA), based on the ICDD Powder Diffraction File (2006 Version).

3. Geologic Setting of Astypalaia and Field Observations

To answer the question about the provenance of the raw materials present at Vathy, a detailed bibliographic review was carried out for the in-depth study of the local lithologies. Field observations through extensive sampling and GPS recording provided valuable information for the petrographic study and the determination of the mineralogical composition of the rocks. For this reason, we tried to perform as much as possible a representative sampling (41 samples in total) at Pyrgos-Vathy, Analipsi, and Livadi peninsulas (Figure 4).

Figure 4. Oblique satellite view (Google Earth Pro) of Astypalaia Island showing the sampling sites.

The geologic structure of Astypalaia has been studied by many researchers [28–33] and consists of sedimentary rocks, i.e., Upper Jurassic to Upper Eocene Alpine geologic formations, as well as meta-Alpine lithologies. According to the geologic map (Figure 5) of [32,34], the lithology of the island consists mostly of limestones, whilst the larger part of the Western region of the island consists of Flysch (Upper Eocene–Oligocene), including sandstones, mudstones, and conglomerates (Figure 6a). In Western Astypalaia, Upper Jurassic neritic medium-bedded limestones, approximately 250 m thick, are overthrusted on flysch. The Eastern part of the island comprises Senonian–Maastrichtian, thick-bedded to massive, rudist limestones and Paleocene–Middle Eocene, thick-bedded to massive, black, nummulitic limestones (Figure 6b). Quaternary sedimentary rocks characterize meta-Alpine lithologies. At several sampling sites, dynamic deformation characteristics were also observed (Figure 6c). Finally, in many sites in the Central part of the island Poros occurred, of aeolian origin, consisting of calcarenites as well as marine and continental fauna (Figure 6d).

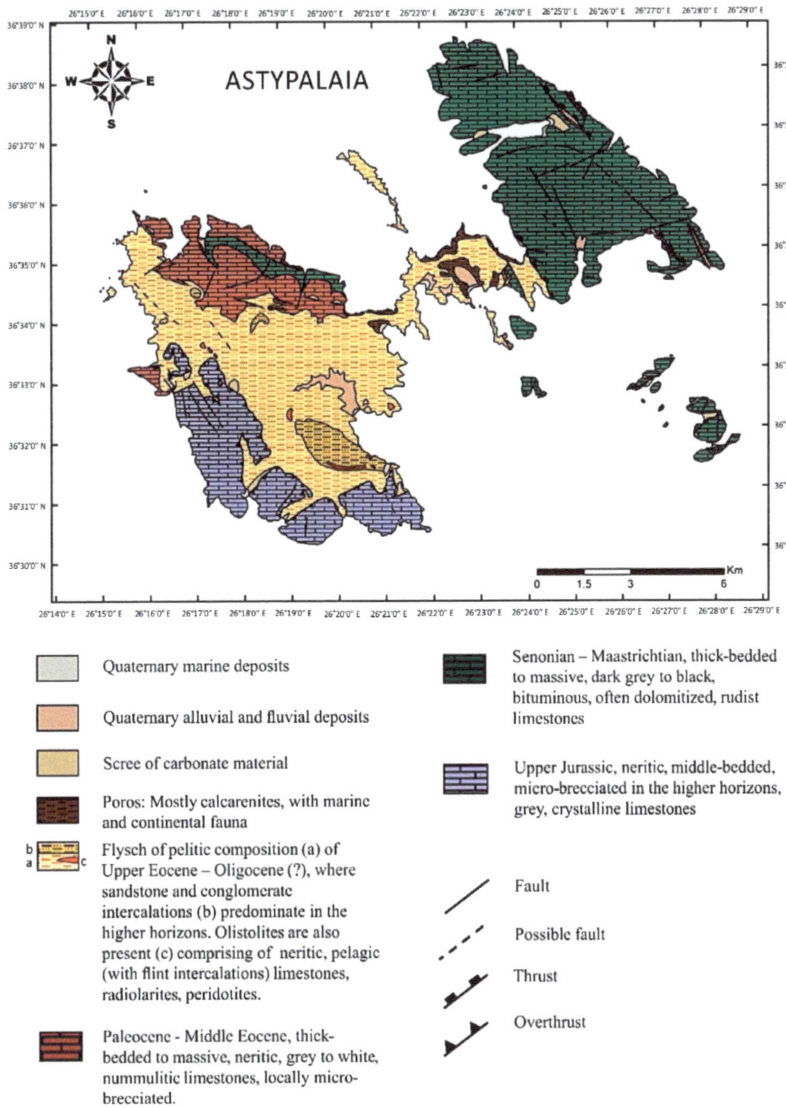

Figure 5. The geological map of Astypalaia (digitized and modified from [34]).

Figure 6. Representative photographs during our field work: (**a**) Sandstone and mudstone intercalations; (**b**) of the Jurassic limestones over thrusting Eocene flysch; (**c**) characteristics of dynamic deformation through the presence of folded sandstones; (**d**) Calcarenite quarry.

4. Results and Discussion

4.1. Microscopic Observation and Spectroscopic Characteristics of the Prehistoric Artefacts

The evaluation of the raw material of the prehistoric artefacts was conducted through microscopic observation and the application of NIRS (Figure 7). The archaeological sample includes: sandstone, shale, marl, limestone and marble, pumice and volcanic rocks of varying chemical composition, steatite, mica schist, obsidian, chalcedony, and bauxite and meta-bauxite. It is important however to study the provenance of the raw materials in order to characterize them as either local, if they belong to the local lithology, or as non-local. This information will clarify the systems of raw material procurement on Astypalaia during the period under study.

As concluded from the application of NIR spectroscopy, the majority of the studied stone implements (grindstones, polishing pebbles, and pounders) are comprised of calcite ($CaCO_3$) (Figure 7a). Calcite has diagnostic features in the near infrared region of the electromagnetic spectrum, displaying a characteristic doublet feature at 1880 nm and 1990 nm approximately, as well as a deep major absorption feature near 2340 nm, shifted to lower wavelength position for more dolomitic ($MgCO_3$) compositions [35,36]. Even if the major absorption feature is distinguishable in all studied spectra indicating the presence of calcite, we observed differences in the reflectance values, the depth and the intensity of the absorption features. The differences are attributed to the textural characteristics of the lithotypes that were used as raw, i.e., the grain size of calcite, the cohesion of the grains, internal grain defects, the presence of impurities, or the presence of inclusions and, if so, the inclusion size [37]. According to [36], powdered carbonate materials have nearly flat spectral characteristics for wavelengths lower than 1600 nm. This parameter seems to be critical for the stone artefacts that display this characteristic, indicating the use of micritic limestone as raw material, i.e., limestone with a grain size of calcite of <4 μm. Other spectra display a negative slope toward higher wavelengths, and even if the major absorption features at 2340 nm approximately are distinguishable, the characteristic doublet is hardly seen and is not clearly developed. As described by [37], even small amounts of organic matter in the mineralogical composition can suppress the spectral bands. Broad, rounded features at 1400 nm and 1900 nm indicate the presence of molecular water either due to hydration or trapping water in the crystal lattice. Furthermore, small absorption features at 1900 nm region are often observed and are attributed to very small water polymers forming inclusions, adhered to the inclusion walls [37,38].

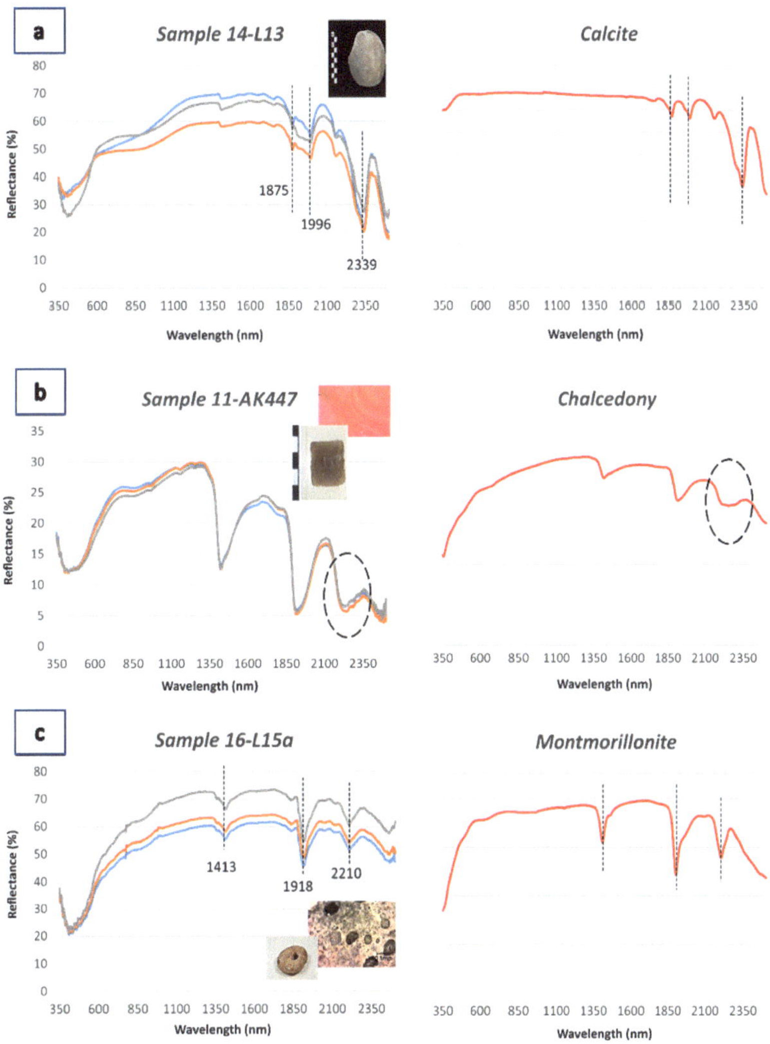

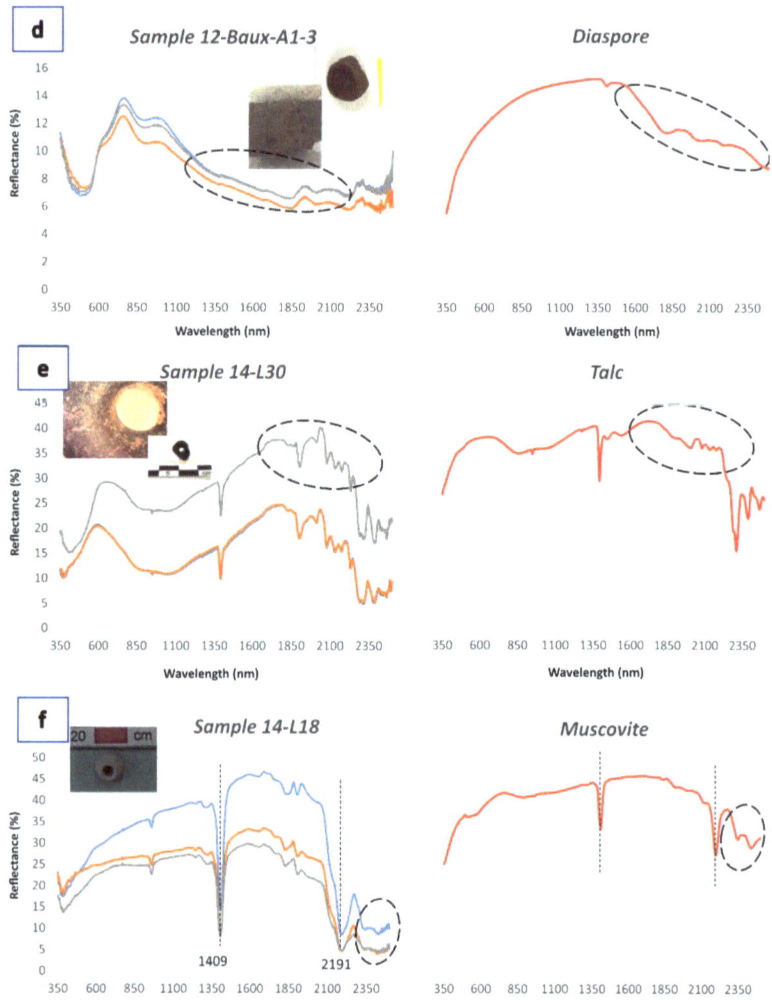

Figure 7. Representative reflectance spectra of the studied stone artefacts (left column), in comparison to the associated laboratory reference, spectra of the USGS spectral library (right column) that were used in the present study [39]: (**a**) pebble; (**b**) bifacial geometric tool; (**c**) floater; (**d**) rubber; (**e**) and (**f**) bead.

We also identified spectra that are characterized by absorption features at approximately 2200 nm and at 2290 nm, except for the absorption bands of calcite, which indicate the presence of more mineralogical components in the raw material. The specific wavelength position of the absorption features is attributed to Al-OH and Mg, Fe-OH molecular vibrations, respectively. The spectral signature of the raw materials, in accordance with the microscopic evaluation of the samples, are related to calcite, as well as clay minerals, i.e., kaolinite, smectite, nontronite, and chlorite. The overlapping features of these minerals make them difficult to identify, and according to the abundance of each mineral, the spectral properties of the raw materials are influenced, i.e., the reflectance values and the depth parameters of the absorption.

Spectra that are characterized by absorption features at 1400 nm and 1900 nm and a rounded absorption feature near 2240 nm due to molecular vibrations of the Si-OH group

(Figure 7b) are typical for chalcedony [27]. The impurities observed through the microscopic study, which are responsible for the color of the artefact varying from brown to white, characterize the spectral properties in the visible region of the electromagnetic spectrum. Sometimes, the spectra are differentiated displaying absorption features near 1000 nm attributed to spin-allowed crystal-field transitions due to the presence of metal cations, with a positive slope toward longer wavelengths. Some samples are not characterized by the rounded absorption feature at 2240 nm, implying the presence of a more crystalline variety of silica, i.e., quartz, rather than chalcedony. The absorption features at 1400 nm and 1900 nm correspond to molecular water H_2O absorbed in crystal grains or incorporated into the silica framework [40].

Pumice artefacts have a light yellowish to beige color, rounded surfaces and high porosity. They comprise amorphous SiO_2, and minor crystals of biotite which were observed through the microscopic study (Figure 7c). The spectral characteristics of some pumice artefacts have a nearly flat reflectance, displaying absorption features at 1900 nm as well as in the visible region of the electromagnetic spectrum at 400 nm. On the contrary, intense, distinguishable absorption features at 1900 and 2200 nm, indicating the presence of hydrous Al-rich minerals, characterize other pumice artefacts. The microscopic study verified the presence locally of altered areas, related to clay minerals, as indicated by the NIRS.

Furthermore, among the lithic artefacts, pebbles of volcanic origin were also studied, whose lithologic characteristics differ from one another, indicating their different chemical composition. Their color, sometimes reddish, while other times a light beige to dark gray color, in combination with their textural characteristics (presence or absence of porosity, shape and network of pores, mineral composition of phenocrysts and fine-grained groundmass, and the presence or absence of amorphous material) indicate different physicochemical conditions for the raw material, raising questions regarding their origin. The phenocrysts identified through microscopic observation are mainly minerals, including feldspars, especially plagioclase, as well as biotite, while the fine-grained groundmass consists of quartz, feldspars, and volcanic glass (amorphous SiO_2).

Some artefacts are comprised of compact material, with a reddish to dark greyish color, indicating a geo-material that stands out from the local lithology of the island. Through microscopic analysis, we observed dark grey color areas forming spots locally, in the reddish material, whereas rarely, some artefacts had a dark gray color uniformly. The spectral characteristics of these samples displayed absorption features at 480 nm and 840 nm, as well as a steep, negative slope from 740 nm toward higher wavelengths. A not intense but broad absorption feature is also identified at 1830 nm (Figure 7d). The microscopic study as well as the spectroscopic analysis of the artefacts indicate the presence of hematite and diaspore in the mineralogy of the raw material, which are indicative mineralogical components of bauxite and meta-bauxite. Hematite (Fe_2O_3) is an iron oxide, common in many geologic environments, indicating oxidizing conditions. Diaspore is an Al-rich mineral (α-AlO(OH)), comprising the major component of bauxite, along with gibbsite and boehmite.

The absorption features observed through the application of NIRS, at the spectra of the green bead at 1390 nm, 1900 nm, 2080 nm, 2134 nm, 2174 nm, 2232 nm, 2320 nm, 2384 nm, and 2465 nm indicate the presence of talc in the raw material (Figure 7e). Talc is a hydrated magnesium silicate ($Mg_3Si_4O_{10}(OH)_2$), occurring as an alteration product of Mg-rich silicate minerals, e.g., olivine, pyroxene, and amphiboles. The broad absorption feature at 830 nm, as well as the positive slope of the reflectance spectra from approximately 1400 nm to 2000 nm, also indicate the presence of a Fe-rich mineral phase, the absorption features of which are probably overlapped from the absorption features of talc, hence it is difficult to be determined. However, a small absorption feature at 2290 nm is attributed to chlorite. The spectroscopic characteristics of the green bead, in contribution with the textural characteristics of the samples, indicate steatite as a raw material.

The other studied bead, similar in size with the aforementioned, has a nearly light grey color and the spectral characteristics of the sample indicate the presence of Al-rich minerals, due to the characteristic absorption features at 2190 nm, indicative of Al-OH molecular vibrations (Figure 7f). These spectral characteristics, in combination with the double absorption feature at 2343 nm and 2431 nm, are typical spectral signatures for mica minerals [27,41], i.e., muscovite, paragonite, lepidolite, phengite, etc. [42]. The wavelength position of the absorption feature at ~2200 nm is used to evaluate the compositional differentiation of these minerals. Paragonite and Al-rich white mica have absorption features smaller than 2195 nm [43], whereas muscovite's absorption feature is located at approximately 2200 nm and phengite's is at up to 2235 nm [44]. Due to the white color of this artefact as well as the diagnostic feature at 2190 nm, we can limit the composition of the raw material to paragonite or Al-rich white mica.

4.2. Petrographic and Mineralogical Analysis of Astypalaian Lithotypes

The rock samples considered in the present study were characterized through petrographic examination (Figure 8) and the mineralogical composition was determined through mineralogical analysis (Figure 9, Table 3).

The petrographic investigation of the sandstone specimens (samples AST3, AST5, AST6, AST7, AST10, AST24, AST28) obtained from Astypalaia revealed that they are moderate to poorly sorted, often characterized by a cataclastic texture. The samples include medium- to fine-grained particles, comprising sub-angular to sub-rounded grains of quartz and in less abundance feldspars (Figure 8a). Subhedral to anhedral calcite, as well as chlorite and illite, comprise the cementing material. Opaque minerals as well as clay minerals are rarely identified displaying higher abundance, observed even macroscopically through light to dark reddish areas (Figure 8b). Quartz is the most abundant minerals, forming either monocrystalline or polycrystalline constituents. The presence of polycrystalline grains, which are not fully separated, indicate small transportation, thus the immature nature of the specimens. The feldspars often display characteristic twinning and are altered to calcite or sericite, i.e., fine-grained mica. Tectonic processes due to diagenesis are imprinted in the textural characteristics of the samples, i.e., the undulose extinction of quartz, the fragmentation of the crystals, and the presence of foliation. Sometimes, the samples are also characterized by a net of micro-veins, filled with fine- to coarse-grained calcite. Patches of yellowish areas that were observed through the petrographic study are attributed to small cavities that are filled by clay minerals. Minor rock fragments that were also identified in some samples include sub-rounded and sub-angular mica schist and chert.

The textural characteristics of the samples, AST8, AST13, AST14, AST22A and AST25 indicate the abundance of calcite, varying in grain size. The samples are often characterized by the presence of micro-veins, filled with more coarse-grained calcite (200–750 μm), showing in general a specific orientation but irregularity in their thickness (Figure 8c). Samples AST8 and AST13 are characterized by the presence of a grumeleuse texture, i.e., micritic clots that are surrounded by microsparitic calcite (Figure 8d). A grumeleuse texture is typical for recrystallization processes, where coarser calcite crystals are developed against smaller crystals. Samples AST21 and AST27 have an orange-beige color and comprise large calcite grains (2–4 mm). As concluded from the mineralogical analysis, the sample AST26 consists of dolomite (Figure 8e), with calcite in trace amounts. Calcite and dolomite differentiation cannot be achieved only through petrographic study, thus the mineralogical analysis by means of XRPD made a decisive contribution. The sample consist of inequigranural polygonal calcite grains, with an almost isometric growth in size, indicating a well-crystallized, more cohesive geologic formation compared to the aforementioned. Furthermore, micro-brecciated zones are observed, imprinting the tectonic processes involved in the formation of the lithotype, where the fragmentation of calcite occur.

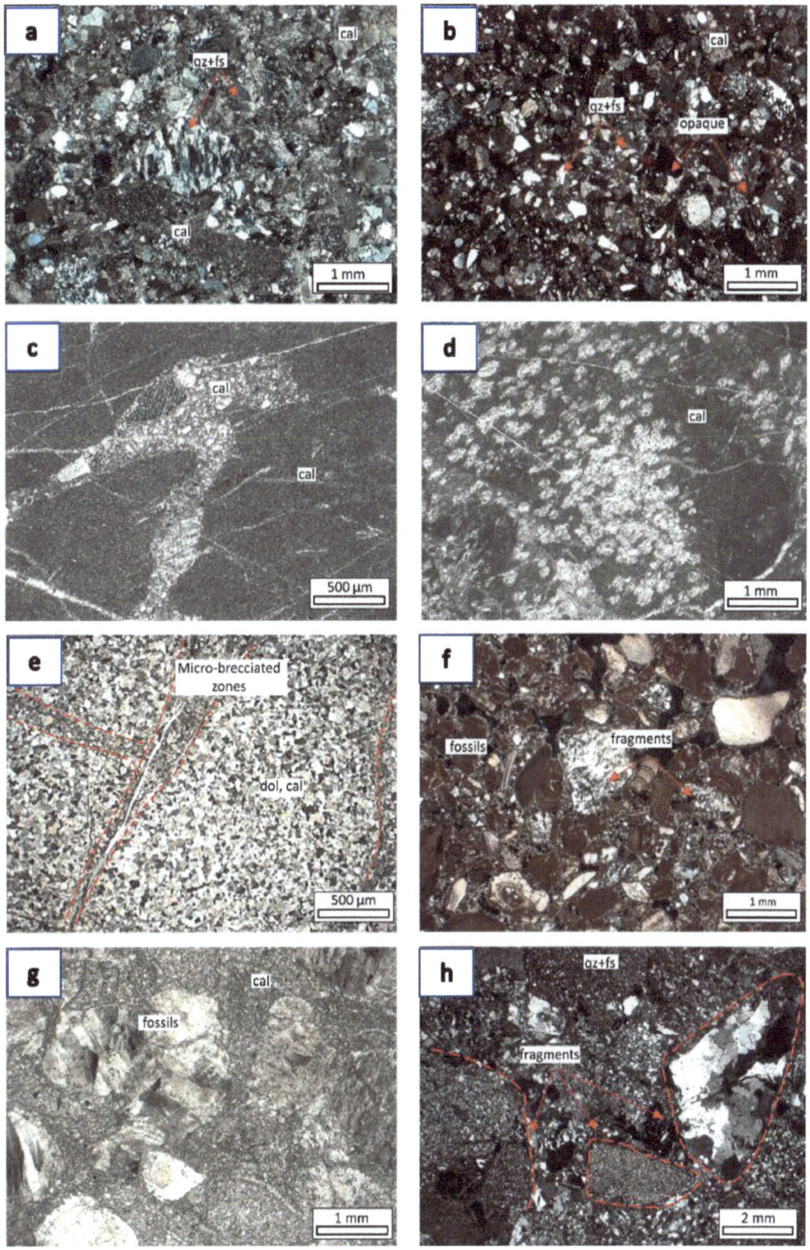

Figure 8. Representative photomicrographs of the studied samples: (**a**,**b**) sandstone specimens (samples AST6, AST7); (**c**) coarse-grained calcite crystals filling micro-veins in micritic limestone (sample AST25); (**d**) micritic limestone with the characteristic grumeleuse texture (sample AST13); (**e**) brecciated zones in coarse-grained limestone (sample AST26); (**f**) biomicrite comprising a variety of fossils and clastic fragments (sample AST9A2); (**g**) biosparite, comprising fossils and sparitic calcite (sample AST4B); (**h**) breccia lithotype characterized by subangular fragments, in fine-grained groundmass of quartz and feldspars (sample AST4A). Mineral abbreviations are according to [44]: calcite (cal); dolomite (dol); feldspars (fs); quartz (qz).

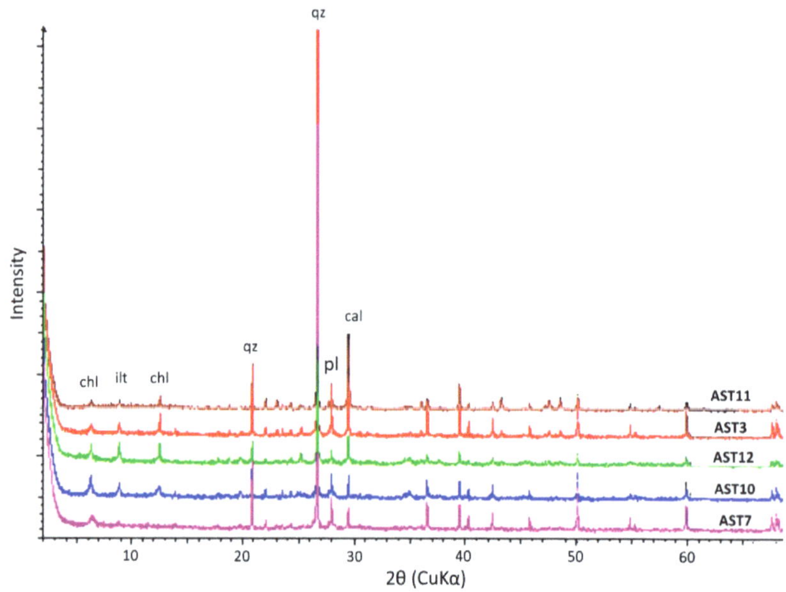

Figure 9. Representative X-ray powder diffractograms of sandstones.Mineral abbreviations are according to [44]: calcite (cal); chlorite (chl); plagioclase (pl); illite (ilt); quartz (qz).

Table 3. The mineralogy of the samples as resulted from the evaluation of the X-ray powder diffractograms. Mineral abbreviations are according to [44]: alkali-feldspars (afs); aragonite (arg); calcite (cal); chlorite (chl); dolomite (dol); illite (ilt); muscovite (ms); plagioclase (pl); quartz (qz).

Sample	XRD	Lithotype
AST-3	qz, pl, cal, ms/ilt, chl	Calcitic sandstone
AST-4B	cal	Biosparite
AST-7	qz, pl, cal, ms/ilt	Calcitic sandstone
AST-9A1	qz, pl, afs, cal, arg	Biomicrite
AST-9A2	qz, pl, afs, cal, arg	Biomicrite
AST-9B	qz, pl, afs, cal, arg	Biomicrite
AST-10	qz, pl, cal, ms/ilt, chl	Calcitic sandstone
AST-11	qz, pl, cal, ms/ilt, chl	Calcitic sandstone
AST-13	cal	Limestone
AST-16B	cal, qz	Calcitic olistholite
AST-21	cal	Calcarenite
AST-22A	cal, arg	Limestone
AST-22B	cal	Limestone
AST-25	cal, dol	Limestone
AST-26	dol, cal	Crystalline limestone/Marble?
AST-27	cal	Calcarenite
AST-28	qz, pl, cal, ms/ilt, chl, clay minerals	Calcitic sandstone

Samples AST9A1, AST9A2, and AST9B display channel-like porous, typical characteristic for loose, non-cohesive lithologies and comprise micritic calcite, bioclasts, and endoclasts (Figure 8f). The textural characteristics of the samples are indicative to biomicrites, coinciding to deposition in shallow hot water environments. The clastic fragments are in agreement with the lithologic formations of Astypalaia, comprising mostly mudstone and sandstone. In the mineralogical analysis of the samples, aragonite was also identified.

On the other hand, samples AST4B and AST15 comprise microsparitic calcite and bioclasts, indicating a fossiliferous origin typical for biosparites, differing though in the grain size of the carbonate constituents (Figure 8g). The structure of the lithotype is more cohesive and no porosity was observed macroscopically, or even microscopically. Furthermore, we also observed micro-zones filled with brecciated calcite, coinciding with the tectonic processes involved.

Sample AST16B represents a calcitic olistholith, abundant in euhedral to subhedral coarse-grained calcite crystals (250 μm–2 mm), in a fine-grained quartzitic groundmass. Samples AST4A and AST4C comprises fragments of carbonate rocks, sandstone and mudstone, quartzitic matrix, as well as calcite and minor clay mineral as cementing material, indicative for breccias (Figure 8h). The samples are moderate to poorly sorted, with sub-angular to sub-rounded constituents. The fragments are usually cracked and cut by micro-veins, rich in calcite and silicic material. Minor opaque minerals also occur alongside the cracks and the micro-veins. Quartz and feldspars are usually characterized by undulose extinction and fragmentation.

4.3. Preliminary Results on the Provenance of the Raw Materials

Our preliminary results for the main mineral raw materials are summarized in Table 4. The stone artefacts characterized as local are related to geo-materials whose lithological characteristics are consistent with the lithology of Astypalaia Island. These include sandstone, limestone/marble, marl, and shale.

Table 4. Classification of the stone artefacts according to lithological characteristics of the raw material, in local and non-local.

Local	Non-Local
Calcitic sandstone	Chalcedony
Shale	Bauxite/Meta-bauxite
Marl	Steatite
Limestone/marble	Paragonite
Pumice and volcanic rocks (rhyodacites to andesites, tuffs)	

Geo-materials characterized as non-local include chalcedony, bauxite/meta-bauxite, steatite, and mica (paragonite most probably), the closest to Astypalaia occurrences of which are illustrated in Figure 10.

Steatite, the so-called 'soapstone' due to its texture, is a magnesium-rich rock comprising mostly talc and to a lesser extent other minerals including amphiboles, chlorite and serpentine. The application of the NIRS at the steatite beads indicated the presence of talc, displaying the related characteristic absorption features at the spectra. In the Aegean Sea, steatite occurs at Tenos, Naxos, Sifnos, and Crete. According to [45], the mineralogical composition of steatite in Crete varies, exhibiting mineralogical heterogeneity, related to the complex geologic structure of the island and the associated host rocks. The main mineral phase comprises mostly of antigorite or tremolite, and other minerals related to the protolith. We can conclude that the origin of the steatite beads is not connected with the occurrences from Crete, due to the presence of talc, identified through the NIRS, thus most possibly the provenance of steatite is to be linked with Tenos, Naxos, or Sifnos.

Bauxites are sedimentary rocks which are characterized by the presence of Al-rich minerals, i.e., gibbsite [$Al(OH)_3$], boehmite [γ-AlO(OH)], and diaspore [α-AlO(OH)], thus their high Al content. Bauxites that have undergone metamorphic processes, processes related to changes in the pressure and temperature that contribute to differentiation in the mineralogy and the textural characteristics of the rock, are called 'meta-bauxites', and such occurrences are extremely rare. In the Aegean Sea islands, bauxite occur at Amorgos, whereas meta-bauxites occur at Naxos [46].

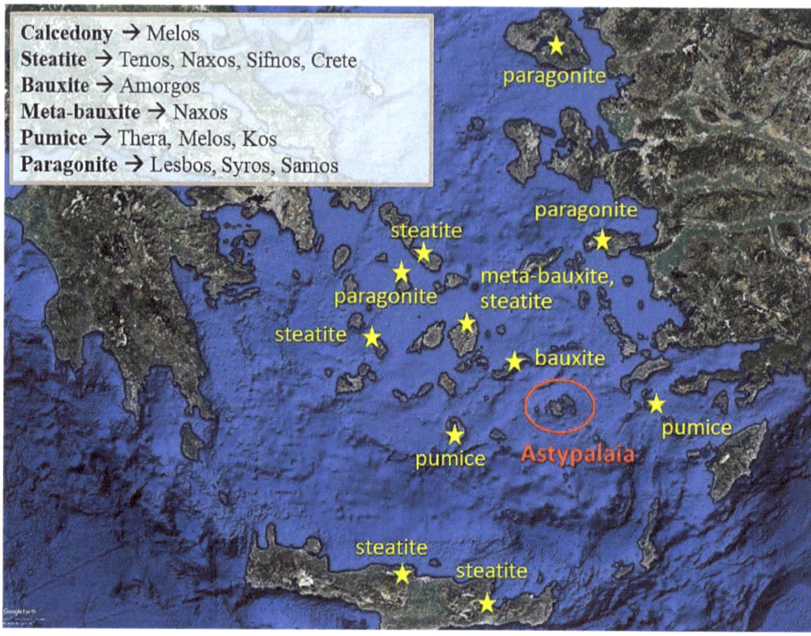

Figure 10. Oblique satellite image (Google Earth Pro) of the islands in the Aegean Sea, with the main occurrences of chalcedony, steatite, bauxite, meta-bauxite, pumice, and paragonite.

On the other hand, chalcedony is a cryptocrystalline variety of silica, mainly consisting of quartz and moganite, i.e., fine-grained crystalline aggregates undiscernible even through microscopic study. Moganite is also a silica mineral comprised of SiO_2, differing though in its crystalline structure, which is trigonal for quartz and monoclinic for moganite. The nearest to Astypalaia occurrence of chalcedony is located at Melos island [47].

Paragonite and white mica are typical minerals occurring in medium to low grade metamorphic environments. Even though these minerals participate in a variety of geologic formations in many islands of the Aegean Sea, the most known occurrences, especially of paragonite are located in the Lesbos [48], Syros [49], and Samos [50] islands.

Regarding the artefacts that are made of pumice, they are an interesting case since although pumice does not belong to the local lithology of the island, abundant cobbles of pumice along with volcanic rocks of varying chemical composition were found in large quantities in Gialoudi Bay (Figure 11) during surface research. In general, the eruptive products are highly influenced by the physicochemical properties of the erupting magma, concerning the chemical composition, viscosity, and gas contents, etc. Furthermore, their detailed study provides valuable information about the eruptive conditions of pre-existing volcanic activities or even contributes to addressing a possible future volcanic danger. The nearest to Astypalaia locations, known for their volcanic activity, are Thera, Melos, and Kos. The conditions of transportation of pyroclastic material at extremely long distances (tens of km) are studied by various researchers [51–53] and is still controversial.

Figure 11. (**a**) View of Gialoudi bay towards the southeast; (**b**) Abundant pumice cobbles in the area of Gialoudi bay; (**c**) representative cobble of volcanic rock in the same area.

5. Conclusions

Non-invasive techniques, including use of a portable microscope and NIR spectroscopy, successfully helped us to evaluate the kind of the raw material that was used for the archaeological artefacts. The geochemical composition in the majority of the stone tools was compared to the available spectral library and successfully identified by the interpretation of the retrieved spectra. The local geo-materials include limestone, marl, sandstone, and shale. The variation on the spectral properties of the raw material, which consists of limestone, indicate the exploitation and extensive usage of similar lithologies relative to their mineralogical component from many and different sites in the island. Diagnostic absorption features in specific wavelength positions of the electromagnetic spectrum helped us to distinguish materials that indicate an allochonous nature, not compatible with the local lithologic formations of Astypalaia, including chalcedony, bauxite/meta-bauxite, steatite, and paragonite.

The materials of volcanic provenance pose a more intricate challenge: although they do not belong to the geological "profile" of Astypalaia, they are in a sense "local", having been deposited in the area tens of thousands of years before the arrival of the people who used them. The study of volcanic pebbles from Gialoudi Bay and the search for similar locations of occurrences in other coastal areas of the island is still under research.

The majority of the artefacts collected during the 2011–2015 surveys at Vathy are made of locally available rocks. The larger and more numerous artefacts—discs, grinding slabs, the pebbles used for grinding/pounding—are of local origin, while the adze, beads (personal items), artefacts with geometric shape, and some of the sturdier pounders are of exogenous provenance. The inhabitants of Vathy in the FNL/EBA exploited both local (including the intrusive volcanic deposits at Gialoudi) and non-local raw materials, taking advantage of their properties according to the task at hand and following long-established practices. These initial results are indications that Astypalaia, although distanced, was not isolated even at these early times, and they encourage us to apply the process of geological recognition and NIRS analysis on the assemblage from the excavations as well.

It cannot be deduced, by surface finds alone, whether the exogenous materials were brought to Vathy as personal items of people from the Cyclades or whether they formed part of an exchange network that included obsidian and pottery as well. Nor can the affinities of a past society with specific areas be pin-pointed through laboratory analyses and scientific models alone, including the study of the quantitative and qualitative characteristics of the larger assemblage (to date some 750 artefacts and utensils have been recovered (E.A. personal communication). of ground tools and implements as it emerges from the on-going excavations at Vathy, and their comparison with similar assemblages from neighboring islands (E.A. has undertaken a comparative study from similar artefacts from Leros, Nissyros, and Rhodes.) will determine any similarities and/or differences.

Due to our interdisciplinary synergy, we are better prepared to understand the activities of the inhabitants of Vathy in the Early Bronze Age as these emerge from the ongoing excavations. The complexity of the features and structures that are being located (infant jar burials, petroglyphs, defensive structures [23], and the increasing number of ground stone artefacts [24] made on various materials) completely justify the analysis undertaken and described in the course of this paper.

Author Contributions: Conceptualization, I.I. and M.K.; methodology, I.I. and M.K.; software, M.K.; validation, I.I., A.V., M.K. and E.A.; formal analysis, M.K.; investigation, I.I. and M.K.; resources, M.K.; data curation, M.K.; writing—original draft preparation, M.K.; writing—review and editing, M.K., I.I., A.V. and E.A.; visualization, M.K.; supervision, I.I. and A.V.; project administration, A.V. and I.I. All authors have read and agreed to the published version of the manuscript.

Funding: This research received no external funding.

Data Availability Statement: Not applicable.

Conflicts of Interest: The authors declare no conflict of interest.

References

1. Biró, K.T. Non-destructive research in archaeology. *J. Radioanal. Nucl. Chem.* **2005**, *265*, 235–240. [CrossRef]
2. Anderson, C.; Törnberg, A.; Törnberg, P. An Evolutionary Developmental Approach to Cultural Evolution. *Curr. Anthropol.* **2014**, *55*, 154–174. [CrossRef]
3. Kintigh, K.; Altschul, J.; Beaudry, M.; Drennan, R.; Kinzig, A.; Kohler, T.; Limp, W.F.; Maschner, H.; Michener, W.; Pauketat, T.; et al. Grand Challenges for Archaeology. *Am. Antiq.* **2014**, *79*, 5–24. [CrossRef]
4. Liritzis, I. Twelve thousand years of non-linear cultural evolution: The physics of chaos in Archaeology. *Synesis J. Sci. Technol. Ethics Policy* **2013**, *4*, G19–G31.
5. Liritzis, I.; Laskaris, N.; Vafiadou, A.; Karapanagiotis, I.; Volonakis, P.; Papageorgopoulou, C.; Bratitsi, M. Archaeometry: An overview. *Sci. Cult.* **2020**, *6*, 49–98. [CrossRef]
6. Philokyprou, M. The initial appearance of ashlar stone in Cyprus. Issues of provenance and use. *Mediterr. Archaeol. Archaeom.* **2011**, *2011*, 37–53.
7. Columbu, S.; Piras, G.; Sitzia, F.; Pagnotta, S.; Raneri, S.; Legnaioli, S.; Palleschi, V.; Lezzerini, M.; Giamello, M. Petrographic and mineralogical characterization of volcanic rocks and surface-depositions on Romanesque Monuments. *Mediterr. Archaeol. Archaeom.* **2018**, *18*, 37–64. [CrossRef]
8. Cau Ontiveros, M.A.; Martinez Farreras, V.; Pecci, A.; Mas Florit, C.; Fantuzzi, L. Archaeometric analysis for provenance and content of roman amphorae from the site of Sa Mesquida (Mallorca, Spain). *Mediterr. Archaeol. Archaeom.* **2018**, *18*, 87–105. [CrossRef]
9. Glascock, M.D. A systematic approach to geochemical sourcing of obsidian artifacts. *Sci. Cult.* **2020**, *6*, 35–47. [CrossRef]
10. Lara, C.; Iliopoulos, I. Ceramic technology, ethnic identification and multiethnic contacts: The archaeological example of the Cuyes river valley (Southeastern Ecuadorian highlands). *J. Archaeol. Sci. Rep.* **2020**, *33*, 102557. [CrossRef]
11. Xanthopoulou, V.; Iliopoulos, I.; Liritzis, I. Characterizations techniques of clays for the archaeometric study of ancient ceramics: A review. *Sci. Cult.* **2020**, *6*, 73–86. [CrossRef]
12. Skarpelis, N. Silcrete and chert as source rocks of early prehistoric artifacts: The case of central Evia (Greece). *Mediterr. Archaeol. Archaeom.* **2021**, *21*, 1–19. [CrossRef]
13. Hunt, G.R. Spectral Signatures of Particulate Minerals in the Visible and near Infrared. *Geophysics* **1977**, *42*, 501–513. [CrossRef]
14. Bruni, S. Etruscan Fine Ware Pottery: Near-Infrared (NIR) Spectroscopy as a Tool for the Investigation of Clay Firing Temperature and Atmosphere. *Minerals* **2022**, *12*, 412. [CrossRef]

15. Inagaki, T.; Yonenobu, H.; Tsuchikawa, S. Near-Infrared Spectroscopic Investigation of the Hydrothermal Degradation Mechanism ofWood as an Analogue of Archaeological Objects. Part I: Softwood. *Appl. Spectrosc.* **2008**, *62*, 1209–1215. [CrossRef]
16. Lichtblau, D.; Strlič, M.; Trafela, T.; Kolar, J.; Ander, M. Determination of mechanical properties of historical paper based on NIR spectroscopy and chemometrics—A new instrument. *Appl. Phys.* **2008**, *29*, 19872. [CrossRef]
17. Delaney, J.K.; Thoury, M.; Zeibel, J.G.; Ricciardi, P.; Morales, K.M.; Dooley, K.A. Visible and infrared imaging spectroscopy of paintings and improved reflectography. *Herit. Sci.* **2016**, *4*, 6. [CrossRef]
18. Tanyaş, H.; Dirican, M.; Lütfi Süzen, M.; Türkmenoğlu, A.G.; Kolat, Ç.; Atakuman, Ç. Identification of possible source areas of stone raw materials combining remote sensing and petrography. *Int. J. Remote Sens.* **2017**, *38*, 3923–3942. [CrossRef]
19. Vlachopoulos, A. Archaeological fieldwork at Vathy, Astypalaia, Dodona. History—Archaeology. *ΜΓ-ΜΔ* **2017**, *2014*, 371–411. (In Greek)
20. Vlachopoulos, A. Vathy, Astypalaia. Research and material studies in a diachronic palimpsest of Aegean Archaeology. *Dodekan. Chron. Κθ Rhodes* **2021**, *2021*, 83–131. (In Greek)
21. Tsigkas, G.; Sfikas, G.; Pasialis, A.; Vlachopoulos, A. Nikou, C. Markerless detection of ancient rock carvings in the wild: Rock art in Vathy, Astypalaia. *Pattern Recognit. Lett.* **2020**, *135*, 337–345. [CrossRef]
22. Adam, E. Ground stone tools and utensils from the Vathy peninsula. The surface finds. In *Vathy, Astypalaia. Ten Years of Research in a Diachronic Palimpsest of the Aegean (2011–2020)*; Vlachopoulos, A., Ed.; Hellenic Ministry of Culture: Athens, Greece; University of Ioannina: Ioannina, Greece; Melissa Publishing House: Athens, Greece, 2022. (In Greek; abstracts in English)
23. Adam, E. Ground stone tools and utensils, finds. In *Vathy, Astypalaia. Ten Years of Research in a Diachronic Palimpsest of the Aegean (2011–2022)*; Vlachopoulos, A., Ed.; Hellenic Ministry of Culture: Athens, Greece; University of Ioannina: Ioannina, Greece; Melissa Publishing House: Athens, Greece, 2022. (In Greek; abstracts in English)
24. Metaxas, O. The chipped stone assemblage from the field survey of Vathy, Astypalaia. In *Vathy, Astypalaia. Ten Years of Research in a Diachronic Palimpsest of the Aegean (2011–2022)*; Vlachopoulos, A., Ed.; Hellenic Ministry of Culture: Athens, Greece; University of Ioannina: Ioannina, Greece; Melissa Publishing House: Athens, Greece, 2022. (In Greek; abstracts in English)
25. Vlachopoulos, A.; Angelopoulou, A. Early Cycladic Figurines from Vathy, Astypalaia. In *Early Cycladic Sculpture in Context from Beyond the Cyclades: Mainland Greece, the North and East Aegean*; Marthari, M., Renfrew, C., Boyd, M., Eds.; Oxbow Books: Oxford, PA, USA, 2019; pp. 202–226.
26. Iliopoulos, I.; Kokkaliari, M. Accessing provenance issues of prehistoric stone artefacts from Vathy (Astypalaia, Greece) through the minero-petrographic characterization of local lithologies. In *Vathy, Astypalaia. Ten Years of Research in a Diachronic Palimpsest of the Aegean, (2011–2022)*; Vlachopoulos, A., Ed.; Hellenic Ministry of Culture: Athens, Greece; University of Ioannina: Ioannina, Greece; Melissa Publishing House: Athens, Greece, 2022. (In Greek; abstracts in English)
27. Clark, R.N.; King, T.V.; Klejwa, M.; Swayze, G.A.; Vergo, N. High spectral resolution reflectance spectroscopy of minerals. *J. Geophys. Res.* **1990**, *95*, 12653–12680. [CrossRef]
28. Christodoulou, G. About the geology of Astypalaia Island. *Sci. Yearb. Issued Fac. Nat. Math. Sci.* **1966**, *10*, 169–180.
29. Marnelis, P.; Bonneau, M. Stratigraphie et structure de l'Île d'Astypalea (Dodécanèse, Grèce). In *VI. Colloquium on the Geology of the Aegean Region*; Kallergis, G., Ed.; 6th Aegeiscollathens; Institute of Geological and Mining Research: Yaoundé, Cameroon, 1977; Volume 1, pp. 323–332.
30. Ring, U. Structure and deformation history of Astypalea island, Aegean Sea. *Bull. Geol. Soc. Greece* **2001**, *34*, 329–335. [CrossRef]
31. Drouga, A. The Effect of Volcanic Ash Sedimentation and Dispersion at the Present and Modern Marine Sediments of Kos, Nissyros and Astypalea Islands (SE Volcanic Arc), Aegean Sea, Greece. Ph.D. Dissertation, Aristotle University of Thessaloniki, Thessaloniki, Greece, 2002.
32. Chatzaras, V. Geotraverse across the Hellenides between Western Crete and the Cycladic Islands. Ph.D. Dissertation, University of Patras, Patras, Greece, 2010; 206p.
33. Papoulia, M.; Karymbalis, E.; Gaki-Papanastassiou, K.; Maroukian, H. Assessment of the susceptibility of the coast of Astypalea Island (SE Aegean Sea) to sea-level rise. *Bull. Geol. Soc. Greece* **2013**, *47*, 305–314.
34. IGME. *Geological Map of Greece 1986, 1:50000*; Astipalaia Sheet; IGME: Athens, Greece, 1986.
35. Clark, R.N. Spectroscopy of rocks and minerals, and principles of spectroscopy. In *Remote Sensing for Earth Sciences—Manual of Remote Sensing*, 3rd ed.; Renz, A.N., Ed.; John Wiley and Sons: New York, NY, USA, 1999; pp. 3–58.
36. Gaffey, S.J. Spectral reflectance of carbonate minerals in the visible and near infra-red (0.35–2.55 microns): Calcite, aragonite, and dolomite. *Am. Mineral.* **1986**, *71*, 151–162.
37. Crowley, J.K. Visible and near-infrared spectra of carbonate rocks-reflectance variations related to petrographic texture and impurities. *J. Geophys. Res.* **1986**, *91*, 5001–5012. [CrossRef]
38. Aines, R.D.; Rossman, G.R. Water in minerals? A peak in the infrared. *J. Geophys. Res.* **1984**, *89*, 4059–4407. [CrossRef]
39. Clark, R.N.; Swayze, G.A.; Wise, R.; Livo, E.; Hoefen, T.; Kokaly, R.; Sutley, S.J. *USGS Digital Spectral Library splib06a*; Digital Data Series; U.S. Geological Survey: Denver, CO, USA, 2007; p. 231.
40. Rice, M.S.; Cloutis, E.A.; Bell, J.F., III; Bish, D.L.; Horgan, B.H.; Mertzman, S.A.; Craig, M.A.; Renaut, R.W.; Gautason, B. Mountain, B. Reflectance spectra diversity of silica-rich materials: Sensitivity to environment and implications for detections on Mars. *Icarus* **2013**, *223*, 499–533. [CrossRef]
41. Doublier, M.P.; Roache, A.; Potel, S. Application of SWIR Spectroscopy in very low-grade metamorphic environments: A comparison with XRD methods. *Geol. Surv. West. Aust.* **2010**, *61*, 2992.

42. Fleet, M.E.; Deer, W.A.; Howie, R.A.; Zussman, J. *Rock-Forming Minerals: Micas*, 2nd ed.; Geological Society of London: London, UK, 2003.
43. Doublier, M.P.; Roache, T.; Potel, S. Short-wavelength infrared spectroscopy: A new petrological tool in low-grade to very low-grade pelites. *Geology* **2010**, *38*, 1031–1034. [CrossRef]
44. Whitney, L.; Evans, W. Abbreviations for names of rock forming minerals. *Am. Mineral.* **2010**, *3*, 227. [CrossRef]
45. Jones, R.E.; Kilikoglou, V.; Olive, V.; Bassiakos, Y.; Ellam, R.; Bray, I.S.J.; Sanderson, D.C.W. A new protocol for the chemical characterisation of steatite—Two case studies in Europe: The Shetland Islands and Crete. *J. Archaeol. Sci.* **2007**, *34*, 626–641. [CrossRef]
46. Richards, H.; Mondillo, N.; Boni, M.; Thorne, R.; Tavlan, M. Bauxite and nickel-cobalt lateritic deposits of the Tethyan belt. Bauxite. *Soc. Econ. Geol.* **2016**, *19*, 349–387.
47. Papavasiliou, K.; Voudouris, P.; Kanellopoulos, C.; Alfieris, D.; Xydous, S. Mineralogy and Geochemistry of the Triades-Galana Pb-Zn-Ag-Au intermediate-high sulfidation epithermal deposit, Milos island, Greece. *Bull. Geol. Soc. Greece* **2016**, *50*, 1969–1979. [CrossRef]
48. Hatzipanagiotou, K.; Tsikouras, B.; Migiros, G.; Gartzos, E.; Serelis, K. Origin of rodingites in ultramafic rocks from Lesvos island (NE Aegean, Greece). *Ofioliti* **2003**, *28*, 13–23.
49. Seck, H.A.; Koetz, J.; Okrusch, M.; Seidel, E.; Stosch, H.G. Geochemistry of a meta-ophiolite suite; an associated metagabbros, eclogites and glaucophanites on the island of Syros, Greece. *Eur. J. Mineral.* **1996**, *8*, 607–623. [CrossRef]
50. Ockenga, E.; Yalçin, Ü.; Medenbach, O.; Schreyer, W. Zincohögbomite, a new mineral from eastern Aegean metabauxites. *Eur. J. Mineral.* **1998**, *10*, 1361–1366. [CrossRef]
51. Allen, S.R.; Cas, R.A.F. Transport of pyroclastic flows across the sea during the explosive, rhyolitic eruption of the Kos Plateau Tuff, Greece. *Bull. Volcanol.* **2001**, *62*, 41–456. [CrossRef]
52. Dufek, J.; Bergantz, G.W. Dynamics and deposits generated by the Kos Plateau Tuff eruption: Controls of basal particle loss on pyroclastic flow transport. *Geochem. Geophys. Geosyst.* **2007**, *8*, Q12007. [CrossRef]
53. Dufek, J.; Manga, M.; Staedter, M. Littoral blasts: Pumice-water heat transfer and the conditions for steam explosions when pyroclastic flows enter the ocean. *J. Geophys. Res. Solid Earths* **2007**, *112*, B11201. [CrossRef]

quaternary

Article

Disasters and Society: Comparing the Shang and Mycenaean Response to Natural Phenomena through Text and Archaeology

Alexander Jan Dimitris Westra [1,2,3,*], Changhong Miao [1], Ioannis Liritzis [1] and Manolis Stefanakis [2]

1 Laboratory of Yellow River Cultural Heritage, Key Research Institute of Yellow River Civilization and Sustainable Development & Collaborative Innovation Center on Yellow River Civilization, Henan University, Kaifeng 475001, China; chhmiao@henu.edu.cn (C.M.); liritzis@henu.edu.cn (I.L.)
2 Department of Mediterranean Studies, University of the Aegean, 81100 Mytilene, Greece; stefanakis@aegean.gr
3 Chercheur Associé au Laboratoire ArScAn | Archéologie de l'Asie Centrale-UMR 7041, Université Paris Nanterre, Ministère de la Culture, 92000 Paris, France
* Correspondence: westra@henu.edu.cn

Abstract: Disasters do and have happened throughout human existence. Their traces are found in the environmental record, archaeological evidence, and historical chronicles. Societal responses to these events vary and depend on ecological and cultural constraints and opportunities. These elements are being discovered more and more on a global scale. When looking at disasters in antiquity, restoring the environmental and geographical context on both the macro- and microscale is necessary. The relationships between global climatic processes and microgeographical approaches ought to be understood by examining detailed societal strategies conceived in response to threatening natural phenomena. Architectural designs, human geography, political geography, technological artefacts, and textual testimony are linked to a society's inherited and real sense of natural threats, such as floods, earthquakes, fires, diseases, etc. The Shang and Mycenaean cultures are prime examples, among others, of Bronze Age societies with distinctive geographical, environmental, and cultural features and structures that defined their attitudes and responses to dangerous natural phenomena, such as floods, earthquakes, landslides, and drought. By leaning on two well-documented societies with little to no apparent similarities in environmental and cultural aspects and no credible evidence of contact, diffusion, or exchange, we can examine them free of the onus of diffused intangible and tangible cultural features. Even though some evidence of long-distance networks in the Bronze Age exists, they presumable had no impact on local adaptive strategies. The Aegean Sea and Yellow River cultural landscapes share many similarities and dissimilarities and vast territorial and cultural expansions. They have an apparent contemporaneity, and both recede and collapse at about the same time. Thus, through the microgeography of a few select Shang and Mycenaean sites and their relevant environmental, archaeological, and historical contexts, and through environmental effects on a global scale, we may understand chain events of scattered human societal changes, collapses, and revolutions on a structural level.

Keywords: cultural; tangible; Yellow River; catastrophic; religion; climatic; dynasty; flood; Shang; Mycenae; myth; history

Citation: Westra, A.J.D.; Miao, C.; Liritzis, I.; Stefanakis, M. Disasters and Society: Comparing the Shang and Mycenaean Response to Natural Phenomena through Text and Archaeology. *Quaternary* **2022**, *5*, 33. https://doi.org/10.3390/quat5030033

Academic Editor: José Javier Baena Preysler

Received: 15 June 2022
Accepted: 13 July 2022
Published: 25 July 2022

1. Introduction

The manner of human interaction within a geographical area is determined by a series of cultural processes where the very geography of the location limits the evolution and layers of cultural values in each environmental niche. The layers of effects of structural geology, topography, climate, vegetation, hydrology on the settlements, economies, ideologies, myths, and cultures of the Aegean Sea and Yellow River regions must be included comprehensively. The Mycenaeans and Shang dealt with rivers, wetlands, dry areas, lack and excess of water, unpredictable rainfall and soil erosion, coastal environments with the background of change and fluctuations in sea level, alluvial, and river systems, and other physical phenomena tied to the exploitation of an ecological niche. However, the idea of a unified, homogeneous response to individual environmental processes and common economic strategies would be misplaced.

Natural disasters are part of an ecological niche and include droughts, plagues, storms, volcanic eruptions, landslides, meteor impact and more [1–4]. They are perennial threats to any society whose traces must be sought in the environmental, archaeological, and historical records. Archaeology and environmental archaeology recognise changes in cultural and ecological systems. It is worth noting that in a period of environmental stability and conservation, the continuous habitation of an area for centuries is equally important to recognising changes in vegetation systems, erosion phases, and other natural phenomena [5] (Figure 1). Discussions like the present one link our contemporary ecological challenges and forthcoming disasters. They shine light on the successes and failures in the management of space, the insights and misunderstandings regarding the processes of nature, and the calamities that have consistently beset organised human life, a hard lesson humanity is seemingly obstinately unable to learn, even in the face of incipient and ongoing environmental disasters.

This paper aims to demonstrate some of the key discussions revolving around natural disasters witnessed in environmental data, archaeological evidence, epigraphic evidence, and historical texts. It focuses on primary discussions regarding the Shang and Mycenaean cultures and associated natural disasters. The paper develops from intertwined sections concerning environmental archaeology and natural disasters, physical and textual evidence of destructions, mythical destructions, and disaster management, prehistoric disaster management involving technology and mythology, natural disasters that led to collapse, archaeoseismicity, which is a catalyst for collapse and resilience, the case for China coupled with protoscientific applications, religious world-views, and cosmology supported by epigraphic evidence, texts, and ideas from later periods (Table 1).

Table 1. Table showing basic chronological comparison of Greece and China (dates from Bintliff, 2012; Underhill, 2013) [6,7].

Greece		China	
ca. 3200 BC	Early Bronze Age (EBA)	ca. 2550–ca. 1950 BC	Longshan Culture
ca. 2000/1900–ca. 1800/1700 BC	Middle Cycladic/Middle Helladic	ca. 2300–1500 BC	Qijia Culture
ca. 1800/1700–ca. 1500 BC	Late Cycladic 1/Late Helladic 1	ca. 1750–ca. 1200 BC	Sanxingdui Culture
ca. 1500–ca. 1400 BC	Late Cycladic 2/Late Helladic 2	ca. 1800–ca. 1450 BC	Yueshi Culture
ca. 1400–ca. 1250/1200 BC	Late Helladic 3A-B	ca. 1850–ca. 1550 BC	Erlitou Culture
ca. 1250/1200–ca. 1050 BC	Late Helladic 3C	ca. 1600–ca. 1400 BC	Erligang Culture/ Early Shang Culture
		ca. 1400–ca. 1250 BC	Huanbei Culture/ Middle Shang Culture
		ca. 1250–1046 BC	Yinxu Culture/ Late Shang Culture
		1046–771 BC	Wesstern Zhou Culture

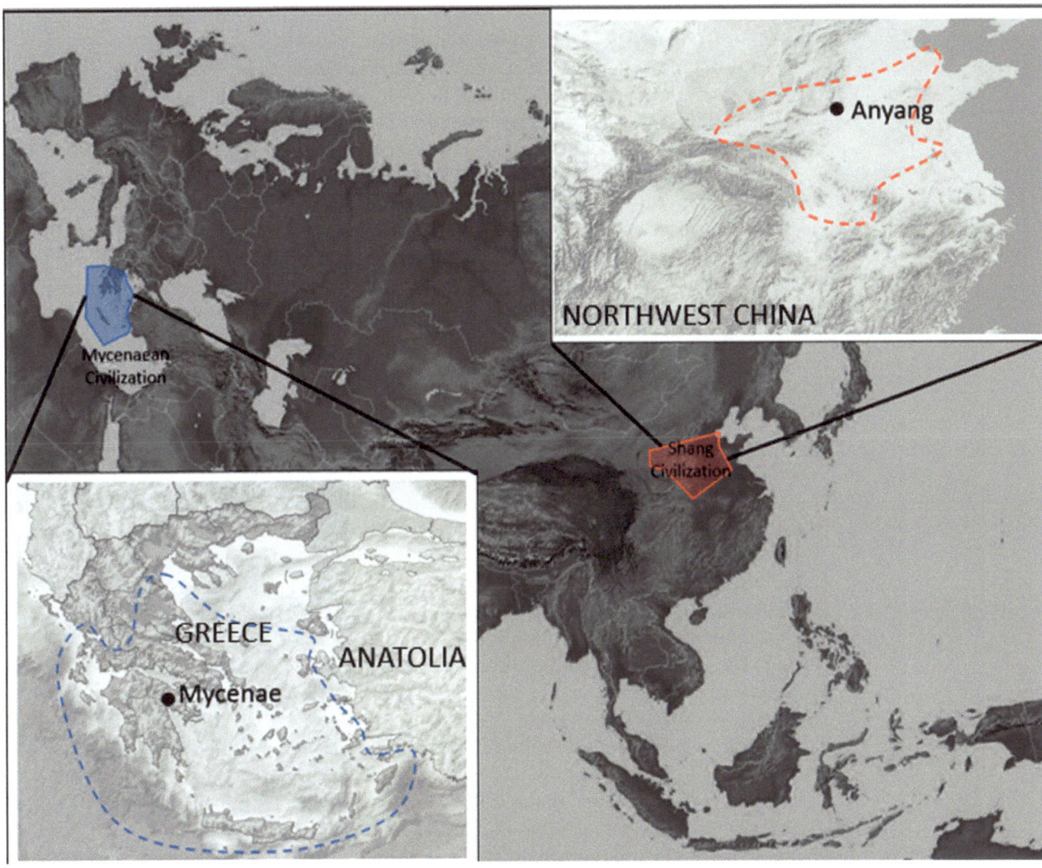

Figure 1. Approximate Mycenaean and Shang territories (Based on [8,9] (basemap free source: https://geology.com/world/asia-physical-map.shtml (access on 10 June 2022))).

2. Disasters through the Environment and Archaeology

2.1. Environmental Archaeology and Natural Disasters

Historical ecology, related to functional anthropology, provides archaeologists with concepts for evaluating cultural evolution [4]. Within cultural zones, technology and lifeways are associated and developed with their environmental context. By extension, the decline of a civilisation can be understood as an ecological disaster caused partly by the misuse of resources and catastrophic natural phenomena. The direct correlation of catastrophic events affecting prehistoric or ancient societies, in this case, the Late Bronze Age Aegean and the middle reach of the Yellow River, with the destruction of the Mycenaean or Shang civilizations, is widely discussed [10,11].

How and who applied environmental knowledge is a crucial question, as the participation of people in environmental change is self-evident and dynamic. Handling and responding to local environments reveal timeless cultures in specific landscape contexts. Environmental processes test social endurance: drought affects the supply of food, ideological and ritual mechanisms are developed to dampen the impact, and a social memory of disasters is created, which results in questioning the ability of the ruling class to appeal to a deity to mitigate threats from the natural world. The practical problem-solving of the Shang and Mycenaeans demonstrates the empirical approach to effectively minimising the effects of earthquakes, floods, and other natural threats with different material and conceptual toolsets, as well as potential societal backlashes.

Distinguishing in the collapse of states and political units, the collapse of civilizations and 'Great Traditions' is not a straightforward matter. [12]. The terms 'collapse' and 'fall' are misleading, because the fragmentation of empires and states into smaller political units does not necessarily translate into a reduction in complexity [10], although in archaeology, it means a reduction in cultural and political complexity. A political collapse, i.e., the collapse or weakening of the state (political hegemony), incurs a chain reaction felt by different strata of society and is visible through archaeological evidence. However, cultures are distinct constellations of material and immaterial phenomena. Collapse does not necessarily lead to an all-out loss of complexity, customs, or the disappearance of the population [10]. Collapse on such a large scale paints historical and ecological collapses with broad strokes and reduces the lived experiences of communities as irrelevant to the continuity of larger abstract entities, such as culture, civilisation, or nations.

2.2. Physical and Textual Evidence of Destructions

For the causes of the Late Bronze Age collapse across the eastern Mediterranean, which spelt the end of the Mycenaean civilisation, a wide array of explanations have been proffered [13–17]. However, linking specific evidence of floods, earthquakes, or volcanic eruptions with prehistoric and legendary records remains complex and contentious, as the impacts and effects of disasters are not well understood [18]. For instance, whether the volcanic eruption on Santorini led to the sudden or gradual collapse of the Minoan civilization has been much debated since Marinatos [19]. Due to the reduced solar radiation and temperatures, the eruption's effects may have had local, regional, and global implications [20] The lack of precise chronologies rends the broader analysis of the relationship between a geological event or catastrophic natural phenomena and broader social and cultural implications hypothetical. For the eruption of the volcano of Thera/Santorini [21,22], the proposed date is circa 1620 BC. A different constellation of natural and political events surround the Thera eruption: a specific catastrophe occur at a precise point in time whose wider environmental and historical consequences are thoroughly investigated. Despite the well-documented effects across the Aegean and Eastern Mediterranean and an increasingly higher chronological resolution, the cause(s) of the Late Bronze Age collapse is unclear and its link to it yet to be reported [23,24].

The Yu and the "Great Flood" myth hypothesis states that an earthquake burst a dam in the Tibetan Plateau, resulting in the cataclysmic flood of the Yellow River basin area circa 2000 BCE, and has been increasingly debated since Wu et al. [11]. The scale and timing of the flood suggests some correspondence with the mythical "Great Flood" of the Yellow River and taming of the waters by the legendary figure of Yu. The story of the hero Yu dredging the water and earning himself the divine right to rule marks the establishment of the legendary Xia dynasty, according to the Shujing (Book of Documents) and the *Shiji* (Records of the Grand Historian Sima Qian). In the traditional Chinese historiographical and archaeological context, moral weight is given to the emergence of the Chinese civilisation from the cataclysmic floods and cultures and societies. Whether natural disasters are a force or catalyst for civilisational destruction, collapse, renewal, and emergence is a historiographical, philosophical, and theological schema and quandary that attributes judgement on ancient destructive natural phenomena. Contrarily to the value judgement placed on terrifying natural phenomena, the environmental and archaeological facts inform us of patterns, changes, shifts, and fluctuations in a scientific setting that ought to be reasonably and relatively void of socio-historical judgement and moralization. The comparison between the interpretations of the cases of the Thera eruption and the 2000 BCE flood, coinciding with the Xia dynasty, the Yellow River flood reveals more about the inherited historiographical, sociological, and disciplinary differences between the Archaeology of Greece and the Archaeology of China than about the prehistoric civilisational mechanisms of collapse or emergence (Figure 2).

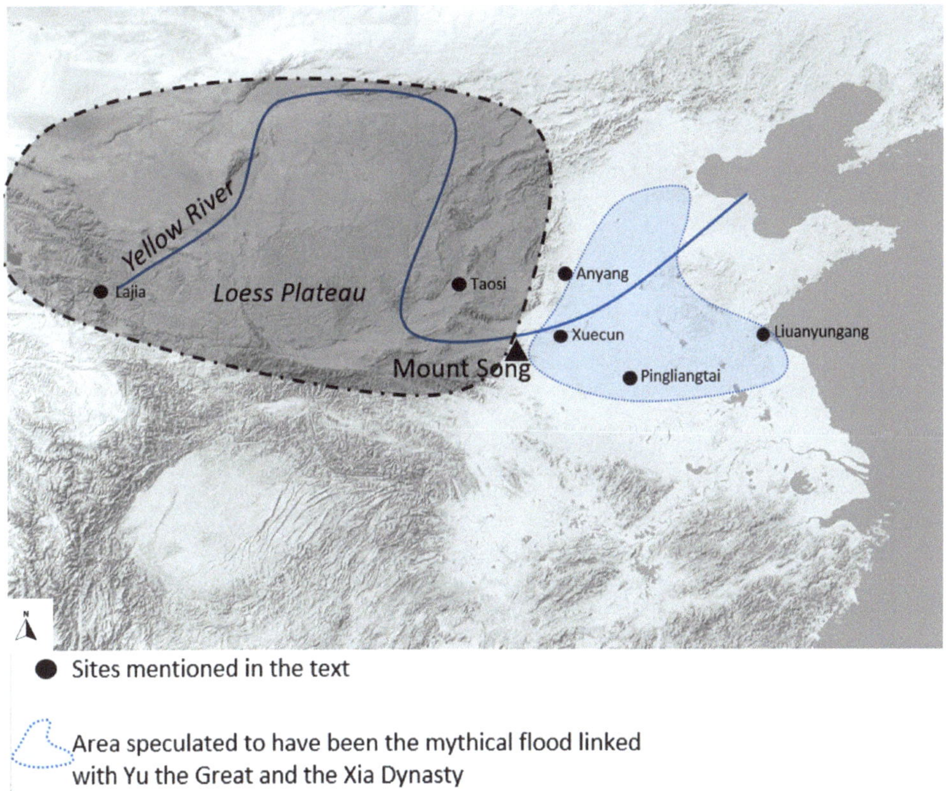

Figure 2. Map showing sites mentioned in the text and associated environmental features around the Yellow River basin such as Mount Song (triangle) (based on the flood dynamics described in [23], under a Creative Commons Attribution License). (basemap: free from Tom Patterson, US National Park Service Natural Earth), distance Taosi-Anyang = 256 km.

3. Early Cultural Engagement with Disasters

3.1. Mythical Destructions and Disaster Management

The experience of earthquakes, volcanic eruptions, and floods is traumatic and incites interest and disquiet in societies. Contemporary analyses and interpretations of past, current, and future destructive natural phenomena follow the historical schemas of ancient historical, philosophical, and theological treatises that considered conflagrations, cataclysms, inundations, and calamities with eschatological claims, moralising virtues, and renewal or renaissance stories, etc. [24]. In archaeology, doomsday is an old-fashioned way of thinking recently reinstated in the face of our contemporary degrading climate [18]. Some search for mythical and legendary catastrophic events as [25] true accounts and historical explanations are affirmed for the rupture and transformations of ancient civilisations, (i.e., Euhemerism). for example, the Biblical Floods, the destruction of Atlantis. and the Yu and Great Flood myth. The flooding of the Yellow River is a common and measurable phenomenon through various methods [26]. The evolution of its course and delta has resulted in large-scale floods in different periods, including the Longshan period [27,28]. One can take issue with the standpoint that considers the Yu and the Great Flood myth a historical *fait accompli* dated to circa 2000 BC (for example, Wang, 2005 [29]) However, the relevant figures of King Yao, Gun, and Yu are not proven historical personages, significant water-control activities worthy of such legend are unavailable, greater water management capabilities of other cultures within the Chinese world are not considered, [30], and other rivers in within the floodplains [31] are omitted. Thus, there are severe scientific issues in taking the historicity of China's Yu and the Great Flood myth as a historical fact.

Such claims are circular in reasoning and controversial in their historical claims and moralisation. Other Euhemeristic practices are found in the minor and nebulous fields of geomythology [32] and archaeomythology [33]. In archaeology, such concerns are found in the long-standing debates concerning the historicity of the Homeric epics or the Xia and Shang dynasties, known as the "Homeric Question" and the "Doubting Antiquity School". Investigating prehistoric and historic societal responses to natural disasters can be fraught with biases and masked by the scientific veneer of environmental science, the venerability of ancient sources, and recent-to-contemporary sociocultural, ideological, and political implications and concerns. Archaeology emerged from an imperialist and colonial perspective and autochthonous (cultural, nationalist) perspectives that can influence contemporary inquiries. No field of study or discipline has emerged from a socio-cultural and ideological vacuum. A thorough comparative archaeological study necessitates a close review of historical and archaeological literature to bring to the surface some of the historical, cultural, and ideological biases surrounding a Sino-Hellenic study [34,35], which is beyond the scope of this paper.

3.2. Prehistoric Disaster Management: Between Technology and Mythology

How did the prehistoric Xia and Shang, and Minoan Mycenaeans engage with their environments? A utilitarian axiomatic claim sees a distinction between our contemporaries and our prehistoric ancestors that lies in tangible and intangible 'tools'. Yet, a prehistoric mind is not inferior (or superior) to ours [36] (paraphrased: 'prehistoric tools, not prehistoric people'). This implies that prehistoric technological developments, infrastructures, and state systems, as well as myths, cults, and religious and literary traditions, are part of a society's practical toolset it may call upon to predict, prevent, or mitigate the effects of inevitably forthcoming natural disasters. Reconstructing these 'toolsets' may render intelligible the 'holy' (spiritual-metaphysical) and the practical, or 'profane', processes (knowledge, materials) that lead to a society's eco-political successes and failures. Historical ecology evaluates environmental expertise and its involvement with the landscape and the environment of the community. The adaptation strategies of groups and developed technologies and lifestyles are not by default repeated in other similar eco-cultural settings [4]. Although Mediterranean cultures have common elements in landscape management, their strategies depend on various timeless historical, cultural, and economic processes. Anti-seismic structures, for example, the experiential knowledge of the Minoans in rebuilding their habitat and mitigation strategies against seismic damage, include building on the bedrock, using wooden frames, posts, and cross-beams set on stone socles, and mud-brick walls. Mud-brick is a global phenomenon, so it cannot be considered a Mediterranean anti-seismic technology. Instead, it sits within a wide array of tools available to each group of prehistoric people.

Technology features in myths, and the archaeological material reflects the visionary ambitions of the Shang and Mycenaean technophiles and the fundamental importance of technology [37] for the 'clans' of the middle reaches of the Yellow River and the Aegean Sea. The legendary achievements of conquest, trade, and arts relied on a fundament of successful agricultural practices. Despite the territorial expansion, vassalization and exchange and diplomacy within and outside the Mycenaean or Shang supremacies, each city and palace relied on locally produced agriculture for wealth and power. Having progressed beyond the irregular Neolithic agricultural practices, the dynastic clans of the Bronze Age, now firmly implanted permanently within palaces and citadels, sought to administer their 'estates'. Hence, reclaiming soils from the water, diverting river systems, draining valleys, irrigating plains, flood control, and storing water became increasingly advanced, efficient, and more significant in scale.

Military might is expressed through the monumental tombs of the citadel at Mycenae, art and material, the extensive cultural expansion across the Aegean, the conflicts with neighbouring larger empires, and the advanced architecture and engineering. Altogether, they testify to the dynamic and aggressive growth of the Mycenaeans [38]. Linear B supports the idea of a centralised administration, but leaves unclear the nature of the Mycenaean state as a kingdom, confederacy, or other. It is mostly discussed through the lens of Homeric epics [39,40]. The rich monumental tombs at Anyang, the burials of mass human sacrifices, large and complex military campaigns, coordinated garrisons, and territorial control, and the incredible and complex cities and productions point to the stratified society and terrifying power of the Shang dynasty. Yet, the oracular literature suggests that the territorial disparity and military might be propped up by allies and subordinates, which sent the ruler on permanent diplomatic missions [41]. More detailed accounts of the Shang state are available elsewhere [6,41–43].

In the archaeological literature, the conscious and ingenious water management capabilities of the Aegean and the Yellow River Bronze Age civilisations are studied to a greater extent than anti-seismic knowledge and engineering. The Greek god Poseidon—in Linear B tablets the 'Earthshaker' (Linear B: e-ne-si-da-o-ne) in Knossos [44]—is evidence of the sociocultural integration of the recurring destructive natural phenomenon into the Minoan and Mycenaean world-view and their associated cultic practices. The building techniques of the Mycenaeans, such as wooden framework wall construction, are found at Mycenae, Tiryns Thebes, and Knossos [45]. Anti-seismic construction techniques are also well attested by the Minoans, with foundations built on bedrock, wall frameworks, symmetrical plans, etc., that could withstand ordinary and extraordinary seismic events [46]. Minoans deployed anti-seismic construction methods that improved over time and extended to monumental buildings and private residences.

4. Disaster as Factor for Human Evolution

4.1. Natural Disaster to Collapse?

Anti-seismic measures are contained within the knowledge and material capabilities of the Shang and Mycenaeans and include both profound and far-reaching understanding and enduring misunderstandings about the inner workings of nature and physics. The Bronze Age societies of the Aegean and Yellow River regions suffered several disasters, leading to the loss of life and material and habitable or cultivable space. Yet, their way of life endured. Consequently, in investigating natural causes for a societal collapse, the question ought to be 'Why did this particular natural disaster prove to be the cause for collapse?'.

The Aegean region is marked by frequent seismic activity [47–49]. Growth in interdisciplinary work in Aegean archaeology is exemplified through archaeoseismology [50,51] and recently in Yellow River archaeology. Natural disasters in an archaeological context are social phenomena [52]. The societal impact of a prehistoric earthquake varies for each community. While they may cause a degree of local decline, or even collapse, they do not cause civilisational collapse [53]. A host of hypotheses exist that range from natural to societal causes, which do not need to be repeated (for a summary, see Middleton, 2020 [11]) (Figure 3).

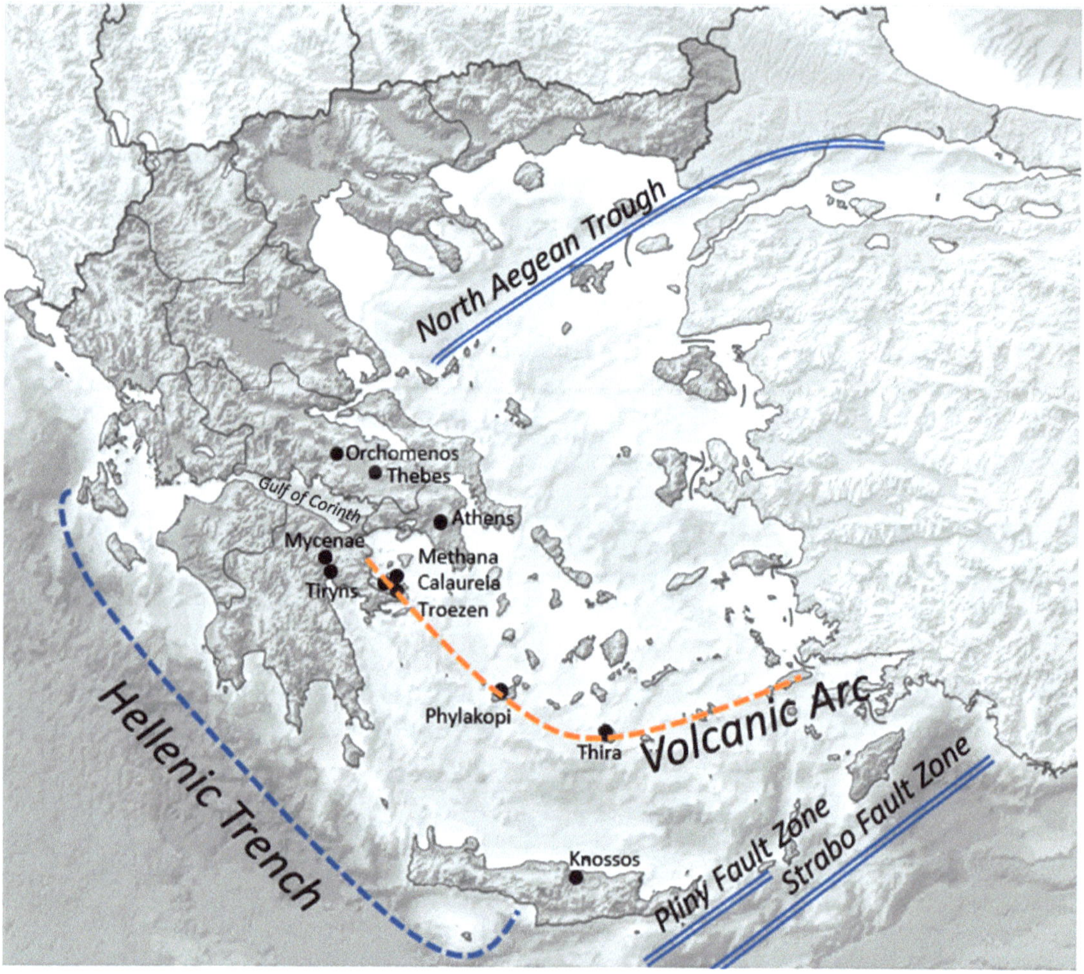

Figure 3. The Volcanic Arc and the Hellenic Trench in Greece with sites mentioned in the text. The Aegean Sea Plate (also called the Hellenic Plate or Aegean Plate) is a minor tectonic plate in the eastern Mediterranean Sea under southern Greece and far western Turkey. Its southern edge is a subduction zone south of Crete, where the African Plate is swept under the Aegean Sea Plate. Seismic and volcanic centers lie along the euro-african subduction zone. The Volcanic Arc (orange dashed line) of the northern Eurasian Plate is a divergent boundary responsible for the formation of the Gulf of Corinth. Note the northern Aegean Trench, the Hellenic Trench, and Pliny and Strabo fault zones in eastern Crete (based on [54,55], scale: Athens-Thebes = 205 km (Basemap: United States National Imagery and Mapping Agency data)).

Destruction of settlements happens, and earthquakes are not the preferred explanation. At first, Evans was reluctant to link the destruction of Knossos palace with seismic activity [51,56]. In addition, society's resilience against natural disasters also indicates human response to them [49,57]. The early Minoan palaces were destroyed between 1900 and 1700 BC by earthquakes [57–59]. The Minoan architectural discipline continuously evolved, incorporating new aesthetic features and practical anti-seismic features [50,60], such as building on bedrock, construction of smaller free-standing blocks, use of wooden frames and cross-beams, façade projections and indentations [61], use of external friezes, lighter upper floors, and use of supportive pillars, among others [62].

Centuries after the Thera eruption, the destruction of the potentially overextended final Mycenaean palaces due to earthquakes or other events (i.e., invasion, Dorian invasion, return of the Heraclidae) do not signify the protracted end of the Mycenaean lifeways a couple of centuries later [63]. The Dorian invasion hypothesis has been rejected [64–66], and the notion of seismic storms causing the Late Bronze Age collapse (see: [57,59] is viewed with increasing scepticism [53]. The Mycenaean collapse is associated with the broader Late Bronze Age collapse of significant kingdoms and cities in the Eastern Mediterranean [67].

Critics reproach the indiscriminate explication of site abandonment or destruction to earthquakes when other factors are unknown [53] and deem it "neocatastrophism" [68,69]. Contrarily, the earthquake hypothesis is preferred, for the sake of parsimony, and stands against a myriad of unproven or unprovable complex hypotheses that can be affixed to the Mycenaean palatial collapse [57]. Subsequently, insufficient evidence and low temporal resolution have led to strained and circular explanations of earthquake-induced collapse or decline [70].

Mythological reports surround the Argolid area in the Peloponnese alleged natural disasters from river flooding [71], including other prehistoric geomythological issues of further deluges (Dardanus, Ogyges, Telchines). In addition, the fourth century BC earthquake and associated tsunami in the Corinthian gulf swallowed the ancient coastal city of Helike [72] and in the Levant [73].

A phenomenon as complex as *civilisational decline* or *collapse* is rife with reductionistic views and historiographical assumptions regarding what and how civilisational genesis, apex, nadir, and necrosis occur and are observable, or are useful as an interpretational paradigm. Adopting an interpretive framework of *civilisational decline* entails a series of complex phenomena of which catastrophic destruction is merely one aspect. In turn, discussing sudden collapse or gradual centuries-long decline also involves the disappearance of enough specific defined civilisational characteristics to render it 'other'. Cultural evolution is thus explained as a non-linear process [74].

4.2. *Archaeoseismicity: A Catalyst for Collapse and Resilience*

Disaster archaeology, in the case of the Mycenaean earthquake hypothesis, invokes several lines of questioning and observation to propose a conjunction of facts and events that risk forcing correlations into causation. An outcome of this discussion is the emergence of the multidisciplinary field of 'archaeoseismology' [53,75]. Archaeoseismology seeks to comprehensively and with interdisciplinarity determine the core seismic culture of an area [76]. Eyewitness accounts and later historical records have historical and sociological value in understanding emotive and pragmatic responses to the knowledge of an eventually forthcoming destructive natural phenomena and its aftermath and fallout. Historic seismicity allows us to extend the earthquake record considerably [77]. Paleoseismology examines the geomorphological and geological evidence for seismic fault activity [78], but its main weakness lies in providing a higher-resolution chronology for seismic events.

Indicators derived from destruction or damage of archaeological sites may provide a higher chronological resolution and the attraction of archaeoseismologists [50,79,80], but ascertaining whether victims, broken pots, and warped structures are the result of earthquakes or other physical pressures and processes is not an unambiguous observation to make [81]. In reality, archaeoseismological simulations of characteristic earthquake damage concluded that it could not be verified [82]. Numerous examples of seismic damage to Minoan and Mycenaean settlements and evidence of repair and reoccupation such as Knossos, Tiryns, and Mycenae exist. Yet, Hinzen et al.'s [83] assessment concluded that the evidence for seismic damage is often inconclusive and that earthquakes causing such widespread destruction to the end of the Mycenaean palatial system is unlikely.

4.3. Disaster Archaeology: The Case of China

The case of China, in comparison, offers different scales of magnitude and geography, and forces different perspectives on the notion of disaster. Earthquakes and floods have caused the demise of vast settled areas. Yet, a coherent culture continued to exist despite the calamities. On February 2nd 1556, an earthquake in Shanxi, Henan, and Shaanxi provinces resulted in the death of circa 800,000 people [84]. China's high population density can lead to tragedies of enormous proportions. However, determining this level of disaster in a prehistoric archaeological context is elusive. Archaeoenvironmental evidence regarding palaeoearthquakes in Xia- and Shang-era China is growing. For example, recent work in Henan Province revealed two faults, two grabens, and ground fissures, indicating a palaeoearthquake at Xuecun. The cultural layers and AMS 14 C from the ash pit date from the palaeoearthquake between 1520 and 1260 BC [85].

The buildings were built with regularly spaced timber posts, strengthened by horizontal cross-beams. The lack of nails and the use of interlock with mortises and tenons gave the building more flexibility and resistance to earthquakes. Earthquakes can also cause other natural disasters, such as floods. Architectural heritages such as *rammed earth* techniques appear independently in China and the Mediterranean regions, and constantly evolve [86]. In China, it is known as *hangtu* and in English as *cob* and is a widespread, simple, and earthquake-resistant building material found along the alluvial plains of north China since the Neolithic Longshan period (c. 3000–1900 BC) [87–89] at sites such as Pingliangtai [90], Lianyungang, Jiangsu, Taosi, and much more [91] as well as Bronze Age sites, such as Erlitou, Longwan, Anyang-Yinxu [92], and later historical periods (Warring States: 475–221 BC) [93]. The *rammed earth* structures (*hangtu*, *cob*, adobe, etc.) are known for their anti-seismic qualities, but whether that was consciously done by the first builder in China or elsewhere is unclear [94] (Figure 4).

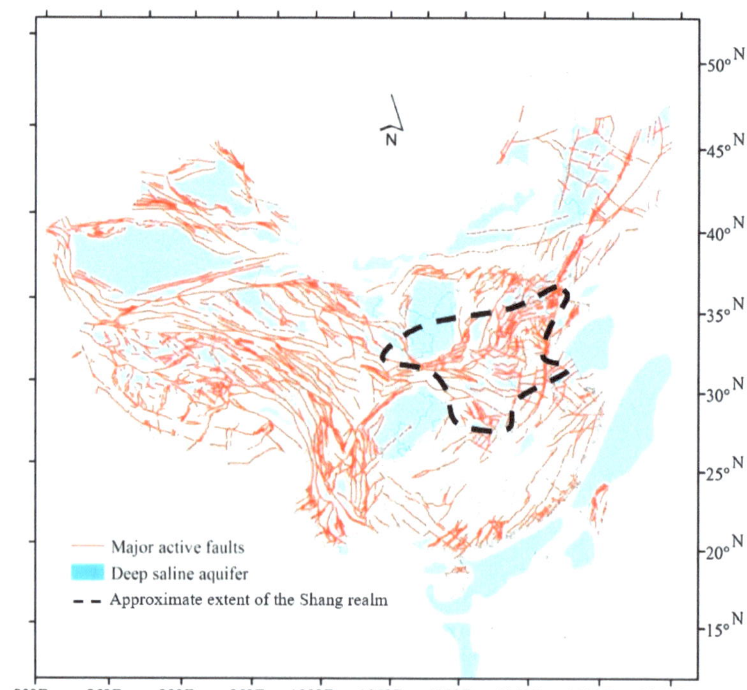

Figure 4. Present-day active seismic faults and deep saline aquifers related to the approximate extent of the Shang culture (dashed line) and beyond (based on source: [95], distances 5° = 555 km).

The Lajia Ruins, for example, are evidence of a natural disaster that destroyed human settlements. In the upper Yellow River, the destruction of the site of Lajia of the Qijia culture has been studied through various palaeoenvironmental means (micromorphology, OSL, paleosol). Around 2200–2000 BC, mudflows caused by rainstorms, flash floods, and earthquakes and the intensive exploitation of the landscape during the Neolithic resulted in the tragedy at Lajia 28, [96–100].

However, the prime suspect for the cause of the demise of the Shang has usually been sought and expressed through traditional historiography that derived from the accounts of Sima Qian (circa 145–86 BC). Nonetheless, the climatic shift in the northern central plains of China in the mid-Holocene is a clear marker of the far-reaching impact climatic change can have on modes of life. The historical connections between climate change and shifting frontiers between farmers and pastoralists of the northern Loess Plateau have received little attention [101]. Recent work has also proposed a connection between the collapse or decline of the Late Shang Dynasty and changing and deteriorating environments in the northern plains. Sedimentary data from the Loess Plateau indicates a sudden increase in aridity circa 1100 BC caused by the shifting maritime monsoon in favour of the continental monsoon. This climatic shift is also recorded in ice cores [102] and confirms its link to the aridity of northwest China [103]. Lake sediments from the Inner Mongolia Plateau show an abrupt reduction in water levels from circa 1100 BC to 400 BC [104,105]. Although not as sudden as an earthquake or flood, the relatively sharp increase in aridity would have destabilised the economy, production, and natural resources of the Shang Dynasty's traditional area of control, resulting in crop failures, livestock failures, mass migrations southwards, and encroachment by northern pastoralists. Altogether, increased aridity, leading to famines, instability, and uprising, could have set the scene for the collapse of the Shang Dynasty and the eventual Zhou takeover of the Shang realm [101]. The Yellow River culture of the Shang gradually expanded southwards in pursuit of copper and other resources, and before its demise, underwent a cultural revival [89,106–108].

Finally, understanding ancient seismicity varies according to each discipline's interest [53]. An exploration of the impact of disasters on the Mycenaean and Shang incorporates seismological, engineering, historical, and anthropological approaches [5]. Unlike sociological observations and conclusions, an archaeological investigation seeks to answer questions that relate to currents of the interests of the discipline, like chronology, and social adaptability, continuity, and discontinuity (Table 1). The placement of the citadel of Mycenae on an active limestone fault scarp (cf. [109], that could have split its cyclopean masonry in half gives us pause to consider the geological knowledge of the Mycenaeans [53] and more broadly the ecological and cultural understanding of the prehistoric Shang and Mycenaeans when faced with inevitable natural dangers. The recovery of Mycenaean and Shang knowledge is inferred from Bronze Age infrastructural projects, epigraphic evidence (Linear B, *Jiaguwen*), and later philosophical treatises. Natural phenomena were keenly observed and a core point of interest. Scientific and religious notions were developed within their own cultural and ideological frameworks and applied using tools, experiences, and knowledge. From later sources, earthquakes were not associated with tectonic plates, but with such causes as excessive Yin and trapped Yang, the winds, and the waters, which caused the world to shake. The overall cosmology of the Mycenaean and Shang visions of the world informed their cultic, religious, societal, and ideological approaches to dealing with natural disasters, while the practical applications visible in the archaeological record are derived from the experiential knowledge of materials and their properties.

5. Early Understanding of Disaster

5.1. Protoscience and Religion

Protoscientific applications, religious world-views, and cosmology can be linked to some extent, but they should not be confused. In other words, prayers, sacrifices, or appeals to Poseidon, Shang Di, or any deity or supernatural mover and properly built infrastructural works, such as dams, dykes, and canals, and disaster-prepared political systems may both aim at avoiding and mitigating future disasters, but their approaches may have little in common.

The repeated occurrences of natural disasters have led the Chinese to compile probably the most reliable catalogue of them: the Guoyu Zhouyu (Historical Stories of States: Zhou), dating to the Spring and Autumn Period (Chunqiu) 770–476 BC. It reflects on how the Yellow River and other rivers drying up caused the collapse of the Shang [110]. Rivers drying up, droughts, famines, dust storms, and so on are mentioned in much of Chinese literature [101] and renders them plausible and the investigation into environmental causes for the collapse of the Shang Dynasties realm a valid hypothesis. The study of historical earthquakes in China is fertile ground, thanks to its extensive and accurate historical record. Since the Shang Dynasty, a variety of natural disasters have been recorded. The Catalogue of Chinese Earthquake, published in 1960, lists 585 historical earthquakes going as far back as 1189 BC. The Taiping Yulan, compiled by Li Fang and dating to the 10th century AD, lists 45 earthquakes between the 11th century BC and 618 AD [111,112]. Compared with the rest of the world, the Chinese records of historical earthquakes are the most complete [113]. The *Shiji* mentions how an earthquake in 780 BC, during the reign of King Yu of Zhou, interrupted the courses of three rivers.

> "In the second year of the reign of King Yu l (of Chou), the western province's three rivers shaken and their beds raised up, Poyang Fu said: 'The dynasty of the Chou is going to perish. It is necessary that the chhi of heaven and earth should not lose their order; if they overstep their order (it is because there When the Yang is hidden and cannot come forth, or when the Yin bars its way and it cannot rise up, then there is what we call an earthquake. Now we see that the three rivers have dried up by this shaking; it is because the Yang has lost its place and the Yin has overburdened it. When the Yang has lost its rank and finds itself (subordinate to) the Yin, the springs become closed, and when this has happened the kingdom must be lost. When water and earth are propitious the people make use of them, when they are not, the people are deprived of what they need. Formerly when the rivers I and Lo dried up, the dynasty of the Hsia perished. When the Ho dried up, the dynasty of the Shang perished.' Now the virtue of the Chou is in the same state as that of these dynasties was in their decline. The Chou will be ruined before ten years are out; so it is written in the cycle of numbers (Sima Qian, *Shiji*, Annals, 36 (Translation, [113] pp. 624–625)."

5.2. Epigraphic Evidence

Poseidon, the "Earthshaker" [*enesidao*], was prolifically worshipped in the Peloponnese. We turn to the prehistoric epigraphic textual evidence from the Linear B tablets and the *Jiaguwen* inscriptions (Oracle Bones). Natural phenomena, like earthquakes and floods, were rationalised by later Greeks and Chinese thinkers. Turning to proto-scientific explanations of natural phenomena, we must rely on contemporary epigraphic text, interpretation of archaeological evidence, and later philosophical treatises to reconstruct the Mycenaean and Shang modes of thinking.

Poseidon, the most commonly found name of a god in the proto-Greek Linear B tablets [114], is known as e-ne-si-da-o-ne ("Earthshaker") [115] from the 14th century BC Tablets in Knossos [114,116] and survives into the classical period as the god of the sea, earthquake, and tsunami [117] (see Figure 3).

The Linear B tablet mentioning Poseidon can also be linked to sanctuaries dedicated to Poseidon at Methana, Calaureia, and Troezen [118]. The cult of Poseidon and likely to a female counterpart by the Mycenaean is well established [119,120] and supported by the appearance of the name po-se-da-o (Tablet PY Es(-) 653 or PY Un(2) 6; also see Duhoux and Davis (2008) [121]), the cult at Phylakopi and the presence of a Hollow Psi figurine, and the pair of human figures driving the Methana chariot models [122]. This is also reinforced by the much later account of Diodorus (15.49.4), who describes the reverence paid by the Peloponnesians to their patron god, Poseidon.

> "That it was Poseidon's wrath that was wreaked upon these cities they allege that clear proofs are at hand: first, it is distinctly conceived that authority over earthquakes and floods belongs to this god, and also it is the ancient belief that the Peloponnese was an habitation of Poseidon; and this country is regarded as sacred in a way to Poseidon, and, speaking generally, all the cities in the Peloponnese pay honour to this god more than to any other of the immortals (15.49.4: *Diodorus Siculus*, tr. 1989 [123])."

Unlike the Mycenaeans, whose writing was mostly accounting, as it derives from the Mesopotamian tradition, the *Jiaguwen* reveals information relating to the divinatory and ancestral cult of the Shang [124,125], there is virtually no surviving secular use of writing [126].

In the Oracle Bones (*Jiaguwen*) inscriptions, commonplace divination is the encounter of some general 'disaster' or 'misfortune'. For instance, the character huò has been subject to much debate about the divination practices of the Shang.

The *Jiaguwen* texts used for divinations have mostly been found at Anyang, in Henan Province, at the site of the last Shang capital known as Yinxu [41,127,128]. After the Shang dynasty was overthrown, scapulomancy and plastromancy were still practised, but held much less importance in state affairs than in the Bronze Age [125,129,130]. The Oracle Bones is written in the form of Chinese that was maybe archaic for the late Shang kings, which would have been the 'language of the ancestors'. Within the corpus are mentioned sacrifices to the Yellow River (he) and the winds (Feng), suggesting an appeal to the spirits of natural elements [125]. The character yu or yuji is believed to mean 'apotropaic rite against disasters', or 'religious ritual performed to prevent and eliminate disasters' [131] such as illnesses, natural disasters, and crop failures:

> 10,152: Crack-making on xinyou, performs yu ritual for flood damage.
> 14,407: Crack-making on xinyou, divining: performs yu ritual for flood damage, and sheep are offered for rituals.
> 72: yu ritual . . . for Shangjia against disasters [131].

The Shang dynasty was established circa 1600 BC and inherited a host of religious ideas, scientific notions, and mythological traditions from previous and unknown cultures, as well as trends towards urbanisation and state formation. One should be careful not to project later historical accounts and modern understanding on such an archaic Bronze Age culture as was the Shang [41]. For a more detailed discussion of the religious landscape of early China, see [132,133]. An impression of the Shang religious beliefs could be summarised as natural phenomena considered as deities, such as the sun, the winds, the earth, the river, etc., whose connection with the Shang ancestral lineage is blurry. *He*, the river deity (spirit—*shen*, or force) and *Yue* (mountain) figure prominently in the Pantheon and are conventionally believed to refer to the Yellow River and *Songshan* (Mt. Song). Sacrifices were made to these *shen* (spirits) [41]. Lastly, the identity of *Di* has been debated. Responsible for the *huo* (catastrophe, disaster, ill omen), he is linked with weather and crops [134]. The *Shang* rulers appealed to *Di* to divine his intentions on projects, wars, and buildings [135].

5.3. Texts and Ideas from Later Periods

The growing interest in Sino-Hellenic comparative historical and philosophical studies [136,137] begins from the early historical periods of China and Greece. During the antiquity of Greece and China, several conceptions about natural disasters and their explanations are formulated. For instance, the *Shiji* quote refers to the Yin and Yang, a prevalent notion during the Han Dynasty that earthquakes could be foretold astrologically. In ideas found in the *I-Ching* (Book of Changes), earthquakes or thunders could be due to constrained Yang or an excess of Yin [113]. Whether such theories can be found with some approximation to periods before the eighth century BC is debatable [113]. The *Meteorologica*, reiterating Anaxagoras, stated that earthquakes were the result of water excess from the upper regions flowing into the hollows of the earth. Democritus considered that earthquakes were the result of soil saturated with water. Anaximenes proposed that seismic shocks result from masses of world falling in caverns. Aristotle put forward the notion of earthquakes being due to instability of the pneuma/vapour [138].

6. Conclusions

Ancient texts dating to periods several centuries after the disasters have sometimes been used to glean some clues and offer general remarks about prehistoric religion and beliefs regarding divination and natural disasters in Greece and China [130], in particular, the sociological implications of divinations as a political and rhetorical tool. Perhaps the comparison of Greek and Chinese mantic practices can be insightful for the archaeological interpretations of Shang and Mycenaean religion and sociological implications of natural events [138].

Yet, the remains of destruction and abandonment of once-thriving settlements are the staple of archaeological research. The causes that led to the sudden or gradual abandonment of settled sites and regions are not clear and must be examined. The catalogue of lost, destroyed, or abandoned community sites is large and incomplete. The fall of a once-thriving world contains the ingredients for many literary, political, religious, ideological, and overall cultural concerns and products. The *Bamboo Annals* and other works [139] describe earthquakes occurring during the time of the mythical emperor Fa (1831 BC) and the Gui emperor (1809 BC). However, their reliability and historicity remain doubtful [85]. Lastly, we may conclude that the Yu and the Great Flood parable, historicised in the Book of Documents (*Shiji*), which occurred in the time of the mythical King Yao, refers to a 'flood' (hongshui) from the mythical primordial sea or waters [140].

If we consider these myths, such as Yu the Great, Deucalion, Ogyges, and so on, as *living* fossils of human culture, then there is still no convincing methodical way of blending archaeological, historical, environmental, and legendary evidence. If there were no prehistoric people, just prehistoric tools, then the experiences of natural dangers and disasters and fears of a cataclysmic collapse of apocalyptical destruction were also woven into the fabric of their world-view, which was then expressed and addressed through practical and spiritual means. The Mycenaeans and the Shang were intelligent observers of their ecology. It can be assumed that no natural (or supernatural) phenomenon, regular or exceptional, was ignored. These occurrences grouped ecological phenomena and societal and spiritual beliefs, which blended into convictions about the triangular influence between deities on nature, nature, and humans (nature–humans–deities) and the ideological legitimacy of archaic rulership. The feats of the Mycenaean and Shang engineers and architects are impressive. Their resilience to floods, droughts, earthquakes, and other catastrophic natural phenomena is understood through practical, intellectual, emotive, ideological, and spiritual engagements with perennial threats.

Author Contributions: Conceptualization, I.L. and A.J.D.W.; methodology, I.L., A.J.D.W., M.S. and C.M.; software, I.L. and A.J.D.W.; resources, A.J.D.W., M.S., I.L. and C.M.; writing—original draft preparation, A.J.D.W. and I.L.; writing—review and editing, A.J.D.W., I.L. and C.M.; visualisation, A.J.D.W., I.L. and C.M.; supervision, I.L., M.S. and C.M.; project administration, I.L. and A.J.D.W. All authors have read and agreed to the published version of the manuscript.

Funding: This work was supported by the National Natural Science Foundation of China (Fund project host: Miao Changhong), grant number 42171186.

Acknowledgments: A.J.D.W expresses special thanks to the École Française d'Athènes 1-month grant for access to its library in Athens, Greece, which made this paper's completion possible. I.L. thanks the Laboratory of Yellow River Cultural Heritage, Key Research Institute of Yellow River Civilization and Sustainable Development & Collaborative Innovation Center on Yellow River Civilization, Henan University, China, for supporting the Sino-Hellenic Academic Project (www.huaxiahellas.com, last accessed 1 July 2022).

Conflicts of Interest: The authors declare no conflict of interest.

References

1. Ahn, S.H. Meteoric activities during the 11th century. *Mon. Not. R. Astron. Soc.* **2005**, *358*, 1105–1115. [CrossRef]
2. Ernstson, K.; Sideris, C.; Liritzis, I.; Neumair, A. The chiemgau meteorite impact signature of the stöttham archaeological site (southeast germany). *Mediterr. Archaeol. Archaeom.* **2012**, *12*, 249–259.
3. Liritzis, I. Pandemics—From ancient times to COVID-19. Some thoughts. *Mediterr. Archaeol. Archaeom.* **2020**, *13*, i–ix.
4. Walsh, K. *The Archaeology of Mediterranean Landscapes: Human-Environment Interaction from the Neolithic to the Roman Period*; Cambridge University Press: Cambridge, UK, 2014.
5. Liritzis, I.; Westra, A.; Changhong, M. Disaster GeoArchaeology and Natural Cataclysms in World Cultural Evolution: An Overview. *J. Coast. Res.* **2019**, *35*, 1307. [CrossRef]
6. Bintliff, J. *The Complete Archaeology of Greece: From Hunter-Gatherers to the 20th Century A.D.*; Wiley-Blackwell: Oxford, UK; Chichester, UK, 2012.
7. Underhill, A.P. *A Companion to Chinese Archaeology*; John Wiley & Sons: Chichester, UK, 2013.
8. Liu, L.; Chen, X. *State Formation in Early China*; Duckworth: London, UK, 2003.
9. Castleden, R. *The Mycenaeans*; Routledge: London, UK, 2005.
10. Middleton, G. Introducing Collapse. In *Understanding Collapse: Ancient History and Modern Myths*; Cambridge University Press: Cambridge, UK, 2017; pp. 1–50.
11. Wu, Q.; Zhao, Z.; Liu, L.; Granger, D.E.; Wang, H.; Cohen, D.J.; Wu, X.; Ye, M.; Bar-Yosef, O.; Lu, B.; et al. Outburst flood at 1920 BCE supports historicity of China's Great Flood and the Xia dynasty. *Science* **2016**, *353*, 579–582. [CrossRef] [PubMed]
12. Yoffee, N.; Cowgill, G.L. (Eds.) *The Collapse of Ancient States and Civilizations*; University of Arizona Press: Tucson, AZ, USA, 1988.
13. Middleton, G.D. (Ed.) *Collapse and Transformation: The Late Bronze Age to Early Iron Age in the Aegean*; Oxbow Books: Oxford, UK, 2020.
14. Renfrew, C. Systems collapse as social transformation. In *Approaches to Social Archaeology*; Harvard University Press: Cambridge, UK, 1984; pp. 366–389.
15. Desborough, V.D.A. *The Last Mycenaeans and Their Successors: An Archaeological Survey, c. 1200-c. 1000 bc*; Wipf and Stock Publishers: Eugene, OR, USA, 2007.
16. Deger-Jalkotzy, S. The Aegean Islands and the Breakdown of the Mycenaean Palaces around 1200 B.C. In *Eastern Mediterranean: Cyprus–Dodecanese–Crete 16th–6th Century B.C.*; Karageorghis, V., Stampolidis, N., Eds.; University of Crete: Heraklion, Greece, 1998; pp. 105–120.
17. Drews, R. *The End of the Bronze Age: Changes in Warfare and the Catastrophe ca. 1200 BC*; Princeton University Press: Princeton, NJ, USA, 1993.
18. Torrence, R.; Grattan, J. The archaeology of disasters: Past and future trends. In *Natural Disasters and Cultural Change*; Routledge: London, UK, 2003; pp. 17–34.
19. Marinatos, S. The Volcanic Destruction of Minoan Crete. *Antiquity* **1939**, *13*, 425–439. [CrossRef]
20. La Moreaux, P.E. Worldwide environmental impacts from the eruption of Thera. *Environ. Geol.* **1995**, *26*, 172–181. [CrossRef]
21. Manning, S.W. *A Test of Time and a Test of Time Revisited: The Volcano of Thera and the Chronology and History of the Aegean and East Mediterranean in the Mid Second Millennium BC*; Oxbow Books: Oxford, UK, 2014.
22. Zielinski, G.A.; Germani, M.S. New Ice-Core Evidence Challenges the 1620s BC age for the Santorini (Minoan) Eruption. *J. Archaeol. Sci.* **1998**, *25*, 279–289. [CrossRef]
23. Chen, Y. Flood dynamics of the lower Yellow River over the last 3000 years: Characteristics and implications for geoarchaeology. *Quat. Int.* **2019**, *521*, 147–157. [CrossRef]
24. Morhange, C.; Marriner, N.; Bony, G.; Carayon, N.; Flaux, C.; Shah-Hosseini, M. Coastal geoarchaeology and neocatastrophism: A dangerous liaison. *Harb. Harb. Cities East. Mediterranean. Byzas* **2014**, *19*, 705–726.
25. Vitaliano, D.B. Geomythology: The impact of geologic events on history and legend with special reference to Atlantis. *J. Folk. Inst.* **1968**, *5*, 5–30. [CrossRef]
26. Yang, D.; Yu, G.; Xie, Y.; Zhan, D.; Li, Z. Sedimentary records of large Holocene floods from the middle reaches of the Yellow River, China. *Geomorphology* **2000**, *33*, 73–88. [CrossRef]
27. Li, K.; Gong, P.; Hu, H.; Jia, W.; Liu, X.; Gao, W. Spatial variability of human subsistence strategies during the longshan period (~4.6–~3.9 ka bp) and its possible physical environmental contexts in the yellow-huai river area, east China. *Sci. Cult.* **2021**, *7*, 105–116. [CrossRef]
28. Lu, P.; Lü, J.; Zhuang, Y.; Chen, P.; Wang, H.; Tian, Y.; Mo, D.; Xu, J.; Gu, W.; Hu, Y.; et al. Evolution of Holocene alluvial landscapes in the northeastern Songshan Region, Central China: Chronology, models and socio-economic impact. *Catena* **2021**, *197*, 104956. [CrossRef]
29. Wang, H. The *Great Flood* in Yao and Shun's Days and the Origin of the Early State in China. *J. Shaanxi Norm. Univ. (Philos. Soc. Ed.)* **2005**, *3*, 76–86.
30. Yuzhu, Z.; Chang, H.C.; Jiangli, P. A luminescence dating study of the sediment stratigraphy of the Lajia Ruins in the upper Yellow River valley. *China J. Asian Earth Sci.* **2014**, *87*, 157–164.
31. Hou, Y.J. Yu the Great Subdueing the Flood. *J. Anc. Books Collat. Stud.* **2008**, *2*, 85–87+71.
32. Vitaliano, D.B. Geomythology: Geological origins of myths and legends. *Geol. Soc. Lond. Spec. Publ.* **2007**, *273*, 1–7. [CrossRef]
33. Marler, J. Introduction to archaeomythology. *ReVision* **2000**, *23*, 2.

34. Beecroft, A. Comparisons of Greece and China. In *Oxford Handbooks Online*; Oxford University Press: New York, NY, USA, 2015.
35. Liritzis, I.; Westra, A.J. The lower Yangtze River and Aegean Sea in the third millennium BC: Parallel cradles of civilizations. *Asian Archaeol.* **2022**, 1–4. [CrossRef]
36. Cornu, V. Qu'est ce que l'architecture? In *Conférence donnée le 7 Octobre 2009 à la Maison de l'architecture*; RAIC: Ottawa, ON, Canada, 2007.
37. Tassios, T.P. Mycenaean Technology. In *Science and Technology in Homeric Epics*; Paipetis, S.A., Ed.; Springer: Dordrecht, The Netherlands, 2008; pp. 3–33.
38. Glassman, R.M. The Expansion of Mycenaean Greek Civilization. In *The Origins of Democracy in Tribes, City-States and Nation-States*; Springer: Cham, Switzerland, 2017; pp. 759–769.
39. Simpson, R.H. Identifying a Mycenaean state. *Annu. Br. Sch. Athens* **1957**, *52*, 231–259. [CrossRef]
40. Nilsson, M. State Organization in Homer and in the Mycenean Age. *Homer Crit. Assess.* **1999**, 126–148.
41. Eno, R. Shang State Religion and The Pantheon Of The Oracle Texts. In *Early Chinese Religion, Part One: Shang through Han (1250 BC-220 AD)*; Brill: Leiden, The Netherlands, 2009.
42. Keightley, D.N. The late Shang state: When, where, and what? In *The Origins of Chinese Civilization*; Keightley, D.N., Ed.; University of California: Berkeley, CA, USA, 1983; pp. 523–564.
43. Yates, R. The city-state in ancient China. In *The Archaeology of City-States: Cross Cultural Approaches*; Deborah, N., Thomas, C., Eds.; Smithsonian Institution Press: Washington, DC, USA, 1997; pp. 71–90.
44. Driessen, J. 'In bulls doth the Earth-Shaker delight'–Introduction to the volume. *Minoan Earthq. Break. Myth. Through Interdiscip.* **2017**, *5*, 19.
45. Schaar, K.W. Traditional Earthquake-Resistant Construction: The Mycenaean Aspect. *J. Soc. Archit. Hist.* **1974**, *33*, 80–81. [CrossRef]
46. Poursoulis, G. Geological factors part in the Minoan society vulnerability process (Crete–Greece). In *The Mediterranean World Environment and History*; Elsevier: Paris, France, 2002; pp. 25–39.
47. Papadopoulos, G.A. *A Seismic History of Crete. The Hellenic Arc and Trench. Earthquakes and Tsunamis 2000 BC–2011 AD*; Ocelotos Publications: Athens, Greece, 2011.
48. Papadopoulos, G.A. Earthquake sources and seismotectonics in the area of Crete. In *Minoan Earthquakes-Breaking the Myth through Interdisciplinarity*; Jusseret, S., Sintubin, M., Eds.; Leuven University Press: Leuven, Belgium, 2017; Chapter 7, pp. 165–190.
49. Poursoulis, G.; Dalongeville, R.; Helly, B. Destruction des edifices minoens et sismicité récurrente en Crète (Grèce). *Géomorphologie Relief Processus Environ.* **2000**, *6*, 253–265. [CrossRef]
50. Stiros, S.C.; Jones, R.E. *Archaeoseismology*; Fitch Laboratory Occasional Paper 7; Institute of Geology and Mineral Exploration and The British School at Athens: Athens, Greece, 1996.
51. Jusseret, S.; Sintubin, M. (Eds.) *Minoan Earthquakes: Breaking the Myth through Interdisciplinarity*; Leuven University Press: Leuven, Belgium, 2017; Volume 5.
52. Shimoyama, S. Basic characteristics of disasters. In *Natural Disasters and Cultural Change. One World Archaeology*; Torrence, R., Grattan, J., Eds.; Routledge: London, UK; New York, NY, USA, 2002; Volume 45, pp. 19–27.
53. Sintubin, M. Archaeoseismology: Past, present and future. *Quat. Int.* **2011**, *242*, 4–10. [CrossRef]
54. Higgins, M.D.; Higgins, R. *A Geological Companion to Greece and the Aegean*; Duckworth Publishers: London, UK, 1996.
55. Meier, T.; Becker, D.; Endrun, B.; Rische, M.; Bohnhoff, M.; Stöckhert, B.; Harjes, H.P. A Model for the Hellenic Subduction Zone in the area of Crete based on seismological investigations. In *The Geodynamics of the Aegean and Anatolia Geological Society*; Taymaz, T., Dilek, Y., Eds.; Geological Society: London, UK, 2007; pp. 183–200.
56. Palyvou, C. *Daidalos at Work. A Phenomenological Approach to the Study of Minoan Architecture*; INSTAP Academic Press: Philadelphia, PA, USA, 2018.
57. Nur, A. *Apocalypse. Earthquakes, Archaeology, and the Wrath of God*; Princeton University Press: Princeton, NJ, USA, 2008.
58. La Rosa, V. A hypothesis on earthquakes and political power in Minoan Crete. *Ann. Geophys.* **1995**, *38*, 881–891. [CrossRef]
59. Nur, A.; Cline, E.H. Poseidon's horses: Plate tectonics and earthquake Storms in the Late Bronze age Aegean and eastern Mediterranean. *J. Archaeol. Sci.* **2000**, *27*, 43–63. [CrossRef]
60. Tsakanika-Theohari, E. The constructional analysis of timber load bearing systems as a tool for interpreting Aegean Bronze Age architecture. In Proceedings of the Symposium Bronze Age Architectural Traditions in the Eastern Mediterranean. Diffusion and Diversity, Munich, Germany, 7–8 May 2008; pp. 127–142.
61. Driessen, J. Earthquake–resistant construction and the wrath of "Earth–shaker". *J. Soc. Archit. Hist.* **1987**, *46*, 171–178. [CrossRef]
62. Alberti, L. Lessons from the past: Some Greek Bronze Age responses to natural disasters and their modern counterparts. *Ann. Geophys.* **2019**, *10*, 1–14. [CrossRef]
63. Vanschoonwinkel, J. Earthquakes and the end of the Mycenaean palaces. *Les Études Class.* **2002**, *70*, 1–2.
64. Andronikos, M. E'dorike Eisvole kai ta archaiologika Euremata. *Hellenika* **1954**, *13*, 221–240.
65. Hooker, J.T. New Reflexions on the Dorian Invasion. *Klio* **1979**, *61*, 353–360. [CrossRef]
66. Thomas, G. The Celts: A Model for the Dorian Invasion? *SMEA* **1980**, *21*, 303–308.
67. Broodbank, C. *The Making of the Middle Sea: A History of the Mediterranean from the Beginning to the Emergence of the Classical World*; Thames & Hudson: London, UK, 2013.
68. Ambraseys, N.N. Archaeoseismology and neocatastrophism. *Seismol. Res. Lett.* **2005**, *76*, 560–564. [CrossRef]

69. Ambraseys, N.N. Earthquakes and archaeology. *J. Archaeol. Sci.* **2006**, *33*, 1008–1016. [CrossRef]
70. Rucker, J.D.; Niemi, T.N. Historical earthquake catalogues and archaeological data: Achieving synthesis without circular reasoning. In *Ancient Earthquakes*; Sintubin, M., Stewart, I.S., Niemi, T.N., Altunel, E., Eds.; Geological Society of America Special Paper; Geological Society of America: Boulder, CO, USA, 2010; Volume 471, pp. 97–106.
71. Liritzis, I.; Raftopoulou, M. Argolid: Connection of prehistoric legends with geoenvironmental and archaeological evidence. *TUBA-AR* **1999**, *II*, 87–99. [CrossRef]
72. Liritzis, I.; Galloway, R.B.; Katsonopoulou, D.; Soters, D. In search of ancient Helike, Golf of Corinth, Greece. *J. Coast. Res.* **2001**, *17*, 118–123.
73. Hirschfeld, Y. The crisis of the sixth century: Climatic change, natural disasters and the plague. *Mediterr. Archaeol. Archaeom.* **2006**, *6*, 19–32.
74. Liritzis, I. Twelve thousand years of non-linear cultural evolution: The physics of chaos in Archaeology. *SYNESIS J. Sci. Technol. Ethics Policy* **2013**, *4*, G19–G31.
75. Noller, J.S. Archaeoseismology. In *Earth Sciences and Archaeology*; Springer: Boston, MA, USA, 2001; pp. 143–170.
76. Sintubin, M.; Stewart, I.S. A logical methodology for archaeoseismology: A proof of concept at the archaeological site of Sagalassos, Southwest Turkey. *Bull. Seismol. Soc. Am.* **2008**, *98*, 2209–2230. [CrossRef]
77. Guidoboni, E. Historical seismology: The long memory of the inhabited world. In *International Handbook of Earthquake and Engineering Seismology*; Lee, W.H.K., Kanamori, H., Jennings, P.C., Kisslinger, C., Eds.; International Geophysics Series 81A; Academic Press: Amsterdam, The Netherlands, 2002; pp. 775–790.
78. Caputo, R.; Pavlides, S.B. Earthquake Geology: Methods and Applications. In *Tectonophysics*; Elsevier: Amsterdam, The Netherlands, 2008; Volume 453.
79. Galadini, F.; Hinzen, K.G.; Stiros, S.C. Archaeoseismology at the beginning of the 21st century. *J. Seismol.* **2006**, *10*, 395–414. [CrossRef]
80. Jones, R.E.; Stiros, S.C. The advent of archaeoseismology in the Mediterranean. In *The Archaeology of Geological Catastrophes*; McGuire, W.G., Griffiths, D.R., Hancock, P.L., Stewart, I.S., Eds.; Special Publications 171; Geological Society: London, UK, 2000; pp. 25–32.
81. Hinzen, K.G.; Schreiber, S.; Yerli, B. The Lycian Sarcophagus of Arttumpara, Pinara, Turkey: Testing Seismogenic and Anthropogenic damage Scenarios. *Bull. Seismol. Soc. Am.* **2010**, *100*, 3148–3164. [CrossRef]
82. Hinzen, K.G. Simulation of Toppling columns in archaeoseismology. *Bull. Seismol. Soc. Am.* **2009**, *99*, 2855–2875. [CrossRef]
83. Hinzen, K.G.; Maran, J.; Hinojosa-Prieto, H.; Damm-Meinhardt, U.; Reamer, S.K.; Tzislakis, J.; Kemna, K.; Schweppe, G.; Fleischer, C.; Demakopoulou, K. Reassessing the Mycenaean earthquake hypothesis: Results of the HERACLES project from Tiryns and Midea, Greece. *Bull. Seismol. Soc. Am.* **2018**, *108*, 1046–1070. [CrossRef]
84. Rao, G.; Cheng, Y.; Lin, A.; Yan, B. Relationship between landslides and active normal faulting in the epicentral area of the AD 1556 M~ 8.5 Huaxian Earthquake, SE Weihe Graben (Central China). *J. Earth Sci.* **2017**, *28*, 545–554. [CrossRef]
85. Xia, Z.; Zhang, X.; Chu, X.; Zhang, J. Discovery and significance of buried paleoearthquake of the early Shang Dynasty (1260–1520 BC) in Xuecun, Xingyang, Henan Province, China. *Chin. Sci. Bull.* **2009**, *55*, 1186–1192. [CrossRef]
86. Jaquin, P.A.; Augarde, C.E.; Gerrard, C.M. Chronological Description of the Spatial Development of Rammed Earth Techniques. *Int. J. Archit. Herit. Conserv. Anal. Restor.* **2008**, *2*, 377–400. [CrossRef]
87. Warren, J. *Earthen Architecture: The Conservation of Brick and Earth Structures. A Handbook*; ICOMOS: Paris, France, 1993.
88. Sui Pheng, L. Construction of dwellings and structures in ancient China. *Struct. Surv.* **2001**, *19*, 262–274. [CrossRef]
89. Qin, Z. Exploring the early anthropocene: Implications from the long-term human–climate interactions in early China. *Mediterr. Archaeol. Archaeom.* **2021**, *21*, 133–148.
90. Bai, Y. On the Early City and the Beginning of the State in Ancient China. In *The Humanities Studies*; Chinese Academy of Social Sciences: Beijing, China, 2003.
91. Liu, L.; Chen, X. *The Archaeology of China: From the Late Paleolithic to the Early Bronze Age*; Cambridge University Press: Cambridge, UK, 2012.
92. Houben, H.; Guillaud, H. *Earthen Architecture: A Comprehensive Guide*; Intermediate Technology Development Group: London, UK, 1994.
93. Shen, C. Early urbanisation in the Eastern Zhou in China (770–221 BC): An archaeological view. *Antiquity* **1994**, *68*, 724–744. [CrossRef]
94. Sui Pheng, L. Techniques for environmental control and structural integrity of buildings in ancient China. *Struct. Surv.* **2004**, *22*, 271–281. [CrossRef]
95. Wei, N.; Li, X.; Wang, Y.; Dahowski, R.T.; Davidson, C.L.; Bromhal, G.S. A preliminary sub-basin scale evaluation framework of site suitability for onshore aquifer-based CO_2 storage in China. *Int. J. Greenh. Gas Control.* **2013**, *12*, 231–246. [CrossRef]
96. Huang, C.C.; Pang, J.; Zhou, Y.; Su, H.; Zhang, Y.; Wang, L. Palaeoenvironmental implications of the prehistorical catastrophes in relation to the Lajia Ruins within the Guanting Basin along the Upper Yellow River, China. *Holocene* **2013**, *23*, 1584–1595. [CrossRef]
97. Wang, H.; Huang, C.C.; Pang, J.; Zhou, Y.; Cuan, Y.; Guo, Y.; Zhang, Y.; Zhou, Q.; Rong, X.; Shang, R. Catastrophic flashflood and mudflow events in the pre-historical Lajia Ruins at the northeast margin of the Chinese Tibetan Plateau. *Quat. Sci. Rev.* **2021**, *251*, 106737. [CrossRef]

98. Zhang, J.; Xia, Z. Deposition Evidences of the 4ka BP Flood Events in Central China Plains. *Quat. Sci.* **2003**, *6*, 667–674.
99. Zhang, Y.; Huang, C.C.; Pang, J.; Guo, Y.; Zhou, Q. Identification of the prehistoric catastrophes at the Lajia Ruins using micromorphological analysis within the Guanting Basin, Minhe County, Qinghai Province. *Archaeol. Anthropol. Sci.* **2018**, *10*, 711–723. [CrossRef]
100. Zhao, H.; Huang, C.C.; Zheng, Z.; Hu, Y.; Zhang, Y.; Guo, Y.; Zhou, Q. New evidence for the catastrophic demise of a prehistoric settlement (the Lajia Ruins) in the Guanting Basin, upper Yellow River, NW China. *J. Asian Earth Sci.* **2017**, *146*, 134–141. [CrossRef]
101. Huang, C.C.; Pang, J.; Li, P. Abruptly increased climatic aridity and its social impact on the Loess Plateau of China at 3100 a BP. *J. Arid. Environ.* **2002**, *52*, 87–99. [CrossRef]
102. O'Brien, S.R.; Mayewski, P.A.; Meeker, L.D.; Meese, D.A.; Twickler, M.S.; Whitlow, S.I. Complexity of Holocene climate as reconstructed from a Greenland ice core. *Science* **1995**, *270*, 1962–1964. [CrossRef]
103. Biscaye, P.E.; Grousset, F.E.; Revel, M.; Van der Gaast, S.; Zielinsky, G.A.; Vaars, A.; Kukla, G. Asian provenance of last glacial maximum dust in the GISP2 ice core, summit, Greenland. *J. Geophys. Res.* **1997**, *102*, 26765–26781. [CrossRef]
104. Wang, S.M.; Feng, M. Environmental change recorded in the Daihai Lake and its relationship with the change in strength of the southeast monsoon. *Sci. China* **1991**, *21*, 759–768. (In Chinese)
105. Yang, Z.Y. Holocene climatic variations recorded in the Diaojiaohaizi Lake of the Daqingshan Mts. *Geogr. Res.* **1998**, *17*, 138–144. (In Chinese)
106. Xu, S.H. Development of Shang culture in Middle Yangtze River region on case study on Panlongcheng. *Jianghan Archaeol.* **2003**, *86*, 40–44. (In Chinese)
107. Dou, H. A discussion on the relationship between the Shang Dynasty remains of Yangtze River Plain in Anhui Province and neighbor areas. *Jianghan Archaeol.* **2012**, *124*, 69–80. (In Chinese)
108. Guo, Y.; Mo, D.; Mao, L.; Jin, Y.; Guo, W.; Mudie, P.J. Settlement distribution and its relationship with environmental changes from the Paleolithic to Shang–Zhou period in Liyang Plain, China. *Quat. Int.* **2014**, *321*, 29–36. [CrossRef]
109. Stewart, I.S.; Hancock, P.L. Normal fault zone evolution and fault scarp degradation in the Aegean region. *Basin Res.* **1988**, *1*, 139–153. [CrossRef]
110. Wang, C.; Wang, S.M. The precipitation change in the Yellow River basin during the last 5000 years. *Sci. China* **1987**, *1*, 104–112. (In Chinese)
111. Wang, J.; Lin, G.; Zhang, Z. Methodology to determine the parameters of historical earthquakes in China. *Geosci. Lett.* **2017**, *4*, 4. [CrossRef]
112. Li, S.; Wu, H.; Guo, Z.; Mei, S.; Min, Z. *The Catalogue of Chinese Earthquakes*; Science and Technology Publish House: Beijing, China, 1960; pp. 1–399. (In Chinese)
113. Needham, J.; Wang, L. *Science and Civilisation in China*; Cambridge University Press: Cambridge, UK, 1995; Volume 3.
114. Palaima, T.; Linear, B.; Cline, E.H. *The Oxford Handbook of the Bronze Age Aegean*; Oxford University Press: Oxford, UK, 2010; pp. 356–372.
115. Rougemont, F. Les noms des dieux dans les tablettes inscrites en Linéaire B. In *Nommer les Dieux: Théonymes, Épithètes, Épiclèses dans L'antiquité*; Bylache, N., Brulé, P., Freyburger, G., Lehmann, Y., Pernot, L., Prost, L., Eds.; Brepols: Rennes, France, 2005; pp. 325–388.
116. Gulizio, J.; Pluta, K.; Palaima, T. Religion in the Room of the Chariot Tablets. In *POTNIA: Deities and Religion in the Aegean Bronze Age*; Laffineur, R., Hägg, R., Eds.; Aegaeum: Göteborg, Sweden, 2001; Volume 22, pp. 453–461.
117. Palaima, T. Appendix One: Linear B sources of Greek religion. In *Anthology of Classical Myth: Primary Sources in Translation*; Trzaskoma, S.M., Smith, R.S., Brunet, S., Eds.; Hackett Publishing Company: Indianapolis, Indiana, 2004; pp. 439–454.
118. Konsolaki, E. A Mycenaean sanctuary on Methana. *Pelopon. Sanctuaries Cults* **2002**, 25–36.
119. Vermeule, E. *Götterkult, Archaeologia Homerica III*; Chapter V: Der mykenische Kult nach den Linear B-Dokumenten; Vandenhoeck & Ruprecht: Göttingen, Germany, 1974; pp. 59–72.
120. Chirassi, I. Poseidon-Enesidaon nel pantheon miceneo. In *Atti e Memorie del Primo Congress Internazionale di Micenologie, 27 Settembre–3 Ottobre 1967*; Incunabula Graeca 25; Ateneo: Rome, Italy, 1974; pp. 945–991. Available online: https://www.maremagnum.com/libri-antichi/atti-e-memorie-del-1-congresso-internazionale-di-micenologia/150946723 (accessed on 10 June 2022).
121. Duhoux, Y.; Davies, A.M. (Eds.) *A Companion to Linear B: Mycenaean Greek Texts and Their World*; Peeters: Louvain-la-Neuve, Belgium, 2008; Volume 1.
122. Konsolaki-Yannopoulou, E. Mycenaean religious architecture: The archaeological evidence from Ayios Konstantinos, Methana. In *Celebrations: Selected papers and Discussions from the Tenth Anniversary Symposium of the Norwegian Institute at Athens, 12–16 May 1999*; Wedde, M., Norske Institutt i Athen for Klassiske Studier, Arkeologi og Kulturhistorie, Eds.; The Norwegian Institute at Athens: Bergen, Norway, 2004.
123. Diodorus, S. *Diodorus of Sicily in Twelve Volumes with an English Translation by C. H. Oldfather*; Harvard University Press: Cambridge, MA, USA; William Heinemann, Ltd.: London, UK, 1989; Volumes 4–8.
124. Schmandt-Besserat, D. *Before Writing*; University of Texas Press: Austin, TX, USA, 1992; Volume I, pp. 184–194.
125. Goldin, P.R.; University of Pennsylvania. Some Shang Antecedents of Later Chinese Ideology and Culture. *J. Am. Orient. Soc.* **2017**, *137*, 121–127. [CrossRef]

126. Bagley, R.W. Anyang Writing and the Origin of the Chinese Writing System. In *The First Writing: Script Invention as History and Process*; Stephen, D.H., Ed.; Cambridge University Press: Cambridge, UK, 2004.
127. Chang, K.-C. *Shang Civilization*; Yale University Press: New Haven, CT, USA, 1980.
128. Keightley, D.N. The religious commitment: Shang theology and the genesis of Chinese political culture. *Hist. Relig.* **1978**, *17*, 211–225. [CrossRef]
129. Kalinowski, M. Diviners and Astrologers under the Eastern Zhou: Transmitted Texts and Recent Archaeological Discoveries. In *Early Chinese Religion, Part One: Shang through Han (1250 BC–220 AD)*; John, L., Marc, K., Handbuch der, O., Eds.; Brill: Leiden, The Netherlands, 2009; pp. 349–354.
130. Raphals, L. *Divination and Prediction in Early China and Ancient Greece*; Cambridge University Press: Cambridge, UK, 2013; pp. 129–131.
131. Eun-hee, K.; Hyeok, K. A study on character forms and sentence samples of "yu" in oracle bone inscriptions. *J. Chin. Writ. Syst.* **2019**, *3*, 11–18. [CrossRef]
132. Keightley, D.N. *The Ancestral Landscape: Time, Space and Community in Late Shang China (ca. 1200–1045 B.C.)*; Institute of East Asian Studies; University of California: Berkley, CA, USA, 2000.
133. Michiharu, I.; Ken-ichi, T. Religion and Society. In *Studies in Early Chinese Civilization: Religion, Society, Language, and Palaeography*; Gary, F.A., Ed.; Kansai Gaidai University: Osaka, Japan, 1996; Volume I, pp. 100–104.
134. Anderson, M. An Investigation of Orthographic Variance in Shang Writing. *Sino-Platonic Pap.* **2011**, *215*, 1–16.
135. Allan, S. On the Identity of Shang Di and the Origin of the Concept of a Celestial Mandate (TIAN MING). *Early China* **2007**, *31*, 1–46. [CrossRef]
136. Lloyd, G.E.R. *Adversaries and Authorities: Investigations into Ancient Greek and Chinese Science*; Cambridge University Press: Cambridge, UK, 1996; Volume 42.
137. Tanner, J. Ancient Greece, early China: Sino-Hellenic studies and comparative approaches to the classical world. A review article. *J. Hell. Stud.* **2009**, *129*, 89–109. [CrossRef]
138. Raphals, L. Which Gods if Any: Gods, Cosmologies, and Their Implications for Chinese and Greek Divination. In *Ancient Divination and Experience*; Oxford University Press: Oxford, UK, 2019; pp. 253–276.
139. Guo, L.; Gao, J.; Yan, W. A Comparative Study of earthquake data in Early World History. *Earthq. Res. Shanxi* **1992**, *2*, 15–28. (In Chinese)
140. Ichirô, K. Rituals for the Earth. In *Early Chinese Religion, Part One: Shang through Han (1250 BC–220 AD)(2 vols.)*; Brill: Leiden, The Netherlands, 2009; pp. 201–234.

quaternary

MDPI

Article

Novel Combined Approach of GIS and Electrical Tomography to Identify Marsh/Lake at Kastrouli Late Mycenaean Settlement (Desfina, Greece)

Ioannis Liritzis [1,2,3,4,*], Niki Evelpidou [5], Ilias Fikos [6], Alexandros Stambolidis [6], Nectaria Diamanti [6], Theano Roussari [6], Maria Tzouxanioti [4], Prodromos Louvaris [6] and Gregorios N. Tsokas [6]

1 Laboratory of Yellow River Cultural Heritage, Key Research Institute of Yellow River Civilization and Sustainable Development, Henan University, Minglun Road 85, Kaifeng 475001, China
2 Collaborative Innovation Center on Yellow River Civilization, Henan University, Minglun Road 85, Kaifeng 475001, China
3 European Academy of Sciences & Arts, St. Peter-Bezirk 10, A-5020 Salzburg, Austria
4 Department of Archaeology, School of History, Classics and Archaeology, College of Arts, Humanities & Social Sciences, Edinburgh University, Edinburgh EH8 9AG, UK; mtzouxanioti@geol.uoa.gr
5 Faculty of Geology and Geoenvironment, National and Kapodistrian University of Athens, Panepistimioupolis, 15784 Athens, Greece; evelpidou@geol.uoa.gr
6 Exploration Geophysics Laboratory, Department of Geology, Division of Geophysics, Aristotle University of Thessaloniki, 54124 Thessaloniki, Greece; ifikos@geo.auth.gr (I.F.); astamp@geo.auth.gr (A.S.); ndiamant@geo.auth.gr (N.D.); roussaria@geo.auth.gr (T.R.); info@geo.auth.gr (P.L.); gtsokas@geo.auth.gr (G.N.T.)
* Correspondence: liritzis@henu.edu.cn

Citation: Liritzis, I.; Evelpidou, N.; Fikos, I.; Stambolidis, A.; Diamanti, N.; Roussari, T.; Tzouxanioti, M.; Louvaris, P.; Tsokas, G.N. Novel Combined Approach of GIS and Electrical Tomography to Identify Marsh/Lake at Kastrouli Late Mycenaean Settlement (Desfina, Greece). *Quaternary* **2022**, *5*, 26. https://doi.org/10.3390/quat5020026

Academic Editor: José Javier Baena Preysler

Received: 4 April 2022
Accepted: 29 April 2022
Published: 4 May 2022

Abstract: The Kastrouli Late Bronze settlement in Phocis province, central Greece, has been proved to have been an important center in the periphery of the Mycenaean palaces. It was reused at least partially and was cultivated until the 20th century. The presence of a flat area off the Kastrouli hill and the seasonal flooding nowadays led to the present investigation, questioning the formation of an ancient lake or marsh/swamp. A methodological approach was applied combining the digital elevation model (DEM) and GIS of the wider and confined area, examining slopes between 0 and 5 degrees (0 and 8.75%), with electrical resistivity tomography (ERT) traverses of around 300 and 500 m, reaching a depth of 100 m. The ERT data were rapidly collected on profiles and provided a cross-sectional (2D) plot. It was found that, in the area, there is a basin with a length of 100 m and a depth of around 40–50 m. The sedimentation process over the millennia has filled the basin, with the upper 5–6 m surface layers of the area having a low resistivity. The presence of two natural sinkholes with apparent engineered hydraulic works is noted to conform to drainage and produce a habitable environment, protecting the cultivated land and avoiding a swamp associated with health issues.

Keywords: marsh; swamp; environment; basin; digital elevation model; GIS; slopes; inclinations; Mycenaean; river; flood

1. Introduction

Landscapes are constantly changing as a result of climatic impact; the formation of streams and rivers; and variations in riversides, flooding, water erosion, and tectonic causes. Combining archaeology with investigations of ancient marshes or pods/lakes is figurative for the resetting of palaeoenvironmental status and our comprehension of anthropic landscape impacts. Studying relict lakes from soils and applying surface geomorphology to geoarchaeological datable surfaces (pedoarchaeology) are distinctly helpful in this process, as well as in establishing a chronology of pedogenesis and dwellings [1].

Settlements in Greece were set on high hills to protect against occasional flooding. However, during the prehistoric eras, settlements were placed near lake-margin marshes and ponds. The abundance of aquatic resources in the lake margin and its vulnerability to short- and long-term fluctuations have had an impact on both subsistence and settlements and are thought to have motivated a para-agricultural economy [2]. In Greece, throughout its long history, water has played an indispensable role in the life of these ancient societies, and Greeks were rigorous in the field of water management, carrying out an impressive variety of hydraulic works [3,4]. Hence, the first lakes may have developed into swamps/marshes, introducing malaria from insects (malaria is caused by parasites such as the plasmodium, and it is transmitted to humans by mosquitoes) and hindering agriculture and livestock; thus, attempts to dry out the lakes may have been made [5,6]. In addition, these waterbodies have records of long environmental variation changes and human resource use, as well as nurturing high rates of endemism and biodiversity, with the freshwater enhancing algae overgrowth, fed by excess nutrients [7].

The presence of an ancient lake and/or marsh can be confirmed using various methodological tools: borehole logging [8], archaeological witness [9], historical accounts, mythological legends (see lake gods), and electrical tomography [10] have been quoted.

The drainage of water from agricultural land has existed for a long time in history, dating back to the earliest civilizations in the Middle East since about (circa, ca.) the fourth millennium before our common era (BCE). In the southeastern Mediterranean, the Minoan and Mycenaean cultures devised techniques and master plans for the drainage of agricultural lands in the ca. 15th century (c.) BCE. Since the decline and fall of the Aegean–Anatolia region and Near and Middle East civilizations, society retrieval and agricultural innovation during the Archaic and Classical periods (ca. 9th–4th c. BCE) encompassed successful hydraulic endeavors, mastering drainage and irrigation techniques. Moreover, Mesoamerica, India, and China have extensive histories of drainage and examples of underground cisterns since prehistoric times and such evidence has been found elsewhere in Greece, e.g., in Mycenae, Athens, Tyrins, Zakros, and Tylissos [11–14]. Earlier engineering works are evident at Lake Kopais, Boeotia, where Late Mycenaeans from Orchomenos, Gla, or Thebes settlements drained the lake [6], as indicated by a GIS-based survey of archaeological datasets [15].

In the present paper, the plain adjacent to the Kastrouli fortified settlement (Desfina village, near the town of Delphi, Phocis province, Central Greece) is investigated concerning drainage activities using a geoarchaeological approach (Figure 1). It is a wetland agrarian area, with the presence of a Late Mycenaean site from 14th c. BCE., which was also inhabited at later times. The site has been systematically excavated since 2016, and an interdisciplinary methodological approach has been developed. Chronological studies using radiocarbon (^{14}C) and luminescence dating, characterization and provenance, geophysical prospection investigations, and coastal geoarchaeology have produced invaluable results regarding the Mycenaean presence in southern Phocis [16–23].

Here, we advance these investigations in the search for the presence of an ancient lake or marsh/swamp by first following a novel approach combining a geographical information system (GIS) with a digital elevation model (DEM) and thematic maps, focusing on the sloping of the terrain, together with electrical resistivity tomography (ERT) traverses, supported by apparent archaeological remains.

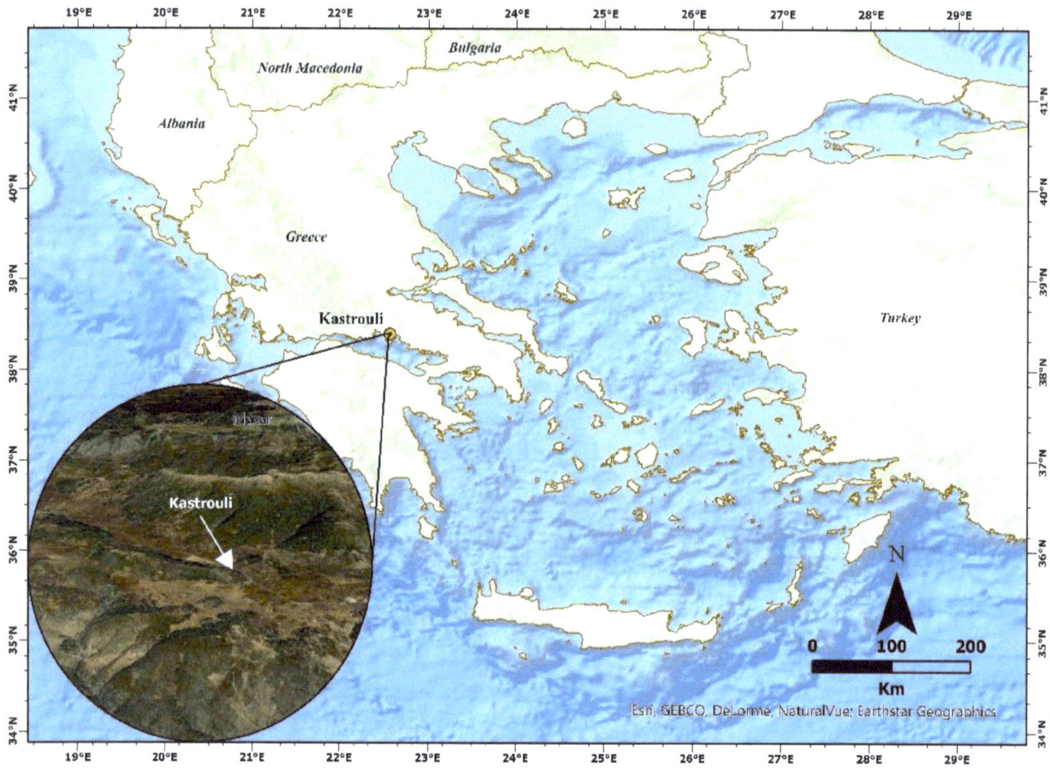

Figure 1. Aerial view of the Kastrouli settlement and environs and the location of the site in the eastern Mediterranean. The traced, shaded landmark on the edge of the plain is noticeable, and it is where water gathers during the winter; this geometric mark indicates plant debris forming a traced boundary of a littoral zone similar to shorelines.

The prompts for this investigation were the contemporary sporadically flooded plain at the Meteles plain off the Kastrouli hillock during winter and the presence of two sinks built with a rim and an entrance leading via a channel to a natural underground fissure. The surface ruins of the two built sinks were apparently engineered to lead seepage water into a natural aquifer present in the limestone bedrock; this observation led to an earlier report on the alleged ancient hydraulic works in the area [22]. These two features suggest the possible existence of a marsh/small lake in the past, as they indicate earlier attempts to drain the formed small lake/marsh. The natural sinks were probably first recognized by the ancient habitants of the area in the nearby prehistoric Kastrouli settlement or its occupants at later times. Thus, the natural outpour was engineered to facilitate drainage. Today, the plains around Desfina village are inundated with long-lasting rainfall during the winter (Figure 2).

Here, we present the first phase of a GIS and electrical tomography combined novel approach to examine the presence of an old swamp/marsh or small lake to confirm the hydraulic works at Kastrouli being an attempt to drain the plain in the past.

The objective of the present paper is to identify and define the boundaries of an ancient marsh/lake using a novel combined approach of a digital elevation model, GIS, and electrical tomography in light of the remains of built surface works, made to reroute vast amounts of flood release into two sinkholes.

Figure 2. A panoramic photo of formation of a seasonal lake in the vicinity of Desfina (on the right), and snowy Mount Parnassus can be seen in the background (© IL).

2. Methods and Instrumentation

2.1. Electrical Resistivity Tomography (ERT)

Electrical resistivity tomography (ERT) is a sophisticated geophysical method that determines the Earth's subsurface resistivity by generally taking measurements on the ground surface [24]. ERT data are rapidly collected on profiles and provide a cross-sectional (2D) plot of the so-called apparent resistivity (Ω·m) versus depth. This is a technical quantity, which is a function of the true resistivities of the formations, constituting the subsurface setting. Next, these data are inverted to yield the distribution of the true resistivity in the subsurface, which is called "electrical resistivity tomography". ERT interpretation is a straightforward task and leads to accurate representations of the geometry and lithology of the subsurface geologic formations.

Clay, metallic oxides, and sulfides are common sedimentary materials that can conduct electrical currents through the material itself. As such, the specific resistance of most near-surface sedimentary materials is indicative of the quantity and chemistry of the pore fluids within the material. The range of resistivity responses for one specific formation depends on the saturation level, ion concentration, the presence of organic fluids (such as non-aqueous-phase liquids), the temperature, the pressure, and the porosity [25,26]. The general principles that ERT is based on are what have been in use by geophysicists for almost a century [27]. Recent developments in field equipment and data processing techniques have enabled rapid two-dimensional routine research and three-dimensional research. Older one-dimensional resistance surveys are still common and useful in many cases but have problems interpreting areas of complex 2D or 3D geology.

Designing the exact parameters of our ERT survey in advance was not possible, mainly due to the lack of information regarding the maximum depth of investigation required. However, since the estimated maximum depth of investigation was of the order of several tens of meters, we employed two custom-built 24-channel cables with 11 m maximum electrode separation. This way, we could modify the spacing and the total length of each ERT from a few meters up to more than 500 m, thus modifying the maximum depth of the investigation and resolution accordingly (Figure 3).

Figure 3. On the site of Meteles plain arranging ERT measurements (© IL). Upper left: four of the co-authors preparing for the ERT measurements in situ, Left: two co-authors lay the cables for ERT, Right: taking a reading in situ.

Apart from the 2 cables, we used the IRIS Syscal Pro, 10—channel system together with 48 steel electrodes. The positioning of each line was determined with differential GPS, ensuring accuracy of a few centimeters.

The first ERT1 was measured with electrode spacing a = 6 m, providing a total length of 282 m and a maximum depth of investigation greater than 60 m. Based on the preliminary results of ERT1, we implemented greater electrode spacing in the second ERT2. Hence, we exploited the maximum possible separation a = 11 m, reaching a maximum depth of investigation of almost 100 m and a total length of 517 m (Figure 4).

The calculation of the true resistivity distribution was performed with the use of DC_2DPro [28], an iterative inversion algorithm that transforms the arrangement of the resistivity data into a geology model that yields the observed distribution of the resistivity values.

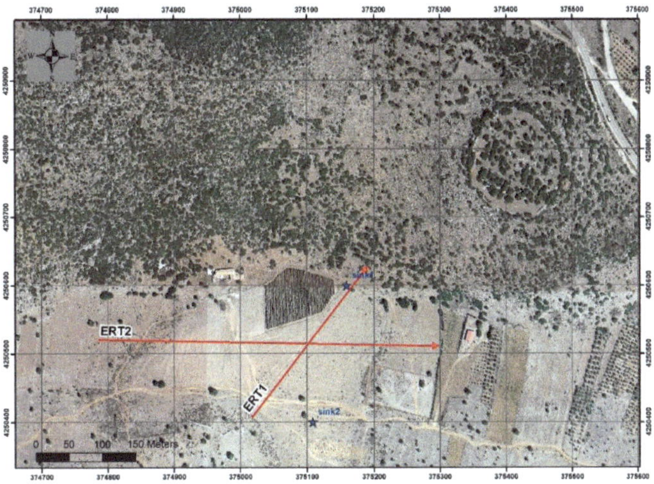

Figure 4. Aerial view of Kastrouli (round circle) and environment with the two ERT traverses (in red). The two sinkholes sink1, sink2 are shown in blue stars. The traced, shaded landmark on the edge of the plain is noticeable, and it is where water gathers during the winter; this geometric mark indicates plant debris forming a traced boundary of a littoral zone. Axis numbers are the HGS (Hellenic Geodetic System) 1987 coordinates in meters.

2.2. GIS, DEM, and Thematic Maps

Geomorphological observations in the Meteles area were made in the field, from existing maps, and from aerial photos derived from an Unmanned Aerial Vehicle (UAV) (DJI Mavic Mini). For the accuracy of field data collection, RTK-GNSS equipment (SPECTRA SP-80) was used. All data were stored in a geographic information system in order to develop a geodatabase for further analysis. Topographic data were simulated in order to produce a digital elevation model (DEM). Based on DEM, a geographical distribution of morphological slopes was developed [29,30], as well as of the drainage system and drainage basins. Analysis of the morphological slopes using several thematic maps helped to investigate, describe, and analyze the changes in the relief.

3. Results

3.1. ERT

Electrical tomography was applied, deploying two traverses along and across the plain (Figure 4). The low conductivity, due to the extremely dry ground during summer (Figure 5), resulted in very high contact resistances, which were compensated for by watering the electrodes. The yielded resistivity imaging of the subsurface indicates a limestone bedrock whose ceiling undulates. At its shallower parts, its depth varies between 0 and 7 m, whereas its depth is ca. 60 m at its deeper parts. The bedrock depth extends beyond 90 m. Figure 6 displays two electrical tomographies. They show variable underground topography with loose sedimentary deposits near the surface (5–10 m). Figure 7 shows a "fence diagram" of the two electrical tomographies. The two sinkholes passing through the natural fissures of the low-resistance limestone bedrock are noticeable. A depression filled with sediments is found along ERT1 at a depth of ca. 60 m, with a locus of loose sediment at 30–40 m of very low resistivity (10–20 ohm-m). The two sinkholes in ERT1 are connected to channels through loose resistance, leading to discharge through the hard rock. In recent times, discharge occurs via natural dryness, but in ancient times, hydraulic works were necessary to remove the water. Since these two engineered sinkholes lie on the present surface, this implies that the boundary of an ancient marsh or small lake could have been close to these sinkholes, while hydraulic drainage works should have been made when the water level was within a meter or less from the present ground surface. The latter delineates the maximum extension of the marsh/lake.

Figure 5. Characteristic crevices and cracks are formed in the ground due to summer dryness (© IL).

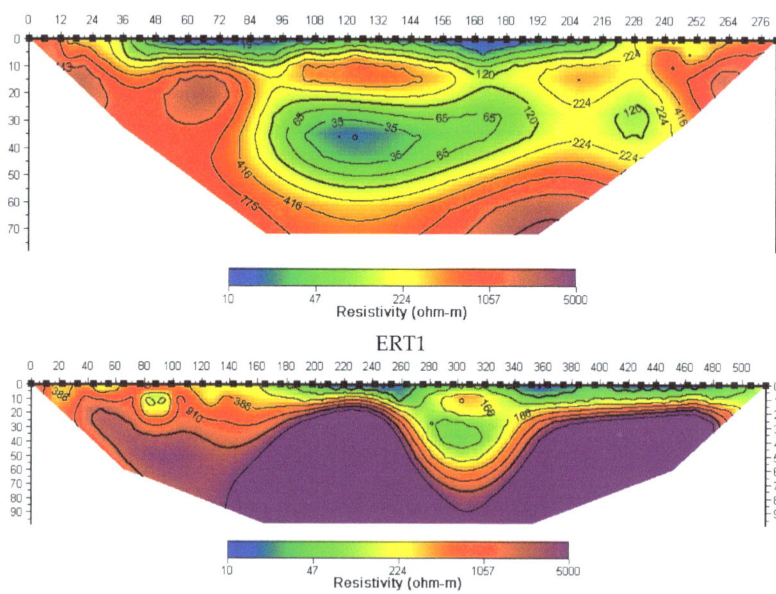

Figure 6. Electrical resistivity tomographies along the profiles ERT1 and ERT2.

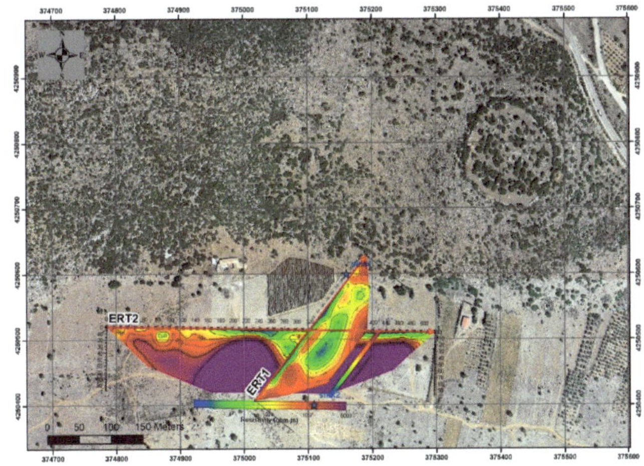

(**A**)

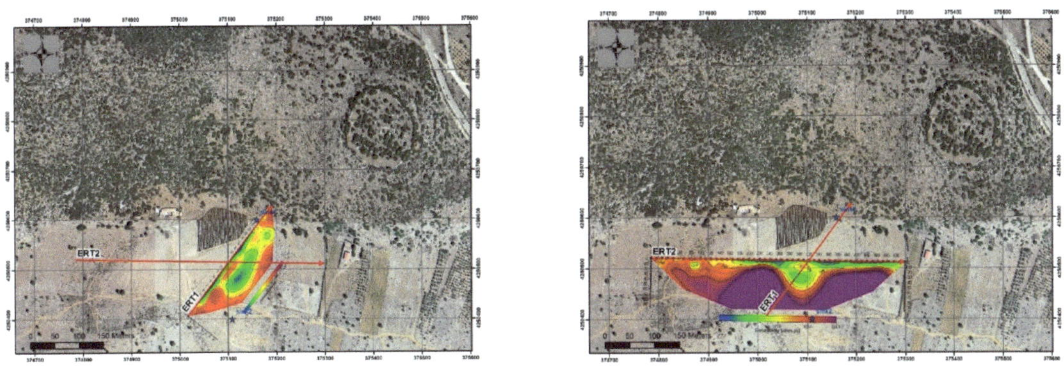

(**B**)

Figure 7. (**A**) Fence diagram of the electrical topographies (ERT1 and ERT2) and (**B**) the ERTs of the Meteles basin shown such that their relation to the basin becomes apparent. Axis numbers are the HGS (Hellenic Geodetic System) 1987 coordinates in meters.

3.2. Slopping via Thematic Maps, GIS, and DEM

It is evident that, nowadays, the geomorphological characteristics of the area of Meteles are not the same as those 3200 years ago. The large amounts of sediment transported during this period must have changed the bottom of a small lake.

For this purpose, using ArcGIS Pro v. 2.9.2. software, we set up a high-resolution DEM 5 × 5 m, which was derived from topographic maps at a scale of 1:5000, spatial data produced from a UAV survey in the main study area [31], and RTK-GNSS data (Figure 8). In the present study, the UAV data provide an accurate metric documentation of the wider archaeological site.

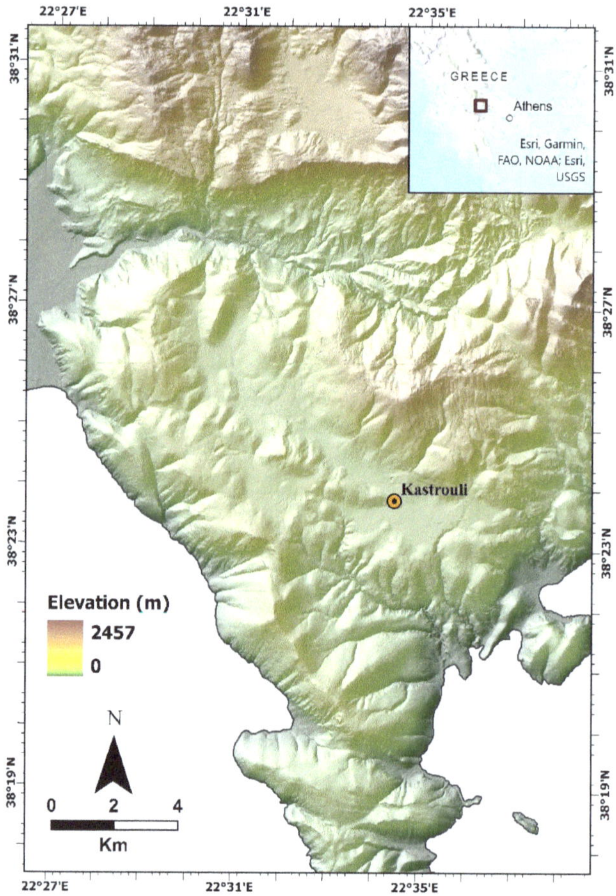

Figure 8. Digital elevation model of Southern Phocis area with Kastrouli fortified settlement.

The DEM was created in order to identify changes in the relief and in the physical characteristics of the study area, e.g., surface drainage information, such as hydrographic network and drainage basins, and to calculate the morphological slopes. The map of the morphological slopes depicts the geographical distribution of the morphological slopes within the drainage basins and, at the same time, reveals where the surface morphology could indicate the existence of sinkholes.

Based on the Food and Agriculture Organization of the United Nations (FAO) slope classification, the analysis indicates a relatively low slope area or flat-to-sloping area of 3.04 Km^2 near the Kastrouli fortified settlement (Figure 9), with inclinations fluctuating between 0 and 5 degrees or 0 and 8.75% (Figure 10). In this area, the authors identified two depressions in the ground, called sinkholes, which are indications of unnatural external surface drainage. Based on the study conducted by Ford and Williams [32,33], sinkholes are produced from different geological processes of endogenous and exogenous origin. Moreover, these sinkholes are located in the relatively low slope area (Figures 10 and 11), which consists mainly of alluvial deposits, limestone formations, and second-generation bauxite ore deposits, indicating both depression and subsidence. Figures 10–12 illustrate variations in the morphological slopes, and the morphological slopes of 1°, 2°, and 3° are delineated to illustrate the area where sinkholes developed.

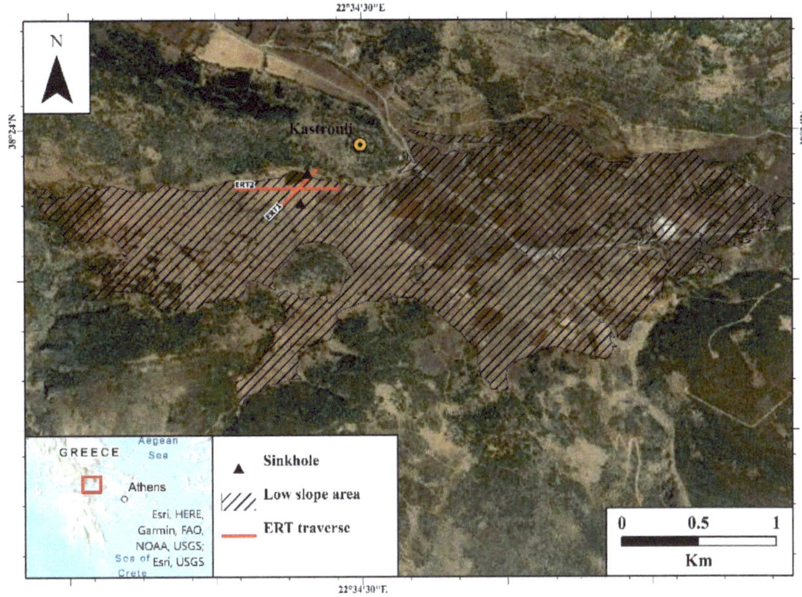

Figure 9. The (northeast) map, derived from the high-resolution digital elevation model, showing the position of the sinkholes, the two ERT traverses (red) and the relatively low slope area.

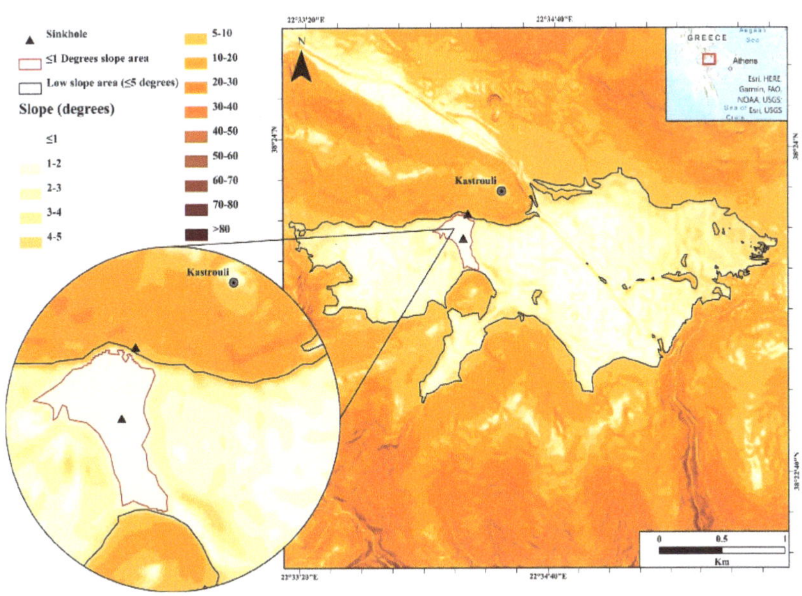

Figure 10. Morphological slopes of Kastrouli area. The black outline identifies the area with inclinations fluctuating between 0 and 5 degrees (0 and 8.75%). On the contrary, the red outline indicates an area of 0.06 Km^2, with inclinations fluctuating between 0 and 1 degrees (0 and 1.75%). Gray triangles are the sinkholes.

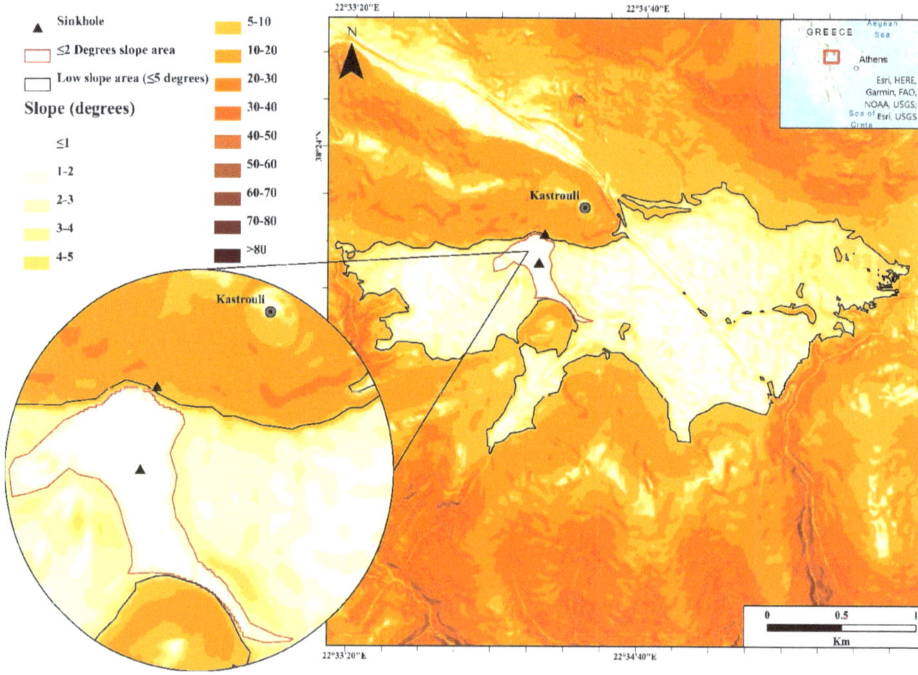

Figure 11. Morphological slopes of Kastrouli area. The black outline identifies the area with inclinations fluctuating between 0 and 5 degrees (0 and 8.75%). On the contrary, the red outline indicates an area of 0.10 Km^2, with inclinations fluctuating between 0 and 2 degrees (0 and 3.49%). Gray triangles are the sinkholes.

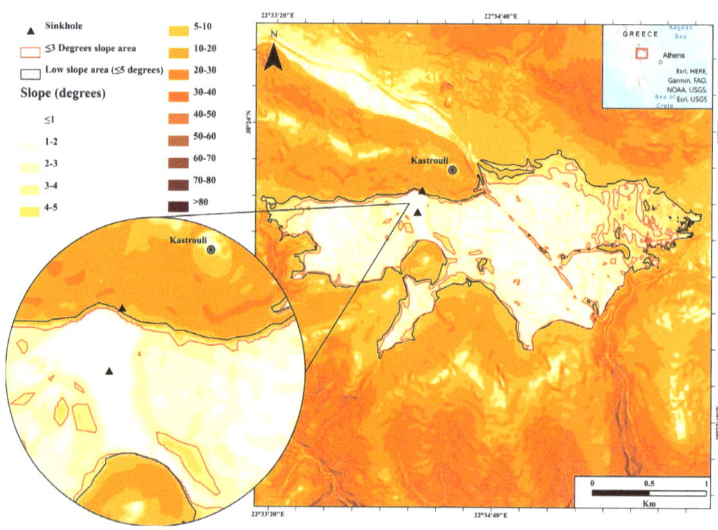

Figure 12. Morphological slopes of Kastrouli area. The black outline identifies the area with inclinations fluctuating between 0 and 5 degrees (0 and 8.75%). On the contrary, the wide red outline indicates an area of 2.33 Km^2, with inclinations fluctuating between 0 and 3 degrees (0 and 5.24%). Gray triangles are the sinkoles.

4. Discussion

The broader geological area of Kastrouli–Desfina consists mostly of Alpine formations of the Parnassus–Ghiona zone and is partly covered by Neogene–Quaternary terrestrial loose deposits. South of Kastrouli, at the coastal zone, there is a small outcrop of the Pindos zone. This Parnassus–Ghiona mountain zone consists of a limestone sequence with bauxite intercalations. The limestones are neritic, middle-to-thick layers of the Triassic to Cretaceous and are covered by flysch, which includes brown-red siltstones, mudstones, and sandstones.

The upper Cretaceous limestones vary in thickness between 150 and 200 m. The basins around Kastrouli and the Desfina valley are filled with Quaternary, mainly Holocene sediments [19,34,35].

At the edges of the Kastrouli–Meteles basin, man-made structural activities on a natural crevice in the bedrock have formed two constructed sinkholes. The southern one has a western opening of around 1.40 m and a width of 0.60 m, and around twelve descending stairs leading to an underground tunnel intentionally dug to divert flood water, reaching a ca. 4 m chamber (Figure 13). To date, no extensive survey has been conducted regarding the ceramic sherds present at the bottom of the tunnel and/or the dating of the stone construction with its mortar. Thus, any dating is possible from the Late Bronze to the medieval period and maybe even modern times.

Figure 13. The southern sinkhole in the center of Meteles plain valley. From the present surface, the top of the interior with steps. Photos are taken from various angles of same sinkhole 2 (© IL).

The second one, north of the plain at the bottom of the foot slope of Kastrouli hill (Figure 14) and built with medium-sized stones, includes a cistern of about 5 by 3 m with a drainage sink, and the drainage location is covered by sediments. This sink was of particular importance, as it has a small-sized arched bridge made with stone block and unpolished stones with mortar, and it is of a larger construction and an obviously engineered channel to drain the water.

Figure 14. The cistern of the northern side sinkhole at Meteles, close up (**left**) and at a distance (**right**) (© IL).

The waters draining into this initially deep basin, made up of variable thicknesses of sedimentary deposits, as shown by ERT, carry with them much of the suspended sediment that has been transported to date by the presence of rivers and streams from the local catchment area. The actions of the current and the waves along the shore of the swamp/small lake are responsible for the additional erosion and sediment deposition, and some material may have been introduced as a result of the action of the wind; furthermore, hydraulic effects dominate in small lakes. Seasonal rivers and streams in the environment transport material of many different sizes, mainly of fine-to-medium grain size, the largest being rolled along torrents. River water enters the Meteles basin during Holocene, and the bed-load transport of the impermeable rock ceases upon reaching today's flat area. The lake basin water outlets seem to define the boundary limits adjacent to the outflow. Because the dynamic processes that keep materials suspended are generally more active near the water inlet to the formed swamp, sediments are usually sorted by size. Rocks, pebbles, and coarse sand are located near the shore, while the finer sand, mud, and silt are, in most cases, in the center of the basin (borehole log work in progress).

Both man-made structures have a cistern, a dug pipe similar to a channel to direct the gathered waters into the openings of the two sinkholes. Drainage was probably made to protect the cultivated land and to avoid a swamp associated with health problems, as it is situated precisely ahead of the district. These are excellent reasons to conclude that the current drainage works in the Meteles plane are later restorations or modifications of a system designed by the Mycenaean residents of the fortified village of Kastrouli [22].

Both ERT measurements, with total lengths of 282 and 517 m, showed that there is indeed a dip in the investigated area with an irregular topography filled with Quaternary sediments at depths ranging between 5–6 m and 60 m for ERT1 and 100 m for ERT2.

However, the high-resolution DEM using ArcGIS derived from topographic maps at a scale of 1:5000, UAV, and RTK-GNSS data identified morphological slopes and other interesting physical characteristics.

This analysis indicated a relatively low slope area of around 3 Km^2, with inclinations fluctuating between 0 and 5 degrees (0 and 8.75%) and sinkholes produced from different geological processes of endogenous and exogenous origin. Depression and subsidence are observed, filled in with alluvial deposits, limestone formations, and second-generation bauxite ore deposits. It is of interest to note that the area covered by the slope of less than 1 degree ($\leq$1.75%) is 0.61 Km^2 (~240 $\times$ 240 m), with two natural outcrop barriers, one being the Kastrouli foothill and the other being close to the southern sink. The increase in the area is profound for slopes higher than 3–5 degrees (5.24–8.75%).

The piedmont in the investigated area gradually increases in elevation at the base of Kastrouli hill in the present upland area. The transition zone between the Meteles plain and the low-relief adjacent hillocks are well defined in the thematic slope maps reinforced by ERT. The traced area between 0 and 5 degrees (0 and 8.75%) results in a dissected plateau of the Desfina landscape and consists of alluvial fans and coalesced alluvial fans supported by the torrent and stream GIS mapping of the region (Figure 15). An area of 2.33 Km^2 makes it a large region, which, in the past, may have suffered from flooding, with dimensions of 3 Km by 1–2 Km.

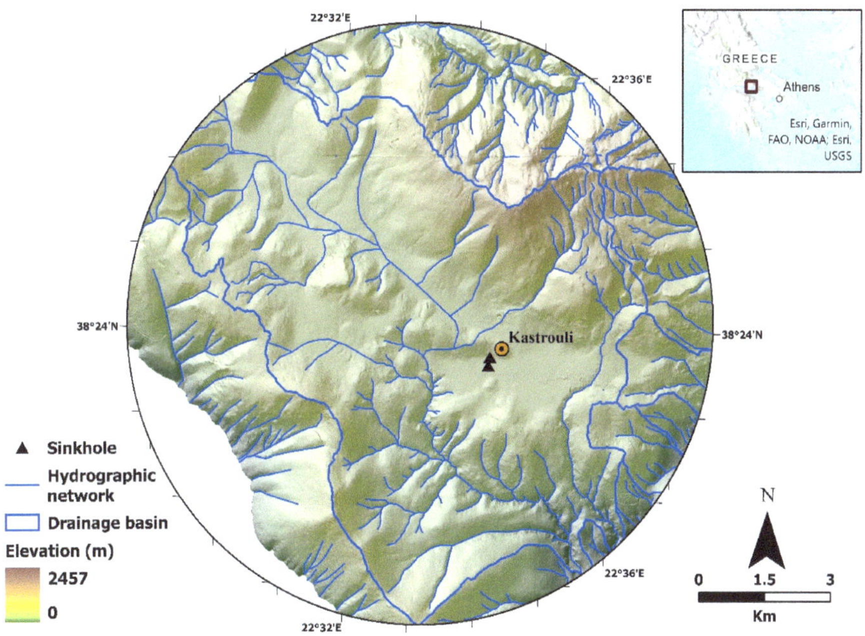

Figure 15. The well-developed hydrographic system of the wider Kastrouli area.

We believe that the drained area confines are most probably at the boundary encircled by a slope of less than 2 degrees ($\leq$3.49%) (less than 0.1 Km^2), whereas the two sinks' positions are relevant and close to the flooded area, where streams and torrents flow downhill and drainage is manageable (Figure 15).

Regarding ERT1, in the first 6 m, there are conductive, loose, clay materials, which are perhaps eroded materials from limestone (Figure 6). At the 60th meter along the tomography, the limestone plunges to a depth of about 35 m. The geological map does not provide any fault indication (as a rift valley) or any bauxite ore body that could explain this image. The basin starts somewhere at about the 95th meter and ends at about 180 m, whereas it seems to reach a depth of 40 m, making the size of this highly conductive layer about 100 m by 30 m. As noted above, in the first 5–6 m, there are conductive materials, while below this formation, there is an antistatic layer of thickness of about 10–20 m. Further down, a relatively conductive formation appears, which may reflect an aquifer.

The existence of the sinkhole may be related to this layer, but our measurements are few, and a (planned) borehole will give more information. There is a sinking in the whole area, but regarding the sinkholes, we note with confidence that the sinks are located at a slightly higher altitude than the basin. This is likely to be affected by the very existence of the zone in which conductive materials accumulate.

In the second section of ERT2 (Figures 6 and 7), the (same) basin is, again, clearly seen, which seems to reach a depth of about 40 m as also noted for ERT1. The basin therefore seems to start from the 260th meter and reaches up to the 340–350th meters along the tomography or about 100 m in length. As in ERT1 and here in the first few meters, we see the conductive material, while deeper, the existence of an antistatic aquifer is most probable.

The ERT and GIS-DEM data sway one's opinion to assume the presence of a cave in that area. That is, the rainwater found a way to dissolute the limestone perhaps together with the existence of a fault. This cave may have collapsed due to the action of a seismic fault, forming the basin and the accumulated deposits.

However, this hypothesis is to be tested if future research verifies the existence of a rift, perhaps conducting a borehole analysis of sediments. However, to date, such information has not been provided by the geological mapping of the area.

5. Conclusions

The set objectives of the present investigation were fulfilled. The sporadic flooding of the Meteles plain that occurs nowadays during winter in the vicinity of the Kastrouli prehistoric settlement (14th to 11th c. BCE) and the presence of two old sinks apparently engineered to lead seepage water into a natural aquifer present in the limestone bedrock were explained. The results of the combined ERT and GIS-DEM geophysical techniques confirmed these observations, indicating the alleged ancient hydraulic works in the area as having been a small lake in the past that developed into a marsh. The surficial entrance to the sinkholes implies that a marsh developed during the last 3500 years or so. The studied area close to the Kastrouli Mycenean settlement, which seems to have been reinhabited at later times, contains a basin with a total length close to 100 m in width and a depth somewhere between 40 and 50 m. It is a small basin with two sinks near it, which drain the water. These natural crevices were reworked by ancient inhabitants and may have continued to last centuries to aid the safe drainage of flooding. In fact, the flooded area and/or the formation of a marsh or swamp, perhaps a small ancient lake, was estimated with DEM and GIS sloping to be around 0.1 km^2, and these two sinkholes were found using ERT, as well as low resistivity values in that part of the bedrock. Future work includes boreholing and the study of the acquired sedimentary cores for dating formation, the origin of the sediments, and a palaeoecological reconstruction of the area.

Author Contributions: Conceptualization, I.L.; methodology, I.L., G.N.T. and N.E.; software, N.E., I.F., A.S., M.T., and P.L.; investigation, I.L., G.N.T., N.E., A.S., I.F., T.R. and N.D.; writing—original draft preparation, I.L.; resources, I.F., A.S., N.E., M.T., P.L. and T.R.; writing—review and editing, I.L., N.E. and G.N.T.; visualization, I.L., N.E., G.N.T., A.S., I.F. and M.T.; supervision, I.L., G.N.T. and N.E. All authors have read and agreed to the published version of the manuscript.

Funding: This research received no external funding.

Institutional Review Board Statement: Not applicable.

Informed Consent Statement: Not applicable.

Conflicts of Interest: The authors declare no conflict of interest.

References

1. Lisetskii, F.N.; Stolba, V.F.; Pichura, V.I. Late-Holocene palaeoenvironments of Southern Crimea: Soils, soil-climate relationship and human impact. *Holocene* **2017**, *27*, 1859–1875. [CrossRef]
2. Hassan, F.A. Holocene lakes and prehistoric settlements of the Western Faiyum, Egypt. *J. Archaeol. Sci.* **1986**, *13*, 483–501. [CrossRef]
3. Mamassis, N.; Moustakas, S.; Zarkadoulas, N. The operation of ancient reclamation works at Lake Copais in Greece. *Water Hist.* **2015**, *7*, 271–287. [CrossRef]
4. Koutsoyiannis, D.; Angelakis, A. Ancient Greece: Agricultural Hydraulic Works. In *Encyclopedia of Water Science*; Trimble, S.W., Ed.; CRC Press: Boca Raton, FL, USA, 2007; pp. 24–27.
5. Viollet, P.-L. *Water Engineering in Ancient Civilizations: 5000 Years of History*; CRC Press: Boca Raton, FL, USA, 2007; ISBN 9789078046059.
6. Kountouri, E.; Petrochilos, N.; Liaros, N.; Oikonomou, V.; Koutsoyiannis, D.; Mamassis, N.; Zarkadoulas, N.; Vött, A.; Hadler, H.; Henning, P.; et al. The Mycenaean drainage works of north Kopais, Greece: A new project incorporating surface surveys, geophysical research and excavation. *Water Supply* **2013**, *13*, 710–718. [CrossRef]
7. Kolding, J.; Van Zwieten, P.A.M.; Marttín, F.; Poulain, F. *Freshwater Small Pelagic Fish and Their Fisheries in Major African Lakes and Reservoirs in Relation to Food Security and Nutrition*; FAO Fisheries and Aquaculture Technical Paper No. 642; Food and Agriculture Organization of the United Nations: Rome, Italy, 2019.
8. Wilke, T.; Wagner, B.; Van Bocxlaer, B.; Albrecht, C.; Ariztegui, D.; Delicado, D.; Francke, A.; Harzhauser, M.; Hauffe, T.; Holtvoeth, J.; et al. Scientific drilling projects in ancient lakes: Integrating geological and biological histories. *Glob. Planet. Change* **2016**, *143*, 118–151. [CrossRef]
9. O'Sullivan, A. Interpreting the Archaeology of Late Bronze Age Lake Settlements. *J. Irish Archaeol.* **1997**, *8*, 115–121.
10. Hoppenbrock, J.; Bücker, M.; Gallistl, J.; Flores Orozco, A.; de la Paz, C.P.; García García, C.E.; Razo Pérez, J.A.; Buckel, J.; Pérez, L. Evaluation of Lake Sediment Thickness from Water-Borne Electrical Resistivity Tomography Data. *Sensors* **2021**, *21*, 8053. [CrossRef]
11. Valipour, M.; Krasilnikof, J.; Yannopoulos, S.; Kumar, R.; Deng, J.; Roccaro, P.; Mays, L.; Grismer, M.E.; Angelakis, A.N. The Evolution of Agricultural Drainage from the Earliest Times to the Present. *Sustainability* **2020**, *12*, 416. [CrossRef]
12. Kollyropoulos, K.; Antoniou, G.P.; Kalavrouziotis, I.K.; Krasilnikoff, J.A.; Koutsoyiannis, D.; Angelakis, A.N. Hydraulic Characteristics of the Drainage Systems of Ancient Hellenic Theatres: Case Study of the Theatre of Dionysus and Its Implications. *J. Irrig. Drain. Eng.* **2015**, *141*, 04015018. [CrossRef]
13. De Feo, G.; Mays, L.W.; Angelakis, A.N. Water and Wastewater Management Technologies in the Ancient Greek and Roman Civilizations. *Treatise Water Sci.* **2011**, *4*, 3–22. [CrossRef]
14. Mays, L.W. *Ancient Water Technologies*; Springer: Dordrecht, The Netherlands, 2010; ISBN 9789048186310.
15. Farinetti, E. *A GIS-Based Study for the Reconstruction and Interpretation of the Archaeological Datasets of Ancient Boeotia*; Archaeopress: Oxford, UK, 2011.
16. Sideris, A.; Liritzis, I.; Liss, B.; Howland, M.D.; Levy, T.E. At-risk cultural heritage: New excavations and finds from the Mycenaean site of Kastrouli, Phokis, Greece. *Mediterr. Archaeol. Archaeom.* **2017**, *17*, 271–285. [CrossRef]
17. Liritzis, I.; Polymeris, G.S.; Vafiadou, A.; Sideris, A.; Levy, T.E. Luminescence dating of stone wall, tomb and ceramics of Kastrouli (Phokis, Greece) Late Helladic settlement: Case study. *J. Cult. Herit.* **2019**, *35*, 76–85. [CrossRef]
18. Liritzis, I.; Xanthopoulou, V.; Palamara, E.; Papageorgiou, I.; Iliopoulos, I.; Zacharias, N.; Vafiadou, A.; Karydas, A.G. Characterization and provenance of ceramic artifacts and local clays from Late Mycenaean Kastrouli (Greece) by means of p-XRF screening and statistical analysis. *J. Cult. Herit.* **2020**, *46*, 61–81. [CrossRef]
19. Xanthopoulou, V.; Iliopoulos, I.; Liritzis, I. Mineralogical and Microstructure Analysis for Characterization and Provenance of Ceramic Artifacts from Late Helladic Kastrouli Settlement, Delphi (Central Greece). *Geosciences* **2021**, *11*, 36. [CrossRef]
20. Levy, T.E.; Sideris, T.; Howland, M.; Liss, B.; Tsokas, G.; Stambolidis, A.; Fikos, E.; Vargemezis, G.; Tsourlos, P.; Georgopoulos, A.; et al. At-Risk World Heritage, Cyber, and Marine Archaeology: The Kastrouli–Antikyra Bay Land and Sea Project, Phokis, Greece. In *Cyber-Archaeology and Grand Narratives: Digital Technology and Deep-Time Perspectives on Culture Change in the Middle East*; Levy, T.E., Jones, I.W.N., Eds.; Springer International Publishing: Cham, Switzerland, 2018; pp. 143–234. ISBN 978-3-319-65693-9.
21. Cantu, K.; Norris, R.; Papatheodorou, G.; Liritzis, I.; Langgut, D.; Geraga, M.; Levy, T. Anthropogenic Erosion from Hellenistic to Recent Times in the Northern Gulf of Corinth, Greece. In *Mediterranean Resilience Collapse and Adaptation in Antique Maritime Societies*; Yasur-Landau, A., Gambash, G., Levy, E., Eds.; Equinox eBooks Publishing: Sheffield, UK, 2022.
22. Liritzis, I. Kastrouli fortified settlement (Desfina, Phokis, Greece): A chronicle of research. *Sci. Cult.* **2021**, *7*, 17–32. [CrossRef]
23. Koh, A.J.; Birney, K.J.; Roy, I.M.; Liritzis, I. The mycenaean citadel and environs of desfina-kastrouli: A transdisciplinary approach to southern phokis. *Mediterr. Archaeol. Archaeom.* **2020**, *20*, 47–73. [CrossRef]
24. Tsourlos, P. *Modeling, Interpretation and Inversion of Multielectrode Resistivity Survey Data*; University of York: York, UK, 1995.
25. Tagg, C. *Earth Resistances*; Pitman Publishing Company: New York, NY, USA, 1964.
26. McNeill, J. Electromagnetic Terrain Conductivity Measurement at Low Induction Numbers. *J. Can.* **1980**, *2*.
27. Schlumberger, C. *Etude sur la Prospection Electrique du Sous-sol*; Gauthier-Villars: Paris, France, 1920.

28. Kim, J. DC2DPro-2D Interpretation System of DC Resistivity Tomography. In *User's Manual and Theory*; KIGAM, Korea Institute of Geoscience and Mineral Resources: Daejeon, Korea, 2009.
29. Coveney, S.; Stewart Fotheringham, A.; Charlton, M.; McCarthy, T. Dual-scale validation of a medium-resolution coastal DEM with terrestrial LiDAR DSM and GPS. *Comput. Geosci.* **2010**, *36*, 489–499. [CrossRef]
30. Schmid, K.A.; Hadley, B.C.; Wijekoon, N. Vertical Accuracy and Use of Topographic LIDAR Data in Coastal Marshes. *J. Coast. Res.* **2011**, *27*, 116–132. [CrossRef]
31. Manfreda, S.; McCabe, M.F.; Miller, P.E.; Lucas, R.; Pajuelo Madrigal, V.; Mallinis, G.; Ben Dor, E.; Helman, D.; Estes, L.; Ciraolo, G.; et al. On the Use of Unmanned Aerial Systems for Environmental Monitoring. *Remote Sens.* **2018**, *10*, 641. [CrossRef]
32. Ford, D.; Williams, P. Karst Hydrogeology. In *Karst Hydrogeology and Geomorphology*; John Wiley & Sons, Ltd.: Hoboken, NJ, USA, 2007; pp. 103–144. ISBN 9781118684986.
33. Ford, D.; Williams, P. Analysis of Karst Drainage Systems. In *Karst Hydrogeology and Geomorphology*; John Wiley & Sons, Ltd.: Hoboken, NJ, USA, 2007; pp. 145–208. ISBN 9781118684986.
34. Marinos, P.; Rondoyanni, T. The Archaeological Site of Delphi, Greece: A Site Vulnerable to Earthquakes, Rockfalls and Landslides. In *Landslides: Risk Analysis and Sustainable Disaster Management*; Sassa, K., Fukuoka, H., Wang, F., Wang, G., Eds.; Springer: Berlin/Heidelberg, Germany, 2005; pp. 241–249. ISBN 978-3-540-28680-6.
35. Celet, P. Quelques aspects de l'hydrogeology des regions calcaires meridionales du Parnasse-Helicon (Grece). *Ann Sci. Pays Hell.* **1971**, *3*, 13–16.

quaternary

Article

The Evolution of an Ancient Coastal Lake (Lerna, Peloponnese, Greece)

Efterpi Koskeridou [1,*], Danae Thivaiou [1,†], Christos Psarras [1,†], Evangelia Rentoumi [1], Niki Evelpidou [2], Giannis Saitis [2], Alexandros Petropoulos [2], Chryssanthi Ioakim [3], George Katopodis [4], Konstantinos Papaspyropoulos [5] and Spyros Plessas [6]

1 Department of Historical Geology and Palaeontology, Faculty of Geology and Geoenvironment, National and Kapodistrian University of Athens, 15774 Athens, Greece; dthivaiou@geol.uoa.gr (D.T.); cpsarras@geol.uoa.gr (C.P.); evrentoumi@geol.uoa.gr (E.R.)
2 Department of Geography and Climatology, Faculty of Geology and Geoenvironment, National and Kapodistrian University of Athens, 15774 Athens, Greece; evelpidou@geol.uoa.gr (N.E.); saitij@geol.uoa.gr (G.S.); alexpetrop@geol.uoa.gr (A.P.)
3 Institute of Geology and Mineral Exploration (IGME), Mitseon Street 4, 11742 Athens, Greece; chryssanthi.e.ioakim@gmail.com
4 Institute of Marine Biological Sciences and Inland Waters, Hellenic Centre for Marine Research, 19013 Anavissos, Greece; gkatopodis@hcmr.gr
5 Department of Dymanic, Tectonic and Applied Geology, Faculty of Geology and Geoenvironment, National and Kapodistrian University of Athens, University Campus, 15774 Athens, Greece; kpapaspy@geol.uoa.gr
6 Independent Researcher, Ethnarchou Makariou 16, 16777 Ellinikon-Argyroupoli, Greece; s.plessas@yahoo.gr
* Correspondence: ekosker@geol.uoa.gr
† These authors contributed equally to this work.

Citation: Koskeridou, E.; Thivaiou, D.; Psarras, C.; Rentoumi, E.; Evelpidou, N.; Saitis, G.; Petropoulos, A.; Ioakim, C.; Katopodis, G.; Papaspyropoulos, K.; et al. The Evolution of an Ancient Coastal Lake (Lerna, Peloponnese, Greece). *Quaternary* **2022**, *5*, 22. https://doi.org/10.3390/quat5020022

Academic Editor: Ioannis Liritzis

Received: 13 January 2022
Accepted: 22 March 2022
Published: 8 April 2022

Abstract: Degradation of coastal environments is an issue that many areas in Europe are facing. In the present work, an ancient coastal lake wetland is investigated, the so-called Lake Lerna in NE Peloponnese, Greece. The area hosted early agricultural populations of modern Greece that started modifying their environment as early as the early–middle Neolithic. Two drill cores in the area of the ancient lake were analysed to establish the sedimentological succession and the depositional environments using sub-fossil assemblages (molluscs and ostracods). Three lithological and faunal units were recovered, the latter being confirmed by the statistical ordination method (non-metric multidimensional scaling). The usage of sub-fossil mollusc species for the first time in the region enriched the dataset and contributed significantly to the delimitation of the faunas. These consist of environments characterised by various levels of humidity (from stagnant waters to freshwater lake) and salinity, with ephemeral intrusions of salt water to the lake, documented by mollusc and ostracod populations. We conclude that the lake and its included fauna and flora were mostly affected by climatic fluctuations rather than human intervention in the area.

Keywords: molluscs; palaeoenvironment; ostracods; Holocene; biodiversity; sedimentology

1. Introduction

Coastal ecosystems are important for human activities [1] because most of human activities are connected to the sea [2]. In the region of modern-day Greece, coastal areas started hosting permanent human settlements as early as the Holocene Climatic Optimum (6000–5000 BP) [3]. In this context, coastal lakes, disconnected from the sea by barriers, together with the relevant ecosystems became affected by human presence [4,5]. As such, coastal lakes and wetlands are pockets of high biodiversity and they often host endemic species [6]. In addition, they constitute environments where non-marine molluscs thrive, mainly freshwater and other gastropods that can live in lakes, bogs and swamps, and can be important environmental indicators [7].

The focus of this work is on the so-called Lake Lerna, present in the Gulf of Argos during part of the Holocene, a region which has been exploited by humans since the Neolithic [8]. The area has been studied in the past through several perspectives, such as the reconstruction of palaeoclimates and the development of agriculture [8–10] or the study of legends that took place in the region (namely the myth of Hercules and the Lernaean Hydra [3]). Ancient Lake Lerna formed at about 8630 ± 100 BP [4], covering the south-western part of the Argolis Plain up until recent years [4]. The latter work was the first to describe in detail the structure of the plain, based on sedimentological results and ostracod assemblages. Another aspect that has been analysed by [11] concerns environmental destruction caused by early inhabitants of the area. Recent works [12] identified the climate changes happening in the area during the past 5000 years.

Here, we aim to understand the evolution of Lake Lerna (Figure 1) to examine the possibility of a human impact on the lake in comparison to climate fluctuations. Through the stratigraphic, sedimentological and palaeobiological analysis of two subfossil-rich shallow drill cores, we investigate the palaeoenvironmental evolution of the lake, regarding the alternation of swamp and shallow lake and terrestrial depositional environments.

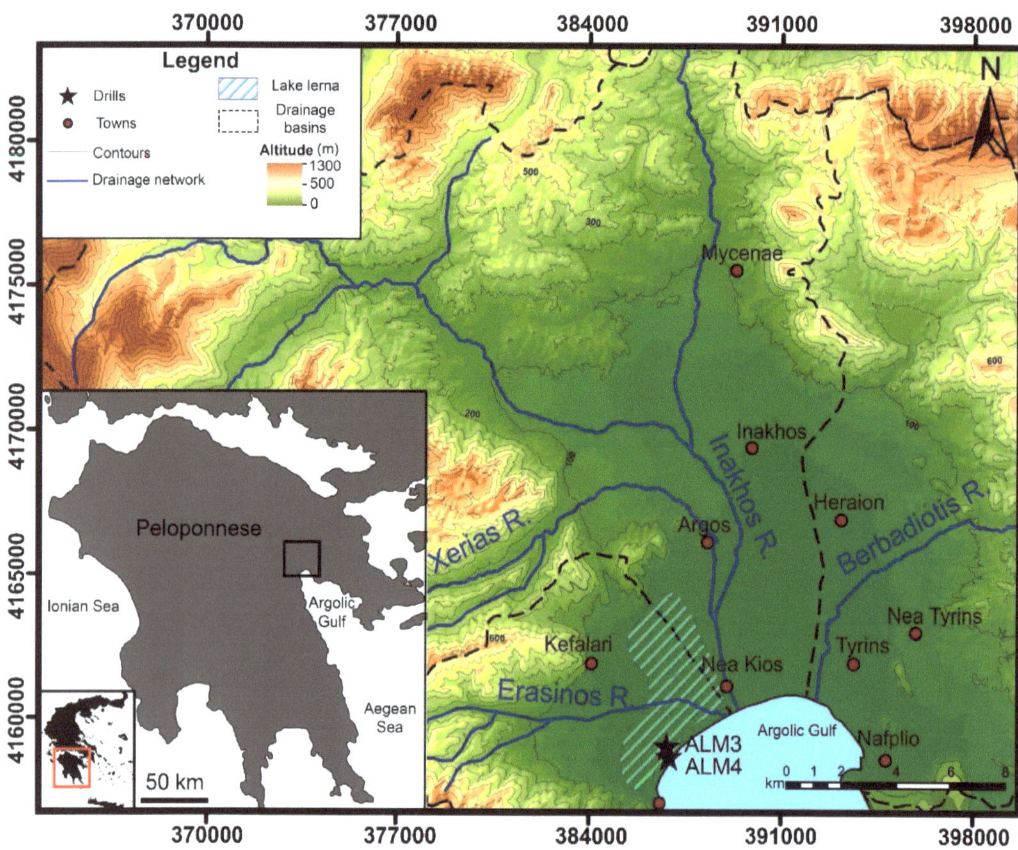

Figure 1. The study area, along with the main morphological structures and the sampling localities in the Argos Plain, Peloponnese, Greece. Made with the geographical coordinate system: Greek Grid, EGSA 87.

Coastal environments are amongst the most disturbed by human activities, especially coastal lakes and wetlands, which are particularly affected by humans [13]. Furthermore, non-marine molluscan faunas have faced a decline in diversity in recent years, as recorded in Europe [14]. The results of the present work contribute to the investigation of coastal environmental and ecological parameters that could potentially affect coastal wetland ecosystems of the past, present and future.

2. Geology and Geomorphology

The wider area of the Argive Plain has been of interest to both archaeologists and natural scientists for the past 50 years [15–23]. Geologists and geoarchaeologists have studied the area focusing on its geomorphological evolution both during the geological time and in prehistorical to historical times.

The Argive Plain consists of alpine formations that appear in its mountainous parts, as well as Upper Pliocene to Upper Pleistocene deposits found in low topographic areas around the Argos basin. The relatively modern Upper Pleistocene-Holocene deposits cover the low elevation and the coastal areas [24–27].

Tectonic action has shaped and still shapes the landscape in the wider area. The western margins of the Argolic Gulf and the Argive Plain are delimited by faults in the direction NW-SE to NW-NNE, in a staggered arrangement (Kiveri-Lerni-Kefalari). The eastern margins of the Argive Plain are also bordered by regular faults, direction NW-SE, which are intersected by faults with a main direction of NE-SW. The arrangement of the Upper Pliocene to Holocene deposits creates a main NW-SE direction of the ruptured structures in the area, with small deviations to the west and east. Regarding the geomorphology, the study area is surrounded by mountain ridges (from W-N-E), with the most important drainage basins being that of Inakhos and Xovriou, which are crossed by the homonymous rivers [28].

The drainage network of the wider area is dendritic and the flow is transient, with small or no runoff quantities most of the hydrological year. Inakhos is an important river, with the highest contribution to the study area. A characteristic landscape often found in the area consists of karstic, dissolving forms, in carbonate formations, seen as dolines and caves. Lake Lerna belongs to a large karstic system that was isolated at about 7000 years BP from the open sea by a beach barrier [4].

3. Materials and Methods

In order to reconstruct the palaeoenvironment, two shallow boreholes were made at the locality of the ancient Lake Lerna, in the area of Almyros, Argive Plain, Peloponnese, for the extraction of two cores: ALM3 (x: 386840.032, y: 4159040.203, 0.862 m.a.s.l.) and ALM4 (x: 386916.843, y: 4158686.257, 0.376 m.a.s.l.) (Figure 1). Sampling was carried out using a portable vibracoring sampler (Cobra) with a diameter of 50 mm, and the sediment was encased inside 1 m closed plastic tubes. The boreholes reached a depth of 400 cm (ALM3) and 398 cm (ALM4).

The cores were analysed at the Laboratory of Physical Geography, Faculty of Geology and Geoenvironment of the National and Kapodistrian University of Athens. The plastic tubes with the cores were split lengthwise, leaving the sediment sequence undisturbed. A stratigraphic analysis was accomplished by studying the sedimentary sequence through visual inspection. Photographs of the retrieved macrofauna were taken under a LEICA M165 C stereoscope, with a LEICA IC90 E camera.

To determine the cores' chronostratigraphy, five samples were prepared for accelerator mass spectrometry (AMS) radiocarbon dating on bulk sediment samples, some containing shell shards and peat. The dating was carried out at Centro di Datazione e Diagnostica dell'Università del Salento (CECAD), Lecce, Italy. The results of AMS dating were calibrated using standard materials supplied by IAEA (International Atomic Energy Agency) and NIST (National Institute of Standard and Technology) (Table 1). The "conventional radiocarbon age" was calculated with a $\delta^{13}C$ correction based on the $^{13}C/^{12}C$ ratio measured directly with the accelerator. For the estimation of the measurement uncertainty (standard deviation), both the radioisotope counting statistics and the scattering of the data have been taken into account. The larger of the two is given as final error (see Table 1). As such, the values of the carbon stable isotopes fractionation term ($\delta^{13}C$) measured by AMS can differ from the natural fractionation and from those measured by IRMS. Moreover, due to the hard water effect or any inwash of detrital organic matter [12], the bulk samples containing low organic material but many shell shards yielded inaccurate dates and were not taken into account.

Table 1. Absolute ages of dated samples.

Sample	Age (BP)	δ13C (‰)	Code	Core Depth	Bulk Sediment Colour
LTL19897A	3200 ± 45	11.3 ± 0.4	ALM3 1–2 3.8α	154 ± 0.5 cm	Brown
LTL20303	1515 ± 45	27.1 ± 0.5	ALM3 190–193.5	190–193.5 cm	Brown-Grey
LTL19748A	5640 ± 45	26.9 ± 0.5	ALM3 2–3 88.5	288 ± 0.5 cm	Grey-Black
LTL20301A	596 ± 45	21.0 ± 0.5	ALM4 203–205	203–205 cm	Brown
LTL20302	4610 ± 45	28.8 ± 0.6	ALM4 328–330	328–330 cm	Black

The stratigraphy of the sedimentary material from the two cores was examined using sedimentological and micropalaeontological techniques. Samples were taken from each core every 7 cm. Parts of the cores that were deemed displaying a change in the sequence were sampled every 2 cm. Each sample (20 g dry-weight) was treated with H_2O_2 (30%) to remove the organic material and agglomerates. The samples were washed using six mesh sizes to separate gravel (>2 mm), sand subdivisions (1 mm, 500 μm, 250 μm, 125 μm, 63 μm) and silt/clay particles (<63 μm). The samples were then dried in an oven at 80 °C.

The acquired data from sieving were used to provide evidence about the depositional environment. The sediments are characterised following Folk and Ward [29]. Sediment texture determination, transport processes and deposition environment were classified using the software GRADISTAT statistical package [30]. The software calculated the sediment distribution and particle size statistics (mean size (Mz), sorting (σ), skewness (Sk), kurtosis (Ku)) using alternative equations expressing values in φ units [29]. The GRATISTAT analysis produces triangular diagrams showing the relative proportion of sand, silt and clay (Figures S1–S8). From the values of one parameter or the combination of many, the depositional environments are determined, and transport and deposition mechanisms are interpreted. Mean size Mz represents the size of the granules of which sediment is composed (fine-grained or coarse-grained), and the type of energy that caused the transport (i.e., water, ice or wind). Grain concentration is measured by σ of the sample around its mean. High σ values mean that the classification in terms of grain size is poor, i.e., during the transport and deposition of the sediment, a limited sorting of its grains took place. Small σ values mean good gradation and therefore good classifications, i.e., the grains of the sediment have been well sorted from the means of transport and storage. The parameter Sk refers to the presence of coarser or finer grains. Curves with excess fine-grained material have a positive value, while those with excess coarse-grained material have negative values. The Ku parameter is a quantitative expression used to describe the deviation from normality. In other words, it measures the ratio between the gradation in the tail of the curve and the gradation in its central part. They are divided into leptokurtic, mesokurtic and platykurtic. Leptokurtic distribution corresponds to a large concentration of grains close to the mean, mesokurtic distribution refers to a normal grain distribution around the mean and platykurtic distribution refers to a large dispersion of grain distribution relative to the mean [31].

All samples were prepared for palaeoenvironmental analyses; molluscs (gastropods and bivalves), ostracods, charophytes and pollen were used for palaeoenvironmental reconstructions. For taxonomic analysis, recovered specimens were separated into morphospecies and were identified to the lowest taxonomic level possible. Selected specimens were photographed under the stereoscope. With an aim of visualising groupings of the data, we performed a non-metric multidimensional scaling (NMDS) using the Bray–Curtis dissimilarity index. All analyses were carried out in RStudio version 3.5.2 and Vegan package 2.5-4, and stratigraphic plots were made with package rioja version 0.9-26. Relative abundances were used for each species in percentages for all stratigraphic plots.

Palynological samples were prepared using standard procedures. Palynomorphs were investigated by extraction with wet sieving through 125 μm and ultrasonic 10 μm mesh sieves. Briefly, samples were treated according to the following procedure: sediment samples underwent carbonate and silicate removal with 10% HCl, they then were decanted three times and 30% HF, neutralised with KOH (10%) and subsequently washed with deionised water and sieved over a precision sieve with 10 μm pore size. Finally, one *Lycopodium* spore tablet was added to each sample to ensure estimation of the absolute abundances and reliability of quantitative data. The residues were mounted in glycerine gel on microscope slides for analysis under a binocular NIKON transmission microscope, and one or two slides of each processed sample were analysed.

4. Results

4.1. Radiocarbon Dating

The results of the radiocarbon dating are presented in Table 1. The ages of the samples vary considerably, most likely due to the hard water effect or any inwash of detrital organic matter [12]. Due to the presence of molluscs along the cores, we are hesitant to take into account of most of our dated samples, except the sample LTL20302 (Table 1), which has enough plant remains (seen as a black-coloured layer) to produce a clearer result (see discussion of [12] for details).

4.2. Cores' Lithostratigraphy

Based on the analysis of both cores, four units of depositional environments were distinguished (units A to D). Combined results of lithostratigraphy, biostratigraphy and dating were acquired from both ALM3 and ALM4 cores. Each sedimentological unit (SU) is described starting at the bottom of each core, following the deposition sequence (Figure 2).

4.2.1. Fluvial Deposits—Sedimentological Unit A (SUA)

Sedimentological Unit A corresponds to the base of the ALM3 core and ranges from 400 cm to 350 cm. It is characterised by brown-grey, slightly gravelly mud sediment (Figure 2); the gravelly fraction represents 3%, the sandy one 7% and the silt/clay 90%. The sediments were characterised as slightly gravelly mud with unimodal distribution, and are moderately well sorted.

The sediment sorting ranges from 0.45φ to 0.7φ, and is classified as moderately well sorted. Skewness (Sk) values range from -0.24φ to -0.36φ, defined as negative to very negative asymmetry. Kurtosis (Ku) values range from 1.4φ to 2.7φ and are classified as leptokurtic to very leptokurtic (Figure S1).

4.2.2. Shallow Lake Swamp Deposits—Sedimentological Unit B (SUB)

This unit covers the largest parts of both ALM3 and ALM4. Characterised by grey and dark brown sediments, with very fine granulometry (silt/clay almost 95%), it is rich in molluscs and organic material. Furthermore, SUB is divided into two sub-units, B1 and B2, with a smooth transition between the two. SUB1 is dominated by grey clay/sandy clay sediments rich in organic material, shells and includes layers of peat and gyttja. A whitish-grey layer of ash was observed at 290 cm SUB2 and is characterised mostly by brown sandy clay (silt/clay 90% and sand 10%).

More specifically, in core ALM3, SUB1 ranges from 350 cm to 250 cm. In this organic-rich, one-meter thick sedimentary sequence, two peat layers were present at 292–290 cm (sample ALM 3-21) and 275–273 cm (sample ALM 3-18) depth. Additionally, at 290–287 cm (sample ALM 3-20), a clay layer, rich in white-grey ash, was observed. Most of the sediments in SUB1 were characterised as slightly gravelly mud to mud [30], with a unimodal distribution, and are well to very well sorted, except for the sample at 275–273 cm depth, which was characterised as sandy mud [30] with a unimodal distribution, and is well to moderately well sorted. SUB2 (250 cm to 180 cm) is characterised by slightly sandy clay sediments, and organic material, such as the peaty layers of B1, are absent. SUB2 is characterised as slightly gravelly mud to sandy mud [30] with a unimodal distribution and is well to moderately sorted.

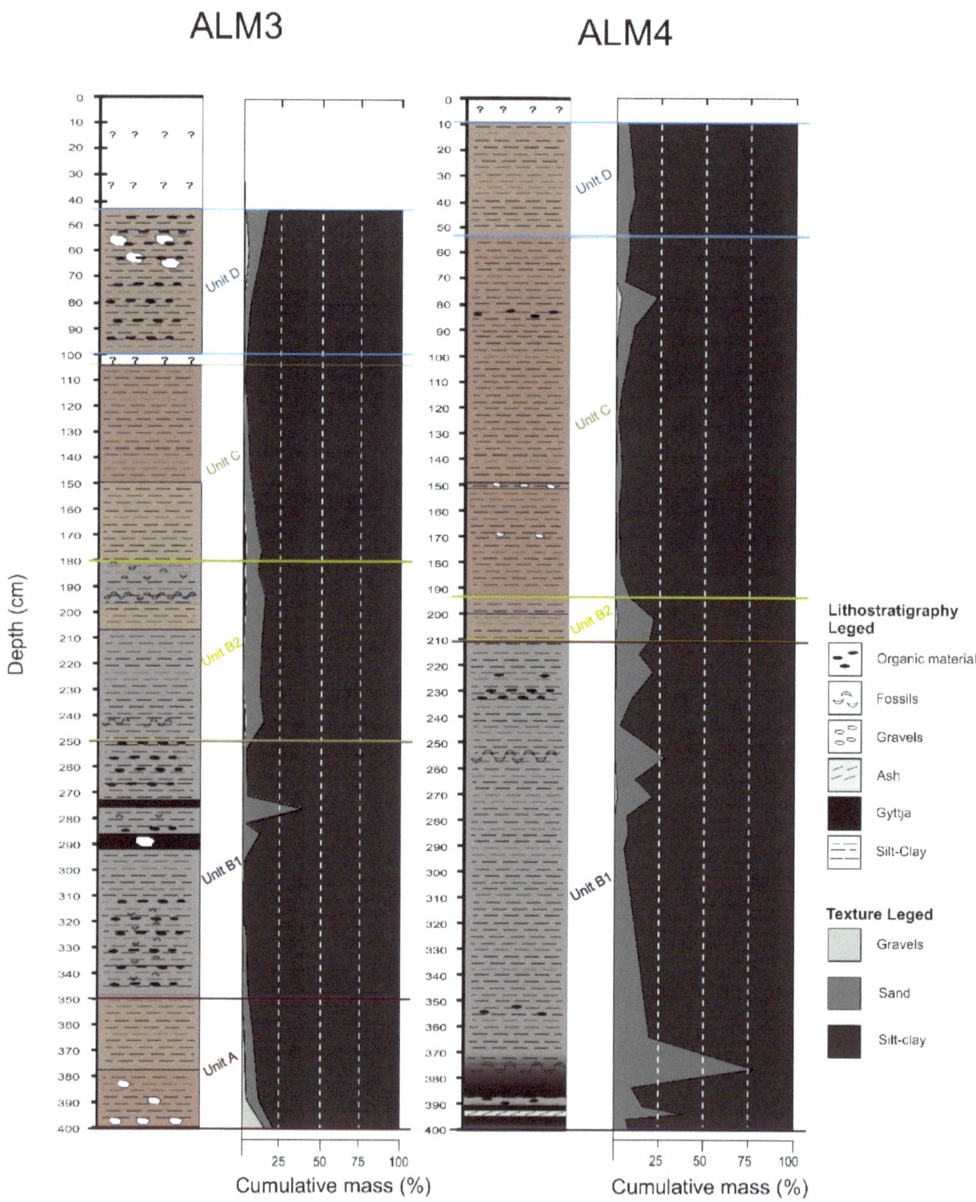

Figure 2. Cores ALM3 and ALM4 lithostratigraphy along with the corresponding sediment texture graphs. The sedimentological units are indicated on each core. The question marks (?) are indicative of disturbed material during coring.

The sediment sorting ranges from 0.3φ to 0.86φ, classified as very well to moderately sorted. Skewness (Sk) values are from −0.38φ to 0φ, which vary from very negative to almost normal asymmetry. Kurtosis (Ku) values range from 0.74φ to 3.16φ, and most of the sediments at sub-unit B1 are classified as platykurtic and as very platykurtic at SUB2 (Figures S1–S3).

In core ALM4, SUB1 was identified between 400 cm and 210 cm. Peat deposits were found at depths of 398–395 cm (sample ALM 4-36) and 385–370 cm (sample ALM 4-31, 32), with a gyttja layer occurring at 392–390 cm (sample ALM 4-34). A light grey material which corresponds to burned organic material (ash) was observed between the gyttja and the lower peat deposit at 395–392 cm (sample ALM 4-35). Further organic material and layers of shells were observed at depths 360–350 cm (sample ALM 4-30), 295–286 cm (sample ALM 4-29, 28), 255 cm (sample ALM 4-24), 245 cm (sample ALM 4-23) and 233–225 cm (sample ALM 4-22, 21). In core ALM4, the sediments of SUB1 are characterised as slightly gravelly sandy mud to mud with a unimodal distribution, and are poorly to very well sorted. Sub-unit B2 ranges from 210 cm to 193 cm depth. This unit is characterised by the smooth transition of the sediment from grey sandy/silty clay to grey-brown silty clay, with the absence of any organic material. Sub-unit B2 is characterised as slightly gravelly sandy mud to mud, with a unimodal distribution, and is well to moderately sorted.

The sediment sorting ranges from 0.3φ to 0.84φ, which is considered as very well sorted to moderately sorted, except for the samples at the depths of 391, 374, 362, 272 and 257 cm, which are characterised as poorly sorted. Skewness (Sk) values range from −0.7φ to 0φ, varying from very negative to almost normal asymmetry. Kurtosis (Ku) values range from 0.74φ to 3φ and most of the sediments of SUB1 and B2 are classified as very leptokurtic, with a few as platykurtic (Figures S5–S7).

Overall, SUB points to a shallow lake swamp depositional environment.

4.2.3. Lake Deposits—Sedimentological Unit C (SUC)

Next in the stratigraphic sequence, SUC consists of grey-brown to brown sandy/silty clays, with some gravel.

In core ALM3, SUC is found between depths of 180 cm and 102 cm. For the first 30 cm, the material consists of grey-brown clay, with a mean percentage of 92% silt/clays and 8% sands. There is a change in colour at 150 cm to brown silty clay, with mean values of silt/clays at 97.5% and sands as low as 2.5%. Granulometric data of SUC are characterised as slightly gravelly mud to mud with a unimodal distribution, and are moderately to very well sorted (Figures S3 and S4).

The same unit is also observed in core ALM4, from 193 cm to 53 cm depth. The dominant material is a brown mud, with a silt/clay value of 98%. Gravels were found occasionally at depths of 170 cm and 150 cm, while brown sandy mud (silt/clays 93% and sands 7%) with layers of organic material was observed from 110 cm up to 53 cm. Granulometric analysis shows that the sediments are characterised as slightly gravelly mud to mud, with mud predominating.

Most values of the statistical parameters of granulometry, as far as sorting is concerned, range from 0.3φ to 0.7φ, characterised as moderately well to very well sorted, except one sample at 84 cm depth of ALM 4, which was characterised as poorly sorted (σ = 1.39φ). Most of the skewness (Sk) values range from −0.3φ to 0φ, varying from negative to almost normal asymmetry, except one sample at 84 cm depth of ALM 4, which is characterised as very negative asymmetry (Sk = −0.71φ). Kurtosis (Ku) values range from 0.74φ to 3.5φ; most of the sediments are classified as platykurtic while the others are classified as very leptokurtic (Figures S7 and S8).

4.2.4. Anthropogenic Deposits—Sedimentological Unit D (SUD)

The last unit of the Almyros cores contains brown-grey silty clays, with the dominating silt/clays reaching a mean value of 90% with the sands at 9% and 1% of gravels. More specifically, in core ALM3, SUD ranges from 100 cm to 43 cm. This sedimentary sequence contains sporadic organic material while gravels were observed at 68 cm until 51 cm.

The granulometric analysis of SUC data characterised the sediments as slightly gravelly sandy mud to slightly gravelly mud [28] with a unimodal distribution, and are moderately to very well sorted.

Regarding core ALM 3 and the samples at depths of 95 and 78 cm, their sorting is 0.3φ, which are very well sorted. Skewness (Sk) values are 0φ, which are considered as almost normal asymmetry. Kurtosis values are 0.74φ and are classified as platykurtic. For samples at depths of 61 and 45 cm, the sorting ranges from 0.87φ to 0.94φ and are characterised as moderately sorted. Skewness (Sk) values range from -0.38φ to -0.4φ, identified as very negative asymmetry. Kurtosis (Ku) values are 3.14φ and 3.7φ, classified as very leptokurtic (Figure S4).

In core ALM4, SUD is observed from the depth of 53 cm up to 10 cm. In contrast to ALM3, the sediment lacked organic material and gravel. The granulometric analysis of SUD in core ALM 4 also shows sediments characterised as slightly gravelly mud to sandy mud by [29], with a unimodal distribution and being very well sorted. The values of the statistical parameters as far as sorting is concerned are 0.45φ, which corresponds to well-sorted material. Skewness (Sk) values are -0.25φ, identified as negative asymmetry, while kurtosis (Ku) values range from 1.3φ to 1.6φ and are classified as leptokurtic to very leptokurtic (Figure S8).

4.3. *Palaeobiological Analysis*

4.3.1. Palaeofauna

Taxonomic analysis yielded 25 species of molluscs (23 gastropod species and two bivalve species) and 24 species of ostracods. All the molluscan fauna (Figure 3) is characteristic of freshwater environments, whereas in the ostracod assemblages, some species that tolerate higher salinities and low oxygen levels [32] are present. Molluscs and ostracods were analysed qualitatively and quantitatively to extract information relative to environmental changes. The analysis of the two cores ALM3 and ALM4 placed the faunas in three different molluscan–ostracod units (MOU) based on the prevailing environmental conditions. These three units reflect similar environments for both mollusc and ostracod faunas (Figures 4–6). The molluscan species and their habitats are presented in Table 2.

The first unit, MOUA, includes samples 18–26 of ALM 3 (Figure 4, 275 to 344 cm) and samples 31–33 of ALM 4 (Figure 5, 374 to 388 cm), and the dominance of a stagnant to slow-moving freshwater environment, where *Gyraulus crista* is the most abundant gastropod. Taxa that inhabit woodland ponds, marshlands and require humid conditions, such as *Hippeutis* sp. [33] and *Carychium* sp., are found in MOUA as smaller variations within the environment. In addition, a taxon that is mostly found in caves or in karstic crevices, *Zospeum* sp., is found in both cores of MOUA in association with other terrestrial molluscs such as *Vallonia* sp. and *Vertigo antivertigo*. The ostracods of MOUA correlate well with the environment that is revealed from the molluscan fauna, and give additional information about the salinity oscillations and the very low levels of dissolved oxygen of the swamp (*Cyprideis torosa, Heterocypris salina*) [34].

The second unit, MOUB, represents a swamp environment found in the stratigraphic sequence (lighter-coloured grey sediments) that includes samples 8–17 of ALM3 (Figure 4, 155 to 270 cm) and 18–30 of ALM4 (Figure 5, 205 to 362 cm). MOUB has the characteristics of a shallow permanent to slow-moving water body marked by the presence of *Planorbis* cf. *atticus*. Dense vegetation is inferred by the increased number of Bithyniidae, whereas elevated salinity is inferred by ostracod assemblages for all units of both cores. A synchronous abundance of the gastropods Bithyniidae and Planorbidae, along with the ostracods *Cyprideis* and *Heterocypris*, is evidence of high-salinity events (Figures 4 and 5). The ostracod fauna of MOUB includes the largest numbers and taxonomic diversity of Candonidae, Cyprididae and Cytherideidae. In MOUB of ALM3 core, a relative reduction in the ostracod population is observed in its upper part (Figure 4, samples 11–8), were

Candona sp. is the dominant taxon. In between this unit, several samples were barren (Figures 4 and 5).

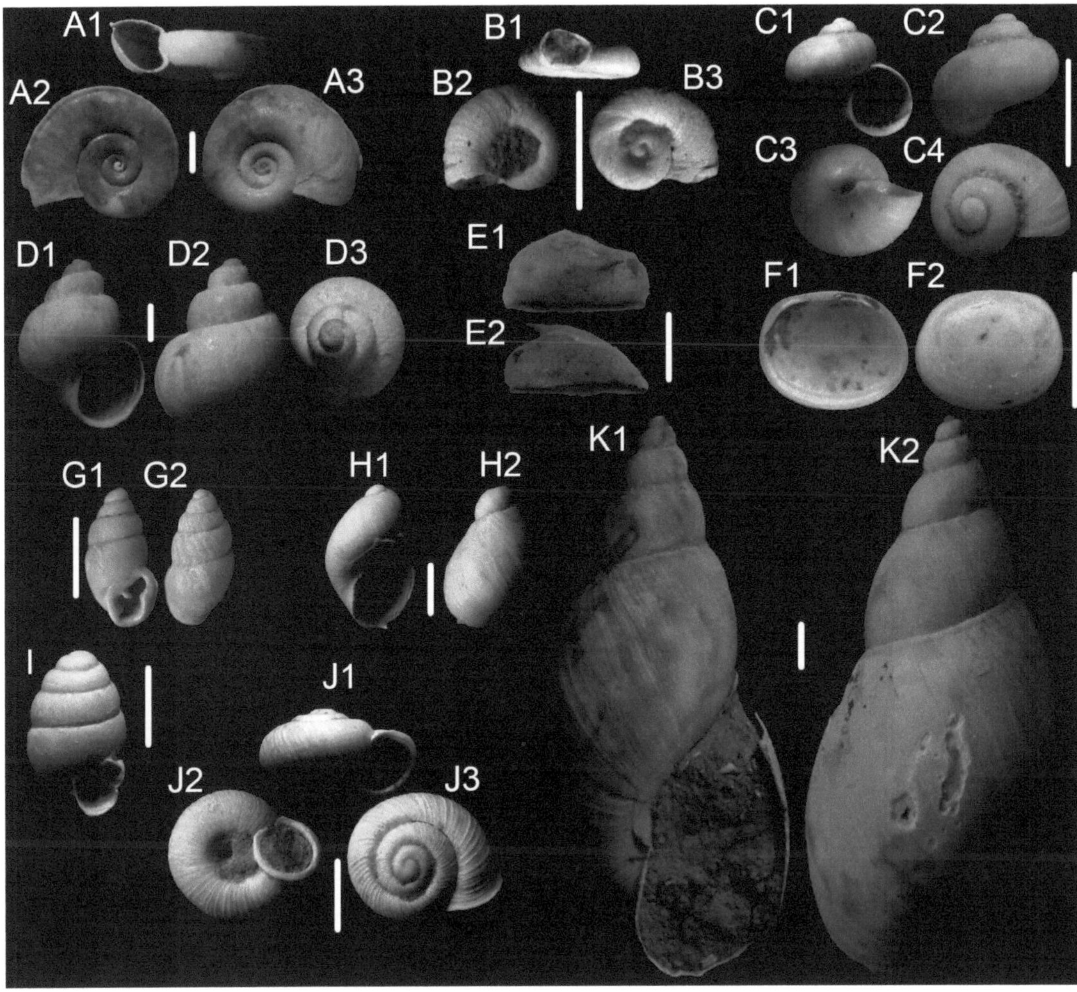

Figure 3. Most abundant molluscs recovered in the ancient Lake Lerna. (**A1–A3**) *Planorbis* cf. *atticus*, (**B1–B3**) *Gyraulus crista*, (**C1–C4**) *Valvata* sp., (**D1–D3**) *Bithynia* sp., (**E1,E2**) *Acroloxus* sp., (**F1,F2**) *Pisidium personatum*, (**G1,G2**) *Carychium* sp., (**H1,H2**) *Oxyloma* sp., (**I**) *Vertigo antivertigo*, (**J1–J3**) *Vallonia* sp., (**K1,K2**) *Stagnicola* sp. Scale bars: 1 mm.

In samples 3–7 of ALM3 (Figure 4, 78 to 145 cm) and 6–17 of ALM4 (Figure 5, 77 to 197 cm), similar environmental conditions prevailed, characterising MOUC (brown-coloured sediments). The absence of molluscs in ALM3 and ALM4, with a few representative specimens of Planorbidae and Bithyniidae, indicate a transition to a freshwater shallow lake with minimal vegetation, where the typical freshwater ostracod species such as *Candona* cf. *candida* and *Ilyocypris bradyi* of both cores are the only present taxa.

The transition from a freshwater lake to a swampy environment is noted in core ALM4, samples 3–5 (Figure 5, 46 to 57 cm), which corresponds to MOUB, with mollusc and ostracod faunas typical of stagnant to slow-moving waters.

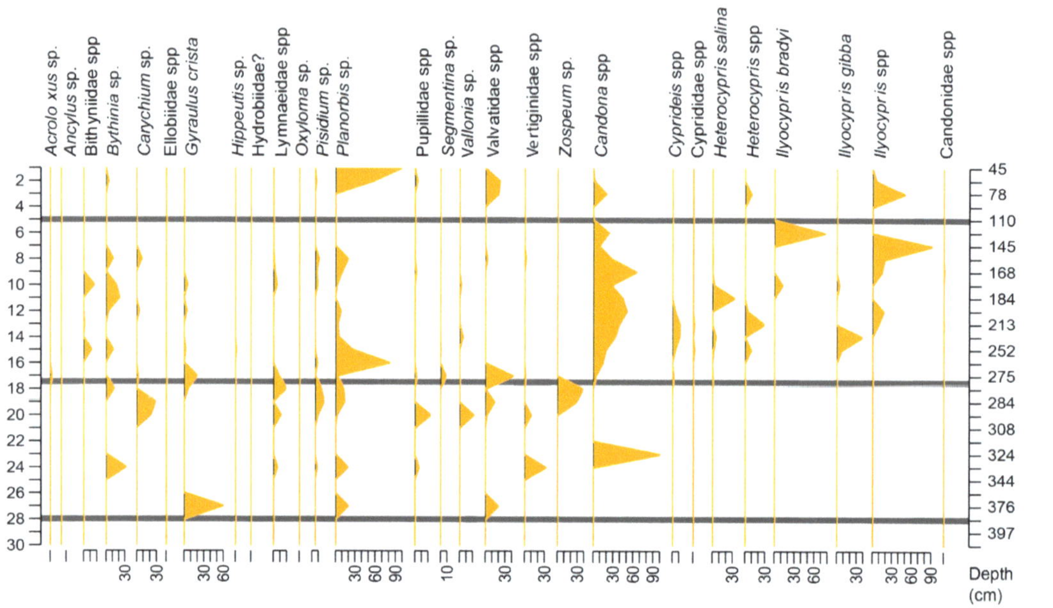

Figure 4. Stratigraphic plot for ALM3, mollusc and ostracod taxa. The horizontal grey lines indicate the limits of molluscan–ostracod units.

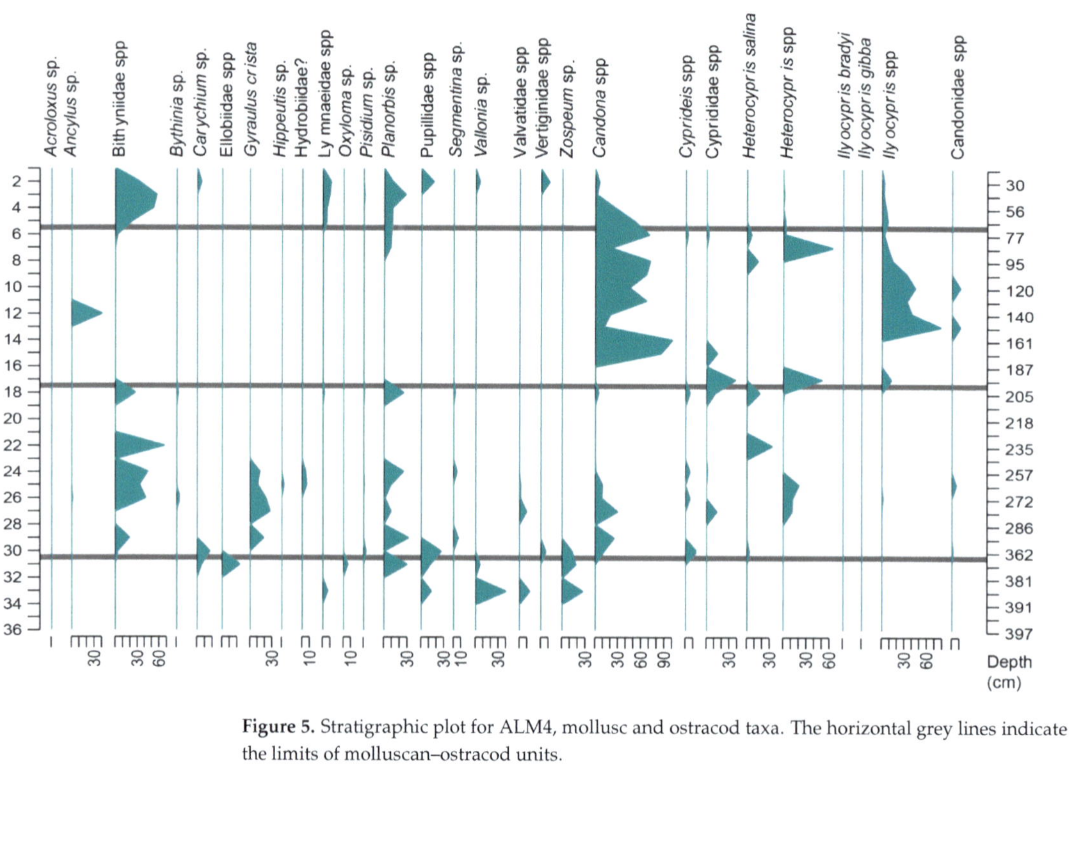

Figure 5. Stratigraphic plot for ALM4, mollusc and ostracod taxa. The horizontal grey lines indicate the limits of molluscan–ostracod units.

Table 2. List of molluscs found in sediments of Lerna and their respective habitats and environments.

Class	Families	Genera	Species	General Habitat	Environment
Gastropoda	Planorbidae	*Planorbis*	cf. *planorbis*	Freshwater	Low energy temporary and permanent ponds, streams, rivers, springs, lakes [32]
		Planorbis	cf. *atticus*		
		Planorbis	sp.		
		Gyraulus	*crista*		
		Hippeutis	sp.		
		Segmentina	sp.		
	Valvatidae	*Valvata*	*cristata*	Freshwater	Cold, clean lakes, rivers, streams [7]
		Valvata	sp.		
	Bithyniidae	*Bithynia*	sp.	Freshwater	Quiet muddy rivers, lakes, ponds, canals, swamps [7]
	Lymnaeidae	*Stagnicola*	sp.	Freshwater	Flowing rivers and streams, lakes to stagnant ponds, swamps [7]
		–	sp.		
	Ellobiidae	*Carychium*	sp.	Permanently wet epigean environments	Aphotic, permanently wet terrestrial biomes [7]
		Zospeum	sp.	Permanently wet subterranean environments	Cave karstic rock crevice dweller [35]
	Vertiginidae	*Vertigo*	*antivertigo*	Terrestrial on water margin vegetation	Wet, unimproved pasture, marshes and tall fen or water margin vegetation [36]
	Pupillidae	*Pupilla*	sp.		Detritus feeder (dead plant remains) [36]
	Valloniidae	*Vallonia*	sp.	Terrestrial moist environment with plant debris	Humid and wet habitats in lowlands, on humid and uncultivated meadows and in calcareous swamps [36]
	Acroloxidae	*Acroloxus*	sp.	Fresh water	Lakes [7]
		Ancylus	sp.		
	Hydrobiidae	*"Hydrobia"*	sp.	Freshwater	Springs, streams, rivers, lakes, groundwater systems, estuarine marshes [36]
	Succineidae	*Oxyloma*	sp.	Terrestrial on water margin vegetation	Feeds on wilting plant parts [36]
Bivalvia	Sphaerioidea	*Pisidium*	*personatum*	Freshwater	Bottom dwelling filter feeder [37]
		Pisidium	sp.		

The ordinations of both mollusc and ostracod faunas that were recovered (Figure 6) show a variation in environments throughout the cores. Three main groups were defined, characterising three environments corresponding to three units that were previously defined: stagnant waters, swamps and freshwater (Figure 6).

4.3.2. Palaeoflora

Sixteen pollen and dinoflagellate cyst samples were analysed for the investigation of local floras, particularly covering the upper part sequence of the core sediments (Figure 7). They were separated in three pollen units, PU1 to PU3, that are characterised by differences in abundances of taxa, although some aquatic taxa seem to appear towards the bottom of the core.

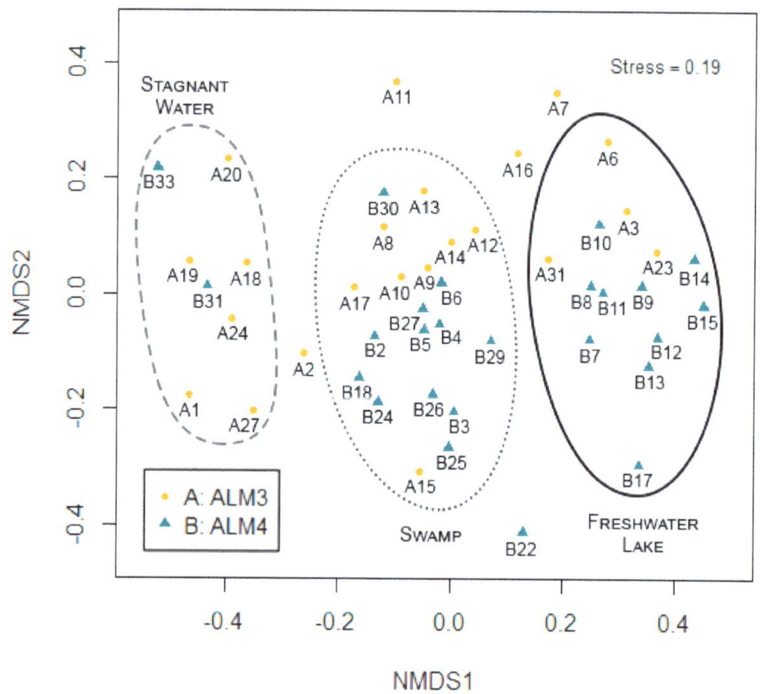

Figure 6. NMDS plot of samples from cores ALM3 and ALM4.

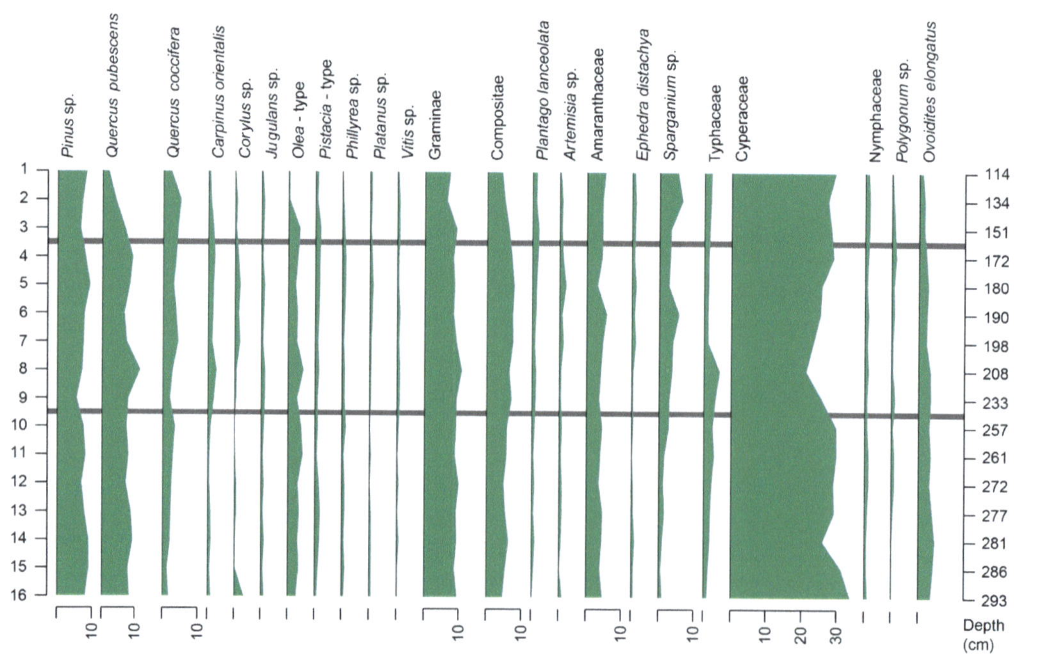

Figure 7. Stratigraphic plot for ALM3 pollen taxa (taxa with abundances higher than 10% are present). The horizontal lines separate the three pollen units.

Pollen Unit 1 (PU1) (Figure 7, 257 to 293 cm) is dominated by angiosperms while presenting a high taxonomic diversity. Woodland vegetation is characterised by mixed vegetation comprising species such as *Juglans*, *Ulmus*, *Ostrya*, *Platanus*, *Acer*, *Castanea*, *Quercus coccifera/ilex* and *Cistus*. Grasses are represented by Gramineae, Asteraceae, Amaranthaceae/Chenopodiaceae, Ericaceae, Caryophyllaceae, Polygonaceae, *Artemisia* and *Ephedra*. They are equally distributed in PU1. What is exceptional in this unit is the presence of aquatic plants such as members of the families Nymphaceae, Typhaceae, Sparganiaceae, Nupharaceae and Cyperaceae. The presence of freshwater phytoplankton genera (*Botryococcus*, *Ascomycetes*, *Ovoidites*) is notable and suggests a freshwater character of the samples.

In PU2 (Figure 7, 172 to 233 cm) there is a progressive increase in coniferous pollen and a more noted presence of xerophytic taxa such as *Quercus coccifera/ilex*. Grasses are present and relatively diverse, dominated by Gramineae and Asteraceae, with *Artemisia* having a stable presence throughout the unit. Aquatic plants are still present, represented by members of Cyperaceae, Nymphaceae and Typhaceae.

Lake deposits of Unit PU3 are defined by three samples, representing a swamp environment that is found in the stratigraphic sequence (lighter-coloured grey sediments) that includes samples 1 to 3 (Figure 7, 114 to 151 cm) of ALM3. They are characterised by the presence of deciduous trees such as *Quercus pubescens* and a stronger presence of aquatic vegetation, mainly represented by the Cyperaceae.

In general, a decrease in non-arboreal pollen (NAP) is observed throughout the core, reaching a maximum in the uppermost sample. Charophyte oogonia are present in samples ALM3, further characterising freshwater settings within PU2.

In total, three units were determined by the aforementioned assemblages. The subunits reflect the variation in the environmental conditions within a greater sedimentary setting. Faunal units coincide significantly with pollen units, with the exception of MUC1, which corresponds to the lower part of PU2.

5. Discussion

All the elements that were analysed in the present work point to a dynamic environment of a coastal wetland. The units defined by the sedimentological and palaeobiological analyses reflect significant changes in the depositional environments of the studied area.

5.1. Sedimentology

The sedimentological analysis revealed four units (SUA to D), separated based on grain size and colouration. The bases of both studied cores are composed of dark-coloured sediments, rich in organic material, leaning towards lighter and brown-coloured sediments in the top part of the cores. Sub-units defined for SUB and SUC indicate smaller-scale changes in the sedimentation, whereas layers of peat mark other events tied to precipitation and oxygen levels in the lake. Deep grey-black sediments indicate anoxic conditions, which show a stagnant lake.

Katrantsiotis and colleagues [12] retrieved a core in the area of Lerna (GreekGrid, EGSA '87: x: 387,952.268, y: 4,159,981.561, 1 m.a.s.l) in order to study environmental changes during the Holocene. The analytical methods included a combination of n-alkane distributions and their hydrogen isotopic composition (δD), as well as sediment optical lithological characterisation, in comparison with the total organic content (TOC). Our results are in agreement with those of [12] (Figure 8). The units with dark-coloured sediment (Table 1) are present in [12] as layers of high organic concentrations, while brown-grey to brown layers correspond to less organic material [12] (Figure 8). However, in [12], all sediments were characterised as gyttja, which in this study is observed only in specific layers rich in organic material.

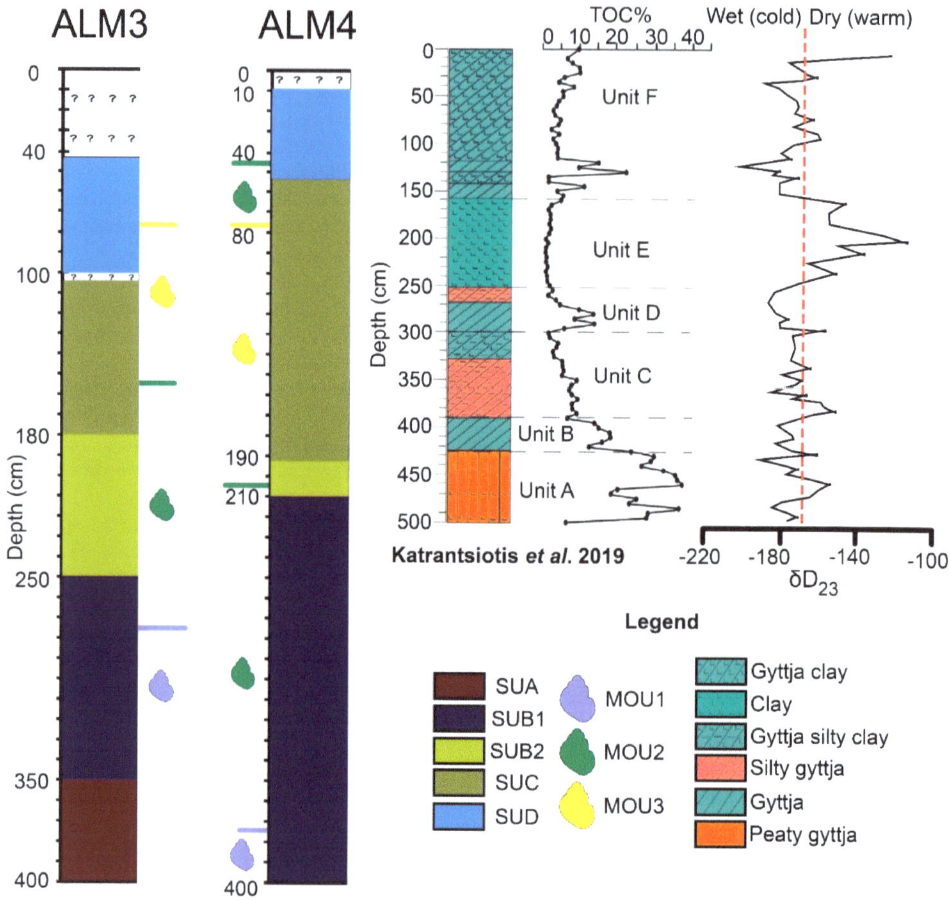

Figure 8. Combination of lithostratigraphic and palaeontological units for the cores ALM3 and ALM4. The question marks (?) are indicative of disturbed material during coring. Comparison with [12] (including TOC and δD_{23} data).

Comparing the present results with [12] yields the following findings.

- Unit B (390–425 cm) of [12], characterised as gyttja with high TOC values, is equivalent to SUB1 of ALM3 (250–350 cm) and ALM4 (350–400 cm) (Figure 8), as these are the levels with the highest concentrations of organic material. The peat deposits seen here between gyttja are rather thin. In [12], Unit B was estimated to have been deposited around 4000 BP.
- The deepest part of Unit C from [12]'s core could be related to the lower SUB2 of ALM3 (208–250 cm) and the middle SUB1 of ALM4 (270–350 cm) (Figure 8). These units have similar silty sediment, as they are characterised by medium to low concentrations in TOC. In [12], the Unit C was dated around 3700 BP.
- The upper Unit C and Unit D of [12] correspond to the upper part of SUB2 (208–180 cm) of ALM3 and the upper part of SUB1 (270–210 cm) and SUB2 of ALM4 (Figure 8). The latter units have similarities in sediment composition as well as in organic material concentration (gyttja silty clay). In [12], the these units were dated around 2500 BP.
- Unit E of [12] is characterised by clay and low TOC, matching with SUC of ALM3 and mid-SUC of ALM4 (100–190 cm) (Figure 8).

- The lower part of Unit F (120–155 cm) of [12] can be compared with lower SUD of ALM3 (75–100 cm) and upper SUC of ALM4 (55–100 cm) (Figure 8). These units are all characterised by clayey sediment with similar concentrations of organic material. The age given for lower Unit F of [12] is approximately 1500 BP.

The most accurate dates of the present study are the ones originating from gyttja, such as the LTL19748A (5640 ± 45 BP) from ALM3, and LTL20302 (4610 ± 45 BP) from ALM4. The two samples are very near the ash layer, present in both cores, as were those of [12]; thus, we expected the dates to be similar, yet they differ by nearly a thousand years. A similar dating result of a seed sample from [12] near the ash layer is around 3970 ± 30 BP, while the "charred" layer itself had a result of 4370 ± 30 BP [12]. The rest of our dating results have considerable uncertainty, as most of them contain mollusc parts. Due to these differences in ages, we are reluctant to discuss age estimations for our samples, and therefore, we mainly compared the cores using sedimentological similarities. We are also concerned about the validity of results of [12] at the upper part of their core (upper 0.5–1 m), since the Argive Plain has been deeply ploughed in recent years, meaning that the upper sedimentary layers are mostly disturbed.

5.2. *Palaeofauna and Flora*

The recovered mollusc and ostracod faunas also correlate with significant environmental changes, mainly by a variation in abundances, which can be useful indicators of the lake's condition at that time. Freshwater taxa are dominant in all samples for both molluscs and ostracods. Nevertheless, some gastropod species, such as the Bithyniidae and Planorbidae, as well as some ostracod species, such as *Cyprideis torosa* and *Heterocypris salina*, can withstand higher salinities [38]. Ordination through NMDS shows groupings of samples coming from different parts of the cores; this reflects the dynamic environments that were successively established in the area of the former lake. The presence of gastropod *Zospeum* sp. in the basal part of both cores suggests that the shells could have been transported through a nearby karstic system and deposited in the lake area.

Pollen data, spaced in the length of core ALM3, depict a trend that agrees with previous findings in the area of the ancient Lake Lerna [39]. Anthropogenic disturbance, from PU2 onwards, is indicated by the increased presence of *Plantago lanceolata* [39]. Since the present work is centred on the freshwater character of the ancient lake, aquatic taxa are mentioned here additionally in PU1 and 2. An element that differs from previous studies, namely [39], is the representation of *Olea*, which is present here in relatively low abundances.

5.3. *Palaeobiological Evolution of the Lake*

Previous studies [4] suggest that the lake formed around 8330 ± 100 BP is a stagnant body of water, accumulating peaty gyttja [12]. The lake was first colonised by Planorbidae, such as *Gyraulus crista*. Moving upwards in the cores, we find gastropod species living in close proximity to lake environments (e.g., Succineidae). This correlates with the wet climate, as suggested by [12] (see also Figure 8, δD_{23} graph), which helped in the accumulation of plants near or inside the lake area, and a subsequent accumulation of detritus and herbivorous gastropods on the lake bottom. The excess organic material caused anoxic events. The absence of an ostracod fauna in the base of the cores further indicates nearly anoxic conditions [34] of the lake (Figures 4 and 5, MOU1). However, the presence of the fully aquatic gastropods Valvatidae is evidence that the water had enough oxygen for these organisms to survive. Furthermore, the transported cave-dwelling gastropod *Zospeum* sp. confirms that the karstic system of the wider area was more active during that wet period [12], and provided water to the lake.

In both cores, we found an ash layer (ALM3 2.95 m, ALM4 3.93 m). This layer might indicate a fire, something that is likely since other low-temperature forest fires were recorded in the Peloponnese around that time [40]. Moreover, this ash layer has been reported as charred material by [12], almost 1 km away to the north east from our sampling sites, highlighting that the fire was widespread in the area.

Sedimentological changes in SUB1 had a drastic effect on the fauna of the lake, with several ostracod species appearing, an indication that the water was no longer anoxic [34]. Furthermore, the gastropod fauna displayed an increase in species diversity and abundance (Figures 4 and 5). The sudden increase in populations, displaying peaks in species of ostracods (Figures 4 and 5) such as *Cyprideis* and *Heterocypris*, indicate swift sea-water influxes into the lake. The simultaneous peaks of gastropods of the families Planorbidae and Bithyniidae can also be explained, as these organisms can tolerate salinity levels up to 16 ppt [38]. The intensity and position of those events is not similar in both cores (Figure 8), suggesting that these influxes were local and sporadic, as the quantity of sea-water was not enough to affect large portions of the lake. The peaks being caused by dry events, which would have resulted in an increase in the salinity, is improbable, since both cores display dissimilar peaks and because this MOU2 corresponds to a wet and cold climate (Figure 8) [12].

After SUB, shifts in sedimentology and aquatic faunas suggest that the marsh transitioned into a clear-water lake environment. This would correspond to the brown-coloured sediment of [12], containing low TOC levels (Figure 8). This period is characterised by a dry and hot climate [9,10,12,41,42] (see also Figure 8, δD_{23} graph), which halted the growth of marginal plants around the lake, thus reducing the deposition of organic material. The absence of molluscs can be a result of various factors, such as the absence of sufficient plants for the molluscs to thrive on. We could argue that the lake was ephemeral at that period of time, but we and [12] did not observe any sedimentological indications of such events, nor did [12] display any abrupt changes in their age model. Therefore, we disregard such a scenario.

Finally, the fauna present in the upper layers of ALM4 (SUC and D) suggests that the lake had sufficient water to sustain both plants and the associated gastropod and ostracod faunas (corresponding to an abrupt increase in TOC [12] seen in Figure 8). This re-establishment of the lake as a permanent body of water with rich fauna and flora partly coincides with the results of [4,18]. It also agrees with the archaeological finds of porotic hyperostosis [43] during the Late Hellenistic Period, as according to the dating results of [12], this wet period started somewhere around 1850 BP. Nevertheless, the reappearance of a swamp would have affected the nearby populations' health.

5.4. *What Is Causing These Environmental Fluctuations?*

Concerning the environmental changes recorded in localities of the Peloponnese [10,12,41,44], there have been alterations of dry and wet conditions throughout the past 6000 years, with human settlements existing in the area from 8500 BP on [42,45]. The results of [12] on the climate fluctuations are interesting as they illustrate that climate can drastically alternate the sedimentology of the lake (Figure 8), thus highlighting the dynamic character of this coastal wetland.

During the Late Bronze age, the population in the Argive Plain was devastated by successive changes in precipitation levels that potentially led to decreased agricultural output around 3200 BP [42,46–48]. In the Late Bronze age (3550–3200 BP), according to [49], the climate was dry, with considerable erosion due to land overuse and deforestation [50], but later studies in the Peloponnese demonstrate a brief wet period between 3300 and 3100 BP [9,10,41]. A prolonged dry period between the wet periods in the Argive Plain [12] (not observable in our samples) might have destabilised the Peloponnesian populations during the end of the Late Bronze Age [41]. After that, the climate was wet, while deforestation and land use, in general, were more intense after the Middle Geometric period (~2750 BP) [51]. Prehistoric degradation of ecosystems has been investigated in the past through vertebrate remains in the area of Lerna during archaeological investigations [11]. It is possible that these changes in the surrounding land were the reason for the lake's prolonged low TOC levels [12] and brown colour seen in our cores. As such, we found clues of indirect human intervention in the lake due to deforestation and land use [51]. For most of the lake's lifetime, however, despite humans potentially interacting with the lake, environmental factors were the key drivers for the water conditions, as well as the fauna and flora living in and around the lake. The evidence for human interaction with the surrounding fauna (e.g., deer, sea shell ornaments) [52,53] as well as the environmental destruction have been investigated through the remains of larger vertebrates in nearby archaeological sites (including birds thriving in lakes and wetlands) [11].

6. Conclusions

Through the study of two cores from the area of ancient Lake Lerna, sedimentological and palaeobiological analyses were carried out to investigate environmental changes, with molluscs being studied and used as palaeoenvironmental indicators for the first time in the region.

Three major units were determined, confirmed by the sedimentological and palaeobiological analyses. They reflect changes in the environments of the area in the following sequence. Fluvial deposits are followed by stagnant waters, then a swamp is formed favouring periodical peat formation; after this, a freshwater lake is characterised by the development of large molluscan and ostracod populations. Occasional storm events could overcome the barrier and contaminate the lake with salt water, evidenced by the molluscan and ostracod faunas, but these events are ephemeral.

The studied material covers a period of the Holocene starting at about 8630 years before the present, during the early to middle Neolithic. This means that all recovered faunas and floras have been potentially affected by the human presence in the area, and no undisturbed conditions were found. Despite that fact, all of our results suggest that the changes happening in the lake were caused mostly by climate fluctuations rather than human intervention.

Concluding, most of the changes happening in the lake are correlated with climatic fluctuations. Parameters such as the vegetation, dead organic material in the water, oxygenation of the water and rainfall in the region can affect the lake's fauna and flora. By investigating these changes, we were able to coordinate our results with other environmental works in the region. Although humans were present, their impact on the lake was minimal in comparison to climatic changes.

The case of Lerna can be used as an example for the evolution of coastal aquatic ecosystems in the Holocene.

Supplementary Materials: The following are available online at https://www.mdpi.com/article/10.3390/quat5020022/s1, Table S1: Dataset used for NMDS analysis and stratigraphic plots, presenting the species and their abundances in the samples of cores ALM3 and ALM4. Table S2: Pollen species recovered in the examined samples of core ALM3. Figures S1–S8: Triangular diagrams of sedimentological texture for both cores (ALM-3 and ALM-4) and their corresponding cumulative particle mass curves created using GRADISTAT statistical package; these include sediment particle size (φ) retained during sieving.

Author Contributions: Conceptualisation, E.K., N.E., D.T. and C.P.; methodology, E.K., C.P., D.T., G.S., A.P. and C.I.; software, D.T., C.P., G.S. and A.P.; validation, E.K., N.E., C.P., D.T., G.S., A.P., E.R., S.P., K.P. and C.I.; formal analysis, D.T., C.P., G.S., A.P., E.R. and C.I.; investigation, E.K., N.E., C.P., D.T., A.P., G.S., E.R., C.I., G.K. and S.P.; resources, E.K. and N.E.; data curation, D.T., C.P., G.S. and A.P.; writing—original draft preparation, C.P., D.T., G.S. and A.P.; writing—review and editing, E.K., N.E., C.P., D.T., A.P., G.S., E.R., S.P. and C.I.; visualisation, C.P., D.T., G.S. and A.P.; supervision, E.K. and N.E.; project administration, E.K.; funding acquisition, E.K. All authors have read and agreed to the published version of the manuscript.

Funding: This research was co-financed by the European Regional Development Fund of the European Union and Greek national funds through the Operational Program Competitiveness, Entrepreneurship and Innovation, under the call RESEARCH-CREATE-INNOVATE (project code: T1EDK-02363).

Acknowledgments: We would like to thank Ilias Mariolakos (NKUoA) for his help during this research. We are grateful to Avgerinos for his help during fieldwork. Thanks to the three anonymous reviewers for their comments that contributed to the final version of the manuscript.

Conflicts of Interest: The authors declare no conflict of interest. The funders had no role in the design of the study; in the collection, analyses or interpretation of data; in the writing of the manuscript, or in the decision to publish the results.

References

1. Economidou, E. The Ecological Value of Coastal Ecosystems. *Ekistics* **1982**, *49*, 98–101.
2. Paine, L. *The Sea and Civilization: A Maritime History of the World*; Knopf Doubleday Publishing Group: New York, NY, USA, 2013.
3. Mariolakos, I.D. Ancient Greece and Water: Climatic Changes, Extreme Events, Water Management, and Rivers in Ancient Greece. *Handb. Environ. Chem.* **2017**, *59*, 3–30. [CrossRef]
4. Zangger, E. Prehistoric Coastal Environments in Greece: The Vanished Landscapes of Dimini Bay and Lake Lema. *J. Field Archaeol.* **1991**, *18*, 1–15. [CrossRef]
5. Oertel, G.F. Coastal Lakes and Lagoons. *Encycl. Earth Sci. Ser.* **2005**, *14*, 263–266. [CrossRef]
6. Barbieri, R.; Zogaris, S.; Kalogianni, E.; Stoumboudi, M.T.; Chatzinikolaou, Y.; Giakoumi, S.; Kapakos, Y.; Kommatas, D.; Koutsikos, N.; Tachos, V.; et al. *Freshwater Fishes and Lampreys of Greece: An Annotated Checklist*; Monographs on Marine Sciences No. 8; Hellenic Centre for Marine Research: Athens, Greece, 2015; p. 130.
7. Strong, E.E.; Gargominy, O.; Ponder, W.F.; Bouchet, P. Global Diversity of Gastropods (Gastropoda; Mollusca) in Freshwater. *Hydrobiologia* **2008**, *595*, 149–166. [CrossRef]
8. Caskey, J.L.; Blackburn, E.T. *Lerna in the Argolid*; American School of Classical Studies at Athens: Athens, Greece, 1997.
9. Weiberg, E.; Unkel, I.; Kouli, K.; Holmgren, K.; Avramidis, P.; Bonnier, A.; Dibble, F.; Finné, M.; Izdebski, A.; Katrantsiotis, C.; et al. The Socio-Environmental History of the Peloponnese during the Holocene: Towards an Integrated Understanding of the Past. *Quat. Sci. Rev.* **2016**, *136*, 40–65. [CrossRef]
10. Norström, E.; Katrantsiotis, C.; Finné, M.; Risberg, J.; Smittenberg, R.H.; Bjursäter, S. Biomarker Hydrogen Isotope Composition (ΔD) as Proxy for Holocene Hydroclimatic Change and Seismic Activity in SW Peloponnese, Greece. *J. Quat. Sci.* **2018**, *33*, 563–574. [CrossRef]
11. Gejvall, N.-G. Prehistoric Environmental Destruction, Lerna (Greece). *Museum Int.* **1973**, *25*, 93–96. [CrossRef]
12. Katrantsiotis, C.; Norström, E.; Smittenberg, R.H.; Finne, M.; Weiberg, E.; Hättestrand, M.; Avramidis, P.; Wastegård, S. Climate Changes in the Eastern Mediterranean over the Last 5000 years and Their Links to the High-Latitude Atmospheric Patterns and Asian Monsoons. *Glob. Planet. Chang.* **2019**, *175*, 36–51. [CrossRef]
13. Janssen, J.; Rodwell, J.; Criado, M.G.; Gubbay, S.; Haynes, T.; Nieto, A.; Sanders, N.; Landucci, F.; Loidi, J.; Ssymank, A.; et al. *European Red List of Habitats Part 2. Terrestrial and Freshwater Habitats*; Publications Office of the European Union: Luxembourg, 2016. [CrossRef]
14. Cuttelod, A.; Seddon, M.; Neubert, E. *European Red List of Non-Marine Molluscs*; Publications Office of the European Union: Luxembourg, 2011. [CrossRef]
15. Kelletat, D.; Kowalczyk, G.; Schröder, B.; Winter, K.-P. A Synoptic View on the Neotectonic Development of the Peloponnesian Coastal Regions. *Z. Dtsch. Geol. Ges.* **1976**, *127*, 447–465. [CrossRef]
16. van Andel, T.H.; Zangger, E.; Demitrack, A. Land Use and Soil Erosion in Prehistoric and Historical Greece. *J. Field Archaeol.* **1990**, *17*, 379. [CrossRef]
17. Finke, E.A. *Landscape Evolution of the Argive Plain, Greece: Paleoecology, Holocene Depositional History, and Coastline Changes*; Stanford University: Stanford, CA, USA, 1988.
18. Zangger, E. The Geoarchaeology of the Argolid. *Am. J. Archaeol.* **1993**, *100*, 636. [CrossRef]
19. Koutsouveli-Nomikou, A. *Contribution a l'Etude Quaternaire de la Région d'Argos (Nord-Est du Péloponnèse, Gréce)*; Université d'Aix: Marseille, France, 1980.
20. Gaki-Papanastassiou, K. *The Geomorphological Evolution in and around the Argive Plain since the Quaternary*; National and Kapodistrian University of Athens: Athens, Greece, 1991. [CrossRef]
21. Seni, A.; Gaki-Papanastassiou, K.; Karymbalis, E.; Zouva, C. Anticipated Sea-Level Rise Vulnerability Assessment Along Part of the Eastern Argolic Gulf and the Argive Plain With the Use of G.I.S. In Proceedings of the 8th Pan-Hellenic Geographical Conference, Athens, Greece, 4–7 October 2007; pp. 146–155.
22. Gaki-Papanastassiou, K.; Maroukian, H. Late Quaternary controls on river behaviour in the eastern part of the Argive plain, eastern Peloponnesus, Greece. In Proceedings of the International Symposium on Mediterranean Quaternary River Environments, Cambridge, UK, 28–29 September 1992; Balkema: Rotterdam, The Netherlands, 1995; pp. 89–95.
23. Liritzis, I.; Raftopoulou, M. Argolid: Connection of Prehistoric Legends with Geoenvironmental and Archaeological Evidence. *Tubaar* **1999**, *2*, 87–99. [CrossRef]
24. De Wever, P. *Etude Géologique des Series Apparaissant en Fenêtre sous l' Allochtone Pindique (Série de Tripolitza et Série Épimetamorphic de Zarouchla), Péloponnèse Septentrional, Gréce*; Université de Lille: Lille, Greece, 1975.
25. Tataris, A.; Kounis, G.; Maragoudakis, N. *Geological Map of Greece on a Scale 1:50,000, "Vagia" Sheet*; IGME: Madrid, Spain, 1970.

26. Exidavelonis, P.; Taktikos, S. *Geological Map of Greece on a Scale 1:50,000, "Tripolis" Sheet*; IGME: Madrid, Spain, 1988.
27. *Geological Map of Greece on a Scale 1:50,000, "Argos" Sheet*; IGME: Madrid, Spain, 1960.
28. Theodoropoulos, D.; Papapetrou-Zamani, A. Geological and Geomorphological Investigations of the Argos Plain. *Ann. Géologiques Pays Hélléniques* **1970**, *22*, 269–294.
29. Folk, R.L.; Ward, W.C. Brazos River Bar [Texas]; a Study in the Significance of Grain Size Parameters. *J. Sediment. Res.* **1957**, *27*, 3–26. [CrossRef]
30. Blott, S.J.; Pye, K. Gradistat: A Grain Size Distribution and Statistics Package for the Analysis of Unconsolidated Sediments. *Earth Surf. Process. Landf.* **2001**, *26*, 1237–1248. [CrossRef]
31. Psilovikos, A.A. Sediment Analysis at Gibraltar Point, Lincolnshire. *East Midl. Geogr.* **1979**, *7*, 128–133.
32. Rossetti, G.; Bartoli, M.; Martens, K. Limnological Characteristics and Recent Ostracods (Crustacea, Ostracoda) of Freshwater Wetlands in the Parco Oglio Sud (Northern Italy). *Ann. Limnol. Int. J. Limnol.* **2004**, *40*, 329–341. [CrossRef]
33. Aneta, S. Woodland Ponds as an Important Habitat of Hippeutis Complanatus (Linnaeus 1758) Occurrence—Effect of Environmental Factors and Habitat Preferences. *Ekológia* **2014**, *33*, 101–115. [CrossRef]
34. Ruiz, F.; Abad, M.; Bodergat, A.M.; Carbonel, P.; Rodríguez-Lázaro, J.; González-Regalado, M.L.; Toscano, A.; García, E.X.; Prenda, J. Freshwater Ostracods as Environmental Tracers. *Int. J. Environ. Sci. Technol.* **2013**, *10*, 1115–1128. [CrossRef]
35. Weigand, A.M.; Jochum, A.; Slapnik, R.; Schnitzler, J.; Zarza, E.; Klussmann-Kolb, A. Evolution of Microgastropods (Ellobioidea, Carychiidae): Integrating Taxonomic, Phylogenetic and Evolutionary Hypotheses. *BMC Evol. Biol.* **2013**, *13*, 18. [CrossRef]
36. Welter-Schultes, F.W. *European Non-Marine Molluscs, a Guide for Species Identification/Bestimmungsbuch für Europäische Land- und Süsswassermollusken*; Planet Poster Editions: Goettingen, Germany, 2012.
37. McMahon, R.F.; Bogan, A.E. 11—MOLLUSCA: BIVALVIA. In *Ecology and Classification of North American Freshwater Invertebrates*; Thorp, J.H., Covich, A.P., Eds.; Academic Press: Cambridge, MA, USA, 2001; pp. 331–429. [CrossRef]
38. Abdelhady, A.A.; Khalil, M.M.; Ismail, E.; Fan, J.; Zhang, S.; Xiao, J. Water Chemistry and Substrate Type as Major Determinants of Molluscan Feeding Habit and Life–Mode in Lagoon Sediments. *Estuar. Coast. Shelf Sci.* **2019**, *220*, 120–130. [CrossRef]
39. Jahns, S. On the Holocene Vegetation History of the Argive Plain (Peloponnese, Southern Greece). *Veg. Hist. Archaeobot.* **1993**, *2*, 187–203. [CrossRef]
40. Norström, E.; West, J.; Kouli, K.; Katrantsiotis, C.; Hättestrand, M.; Smittenberg, R.H. Evaluation of Anhydrosugars as a Molecular Proxy for Paleofire Activity: A Case Study on a Holocene Sediment Core from Agios Floros, Peloponnese, Greece. *Org. Geochem.* **2021**, *153*, 104193. [CrossRef]
41. Finné, M.; Holmgren, K.; Shen, C.C.; Hu, H.M.; Boyd, M.; Stocker, S. Late Bronze Age Climate Change and the Destruction of the Mycenaean Palace of Nestor at Pylos. *PLoS ONE* **2017**, *12*, e0189447. [CrossRef]
42. Weiberg, E.; Hughes, R.E.; Finné, M.; Bonnier, A.; Kaplan, J.O. Mediterranean Land Use Systems from Prehistory to Antiquity: A Case Study from Peloponnese (Greece). *J. Land Use Sci.* **2019**, *14*, 1–20. [CrossRef]
43. Angel, J.L. Ecology and Population in the Eastern Mediterranean. *World Archaeol.* **1972**, *4*, 88–105. [CrossRef]
44. Boyd, M. *Speleothems from Warm Climates: Holocene Records from the Caribbean and Mediterranean Regions*; Department of Physical Geography, Stockholm University: Stockholm, Sweden, 2015.
45. Weiberg, E.; Bevan, A.; Kouli, K.; Katsianis, M.; Woodbridge, J.; Bonnier, A.; Engel, M.; Finné, M.; Fyfe, R.; Maniatis, Y.; et al. Long-Term Trends of Land Use and Demography in Greece: A Comparative Study. *Holocene* **2019**, *29*, 742–760. [CrossRef]
46. Knapp, A.B.; Manning, S.W. Crisis in Context: The End of the Late Bronze Age in the Eastern Mediterranean. *Am. J. Archaeol.* **2016**, *120*, 99–149. [CrossRef]
47. Middleton, G.D. *The Collapse of Palatial Society in LBA Greece and the Postpalatial Period*; University of Michigan Press: Ann Arbor, MI, USA, 2010. [CrossRef]
48. Weiberg, E.; Finné, M. Mind or Matter? People-Environment Interactions and the Demise of Early Helladic II Society in the Northeastern Peloponnese. *Am. J. Archaeol.* **2013**, *117*, 1–31. [CrossRef]
49. Angel, J.L. *The People of Lerna: Analysis of a Prehistoric Aegean Population*; American School of Classical Studies at Athens: Princeton, NJ, USA, 1971; Volume 2.
50. Fuchs, M. An Assessment of Human versus Climatic Impacts on Holocene Soil Erosion in NE Peloponnese, Greece. *Quat. Res.* **2007**, *67*, 349–356. [CrossRef]
51. Weiberg, E.; Bonnier, A.; Finné, M. Land Use, Climate Change and 'Boom-Bust' Sequences in Agricultural Landscapes: Interdisciplinary Perspectives from the Peloponnese (Greece). *J. Anthropol. Archaeol.* **2021**, *63*, 101319. [CrossRef]
52. Reese, D.S. Faunal Remains from Late Helladic Lerna (Argolid, Greece). *Mediterr. Archaeol. Archaeom.* **2008**, *8*, 5–25.
53. Reese, D.S. Faunal Remains from Early Helladic II Lerna (Argolid Greece). *Mediterr. Archaeol. Archaeom.* **2013**, *13*, 289–320.

quaternary

MDPI

Article

Palaeogeographical Reconstruction of Ancient Diolkos Slipway by Using Beachrocks as Proxies, West Corinth Isthmus, Greece

Giannis Saitis *, Anna Karkani, Niki Evelpidou and Hampik Maroukian

Faculty of Geology and Geoenvironment, National and Kapodistrian University of Athens, Panepistimiopolis, 15784 Athens, Greece; ekarkani@geol.uoa.gr (A.K.); evelpidou@geol.uoa.gr (N.E.); maroukian@geol.uoa.gr (H.M.)
* Correspondence: saitij@geol.uoa.gr

Abstract: Beachrocks are well known as significant proxies for paleoenvironmental analysis as they indicate the coastal evolution. The combination of geomorphological and archaeological sea level indicators has a significant contribution to the coastal paleogeographic reconstruction. In this study, we studied a beachrock from the Diolkos area (West Corinth canal, Greece) and remnants of Diolkos slipway to reconstruct the coastal evolution before Diolkos construction until today. We conducted detailed mapping of Diolkos beachrock using DGPS-GNSS, as well as mineralogical analysis and OSL dating of beachrock samples. The results showed that a beachrock slab was preserved before the construction of Diolkos below it, followed by its submergence by a co-seismic event after Diolkos abandonment during 146 B.C. Consequently, a new beachrock was developed on top of the submerged Diolkos around 120 ± 14 A.D. The RSL was stable until 1596 ± 57 A.D. when the beachrock developed even closer to the present-day coastline. After 1596 A.D., it was uplifted by 12 cm until it reached today's condition.

Keywords: Corinth Gulf; geomorphology; coastal evolution; relative sea level; sea level indicator

check for updates

Citation: Saitis, G.; Karkani, A.; Evelpidou, N.; Maroukian, H. Palaeogeographical Reconstruction of Ancient Diolkos Slipway by Using Beachrocks as Proxies, West Corinth Isthmus, Greece. *Quaternary* **2022**, *5*, 7. https://doi.org/10.3390/quat5010007

Academic Editor: Ioannis Liritzis

Received: 8 December 2021
Accepted: 6 January 2022
Published: 18 January 2022

1. Introduction

Research on coastal evolution is an important tool for interpreting future littoral changes in the context of climate change [1,2]. Several recent studies have used a wide variety of coastal sediments in an attempt to reconstruct late Holocene littoral changes [3–8].

Beachrocks are hard littoral deposits, which are composed of a wide variety of beach sediments that are lithified through the precipitation of carbonate cement at the shoreline [9–12]. Typically, the cement is composed of two mineralogies: high magnesium calcite (HMC) and aragonite [13]. The diagenetic environment of beachrocks may be determined by the cement mineralogy and morphology [14]; therefore, the study of these characteristics can enable defining the spatial relationship between the past coastline and the formation zone of the beachrock [10]. In fact, the cement is composed of acicular aragonite with isopachous fringes within the lower intertidal zone, or HMC cements with thin-bladed isopachous crusts, or dark or golden brown, pelletal and micritic, irregular grain coatings and void fillings [15,16].

Beachrocks are frequently used to determine late Holocene shoreline changes for the reconstruction of paleo-shorelines and as sea level indicators for the study of relative sea level changes in various coastal environments [10,12,17–22]. In the absence of other sea level indicators, especially along sandy coastlines, they can provide significant data on relative sea level changes, or even in combination with other sea level markers. Such is the case of the Aegean Sea, where beachrock occurrences are particularly common [12,18,20,22–26].

Amongst the main difficulties in the study of beachrocks, including sea level or paleogeographic indicators, is to obtain a geochronology for their development; commonly, beachrocks are dated through radiocarbon dating using shells or the cement. However, difficulties may arise as, frequently, beachrocks may have undergone more than one diagenetic phase [9], while the cement from one single slab may provide different ages [18]. Dating marine fossils incorporated in the beachrock also facilitates errors as it has been noted that there may be an important time period between the death of the organism and its incorporation into the beachrock. In this latter case, the age obtained should be considered as the maximum [27,28]. During the past decade, luminescence dating has also been used for a variety of coastal sediments [12,29], and it has also been proven useful to obtain a geochronology for beachrock formation in the Mediterranean [12,29–31] and elsewhere [32–34].

In this context, the aim of this study is to investigate the coastal evolution of the Diolkos site at Corinth Gulf, Greece though the morphological mapping of beachrock outcrops, cement investigation, and luminescence dating.

2. Study Area

2.1. Geological and Geotectonic Setting

The study area is located at the Gulf of Corinth (Figure 1), a prominent tectonic and geomorphological structure in the central part of Greece, separating the Peloponnese to the south and central Greece to the north. The Diolkos site lies on the south coasts of the isthmus canal.

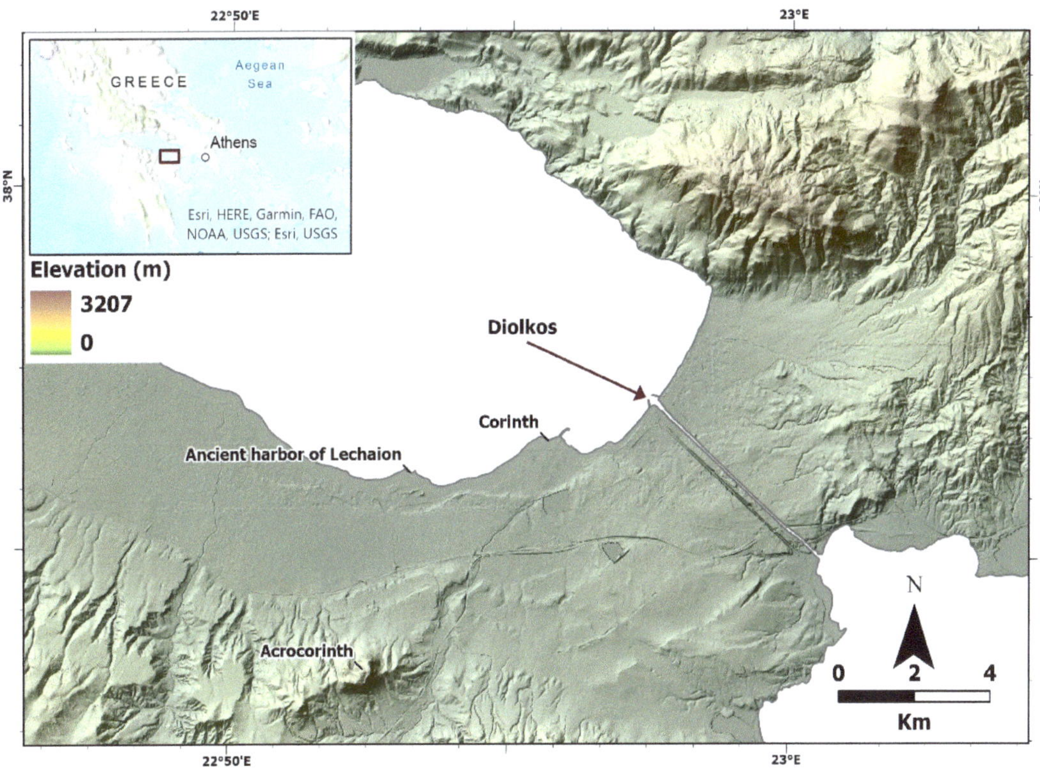

Figure 1. A morphological elevation map of the general area of Corinth. Corinth isthmus (canal) can be observed along with the modern town of Corinth, the Acrocorinth, and the ancient harbor of Lechaion.

The geological basement of Corinth is dominated by rocks of the tectonostratigraphic terrain 3 of the Hellenides [35]. The area of Corinth consists of the rocks of the sub-Pelagonian unit, a part of an old continental margin. This unit is mostly characterized by sedimentary rocks of Triassic to Jurassic grey limestones [35,36]. The topography and geology of this area have been profoundly affected by the fault system, which follows the Central Hellenic Shear Zone extending from the Aegean part to the Ionian Sea and Peloponnese part [37]. This movement along the zone has created many grabens, one of which is the Gulf of Corinth, because of the Pliocene–Quaternary subsidence with marine sediments. The subsidence seized during the middle–lower Pleistocene when the marine sediments were uplifted and eroded.

More specifically, the Corinth canal shows a variety of sedimentary deposits, such as faulted Pliocene–Pleistocene marls, limestones, sandstones, and conglomerates [36]. These deposits can be observed at the canal internal sides [36]. These rocks were formed early in the history of the Gulf of Corinth graben when the southern part was below sea-level and were uplifted relatively recently. To the center of the canal, the oldest exposed rocks are visible. They are a series of pale marls with minor strata of brown sandstones and conglomerates and are Late Pliocene in age [36]. Both these sequences were raised above sea level through an east/west fault system. These steeply inclined normal faults are visible every few hundred meters along the canal [36,37].

From the report of the Hellenic Navy Hydrographic Service, [38] during the period of 1990–2012, the area of Diolkos showed a maximum tide of 0.72 m, a minimum of 0.01 m, and mean tide range of 0.26 m.

2.2. The Ancient Diolkos Slipway

Diolkos was a slipway used for transporting ships from the Saronic to the Corinth Gulf and in reverse. Part of the paved road was excavated by Verdelis [39,40]. The excavations at the western edge of the isthmus revealed that this road was used for transporting ships with the help of a wheeled vehicle, where the ships were moored, exempt from their cargo to the docks of Cenchreae or Lechaion. Diolkos was particularly useful for transporting warships, which had no cargo. The width of the road was 3.5 to 6 m and its length reached 8 km [41]. Diolkos was constructed in the 6th century B.C. in consideration of the need for the prompt crossing of the ships from the Saronic to the Corinth Gulf and is associated with Periander (625–585 B.C.), tyrant of Corinth. However, according to Koutsouba and Nakas [41], the frequent usage of Diolkos had already desisted after the complete destruction of Corinth (146 B.C.).

According to Morhange et al. [42], during 340 B.C., a co-seismic event provoked an uplift of Lechaion harbor by 1.2 m (6.6 km west of Diolkos) and also uplifted the entrance to Diolkos, as shown from the uplifted beachrocks that are formed on the ancient ship slipway. Below Diolkos ancient ship slipway, an older beachrock is present [43].

3. Materials and Methods

3.1. Fieldwork and Laboratory Analysis

For this work, detailed spatial mapping of Diolkos beachrock outcrops was accomplished. A differential global position system (DGPS) with global navigation satellite system (GNSS) system receiver (Spectra SP60) was used for the detailed recording of beachrock elevation/depth (with respect to the mean sea-level), length, and width of the seaward and landward parts with an accuracy of 3 cm. In total, 3 transects (AA′, BB′ and CC′) were accomplished, which also included sampling the front (seaward) and the end (landward) of each beachrock slab [12,29,44] (Figure 2).

Beachrock samples were collected from the top bed of beachrock outcrops, from the front and end slabs. Overall, 4 thin sections were prepared for petrographic and microstratigraphic analyses using a Leica DMLP (Leica Microsystems GmbH, Wetzlar, Germany) petrographic microscope with a digital camera and the corresponding image treatment software. These observations allowed characterizing the constituents, the presence/contribution of bioclasts, as well as the type of the cements and their micro-morphological features.

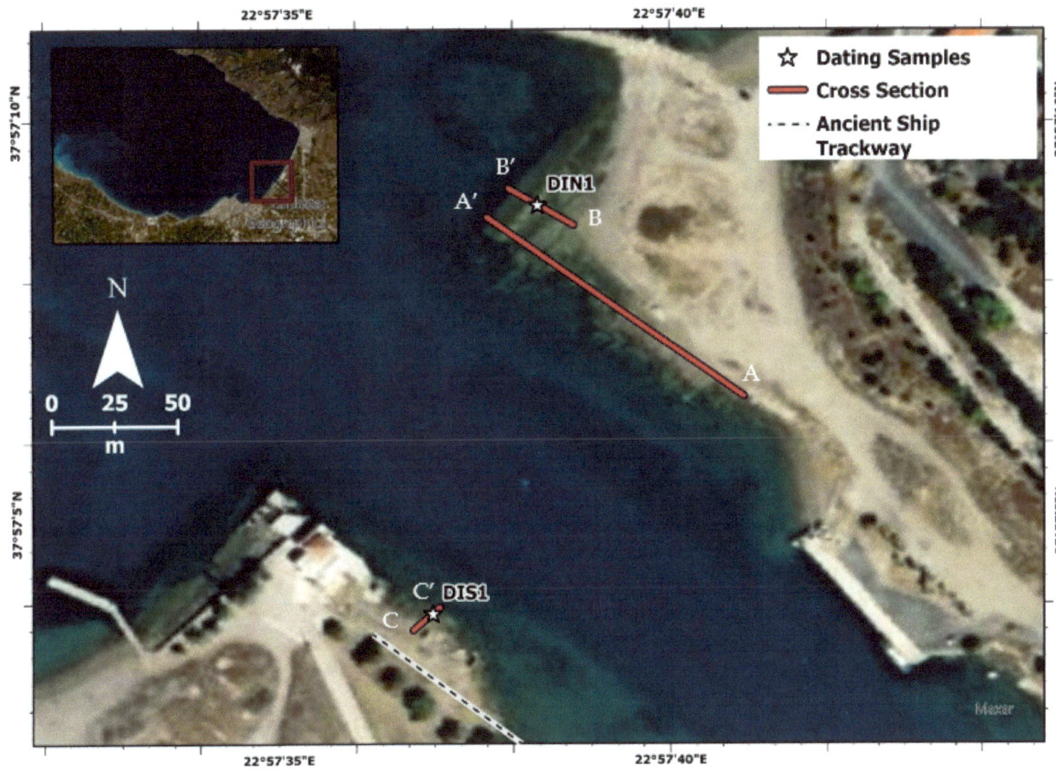

Figure 2. The study area of Diolkos, western part of Corinth canal. In this figure, three cross sections (AA′, BB′, CC′) are illustrated together with the location of OSL samples (DiN1, DiS1). The ancient ship trackway or slipway can be found with the dashed grey/black line.

3.2. Luminescence Dating

The beachrocks were dated by using optically stimulated luminescence (OSL) dating of quartz. Two samples were selected for dating and were processed at the Luminescence Dating Laboratory of the Institute of Physics, Silesian University of Technology, Poland. Quartz grains of 125–200 μm size were selected for use. A germanium spectrometer was used to determine the radioactivity dose rate. The determination of the equivalent dose was measured with the single aliquot regeneration protocol (OSL-SAR).

3.3. Relative Sea Level Reconstruction

For the reconstruction of the relative sea level and the production of relative sea level index points (SLIPs), we used the methodology proposed by Vacchi et al. [45], which has also been used for Mediterranean beachrocks [12,29,46]. SLIPs were produced only for those samples having intertidal formation in terms of cement typology, following the protocol proposed by Mauz et al. [10]. Cement with needles or isopachous fibers of aragonitic cement or isopachous rims and micritic high-magnesium calcite (HMC) cement or HMC cement in stalactitic position and meniscus between grains indicates that the beachrock samples have been formed in the intertidal zone [10].

The altitude of the former sea-level was estimated using the equation of Shennan & Horton [47] for each dated SLIP point. The total vertical error was obtained by adding in quadratic individual errors according to the equation:

$$e_i = (e^2{}_1 + e^2{}_2)^{1/2} \quad (1)$$

where e_1 and e_2 represent the index point error sources while taking into account both indicative range and the additional computational errors associated with the sample altitude [45].

4. Results

4.1. Beachrock Distribution

The detailed spatial mapping of the exposed beachrock of the Diolkos area was performed during summer 2020. The analysis includes the northern and southern parts of the beachrock, which is interrupted by the canal of the Corinth isthmus (Figure 3). The beachrock is composed of pile layers of sediments with plane-parallel seaward bedding. On both sides, the beachrock is highly intervened by human activity (i.e., jetties, Diolkos administration building, submerging road bridge, military installations).

Figure 3. Diolkos area beachrock outcrops. The N corresponds to the northern part and S to the southern part. The Corinth isthmus (canal) can be seen in the middle. The white dashed lines indicate the exposed Diolkos ship slipway. The areas covered by white dotted lines indicate the studied beachrock.

The beachrock, on both sides, has at least 75 cm thickness from its deepest part up to its corresponding exposed outcrop. The largest beachrock width is 105 m on the northern side. The beachrock extends to −75 cm depth and has a maximum elevation of 80 cm ± 5 cm. Due to the complexity, extensive width, and the beachrock distribution, three cross sections were carried out for better data interpretation (Figures 2 and 4).

In more detail, at the northern part, the beachrock lies at a maximum elevation of 68 ± 2 cm (Figure 4a) and maximum depth of −80 ± 2 cm (Figure 4b). In the northern part, a man-made wall structure was present almost at 23 m from the coastline. In Figure 4, the multi-banding morphology of the beachrock is evident. As is evident in Figure 4b, samples CoDi1 and CoDi2 were retrieved for microscopic analysis, as well as the sample

DiN1 for OSL dating. The northern beachrock shows a mean seaward dip of 8° from width 25–95 m, while, from 0–25 m, the mean dip is 12.5°.

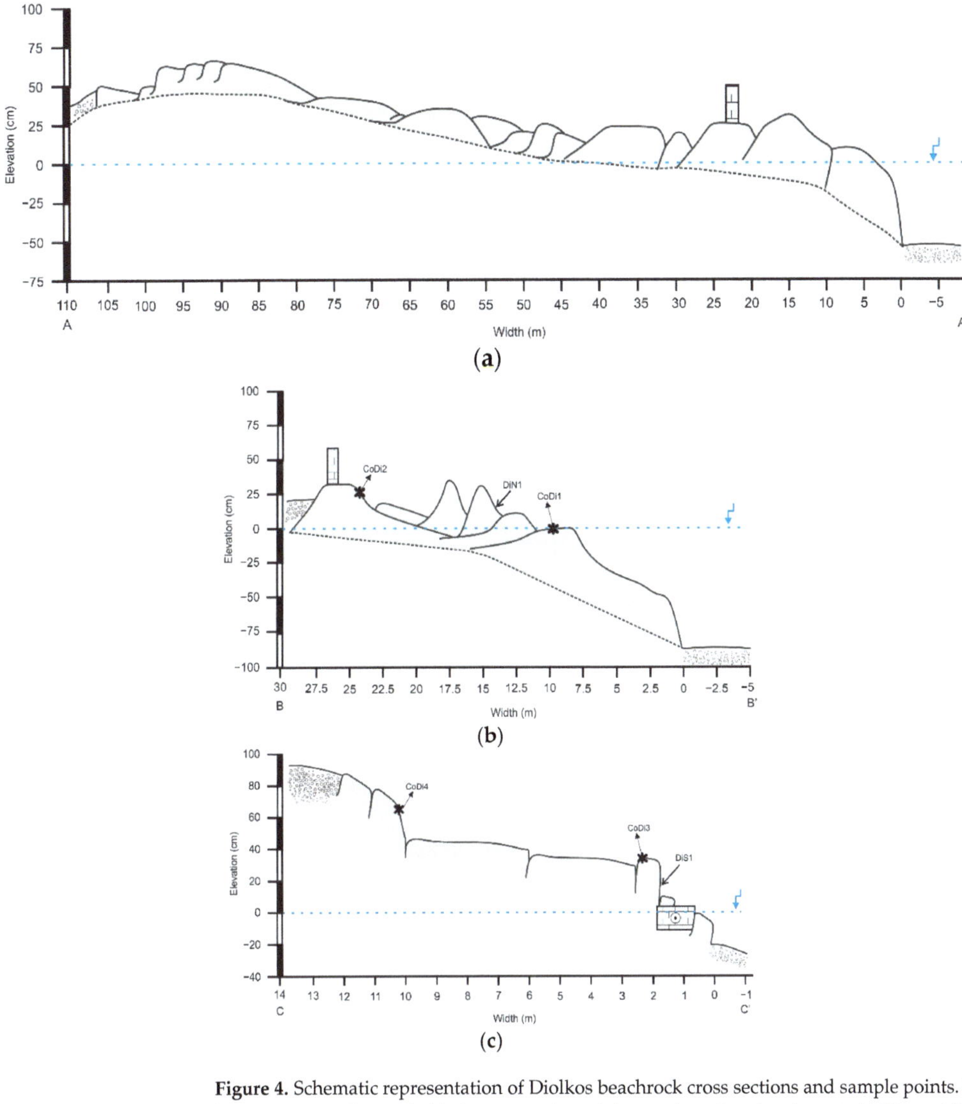

Figure 4. Schematic representation of Diolkos beachrock cross sections and sample points. The blue line represents the m.s.l. (**a**) A cross section of the northern beachrock part. At 22 m width, a wall construction is present. (**b**) A cross section of the northern beachrock part. At 26 m width, a wall construction is present. The star shaped points indicate the samples for microscopic analysis, and the arrow point at the cross section indicates the OSL dating sample. (**c**) A cross section of the southern beachrock part. At 1 m width, a Diolkos remnant is fused with the beachrock. The star shaped points indicate the samples for microscopic analysis, and the arrow point at the cross section indicates the OSL dating sample.

The southern beachrock has a maximum elevation of 83 ± 2 cm and a maximum depth of −20 ± 2 cm. The south–north cross section (Figure 4c) indicates the relationship of the beachrock and the Diolkos remnants, which are located at sea level. Another part of beachrock is submerged and below the Diolkos remnants. At this cross section, samples

CoDi3 and CoDi4 were retrieved for microscopic analysis, as well as sample DiS1 for OSL dating.

4.2. Cement Morphology and Mineralogy

All the observed Diolkos samples show a coherent pattern with sub-rounded and well to medium sorted grains with a general absence of bioclasts (<3%). The lithoclasts mainly consist of quartz, calcite, dolomite, plagioclase, and feldspar (Table 1). The examination of the beachrock cements from Diolkos showed that the micritic high magnesium calcite (HMC) [$(Ca,Mg)CO_3$] cement was the most dominant in all samples (Figure 5). In samples CoDi1 and CoDi2, the cement forms a thin isopachous coating around the sediment grains (Figure 5a,b). In samples CoDi3 and CoDi4, the cement forms a more coherent isopachous coating as well as pellet forms (Figure 5c,d). Furthermore, the CoDi4 sample has brown micritic cement, forming an outer film. A pore filling matrix cement was observed and consisted of very fine sedimentary particles (5–20 μm). Additionally, in sample CoDi4, a meniscus cement formation was noted (Figure 5).

Table 1. Mineral content and textural characteristics of beachrocks derived from the microscopical analysis and association with SLIP and indicative meaning.

Sample	Cement Type	Cement Thickness	Mineral Composition	SLIP	Indicative Meaning
CoDi1	Thin isopachous micritic HMC. No matrix and no bioclasts	<10 μm	Quartz, Mg-Calcite, Dolomite, Plagioclase, clay minerals	Intertidal, undifferentiated	MHW to MLW
CoDi2	Thin isopachous micritic HMC. No matrix and no bioclasts	<10 μm	Quartz, Mg-Calcite, Dolomite, Plagioclase, clay minerals	intertidal, undifferentiated	MHW to MLW
CoDi3	Isopachous micritic HMC and. Matrix infilling and Pellet concentrations.	10–20 μm	Quartz, Mg-Calcite, Dolomite, Plagioclase, clay minerals	Intertidal, undifferentiated	MHW to MLW
CoDi4	Isopachous micritic HMc. brown bio-micritic cement. Pellet concentrations.	10–20 μm	Quartz, Mg-Calcite, Dolomite, Plagioclase, clay minerals	Intertidal, undifferentiated	MHW to MLW

The observed binding material between the grains is mostly upper intertidal cement in samples CoDi1, CoDi2, CoDi3, and CoDi4, while the CoDi4 sample has evident forms and characteristics of freshwater influence. Cements of the upper intertidal zone are <100 μm and they are associated with detrital constituents (rock and shell fragments), which are all present in all the samples. Furthermore, the cement crystals that are forming an isopachous micritic coating serve as another indicator of intertidal zone beachrock formation. In the above samples, there was an absence of meteoric cement.

Combining the microscopic analysis and the detailed field survey, we confirmed that all the samples were beachrock samples with intertidal cement. Thus, the formation zone of the retrieved samples for OSL dating was the intertidal zone.

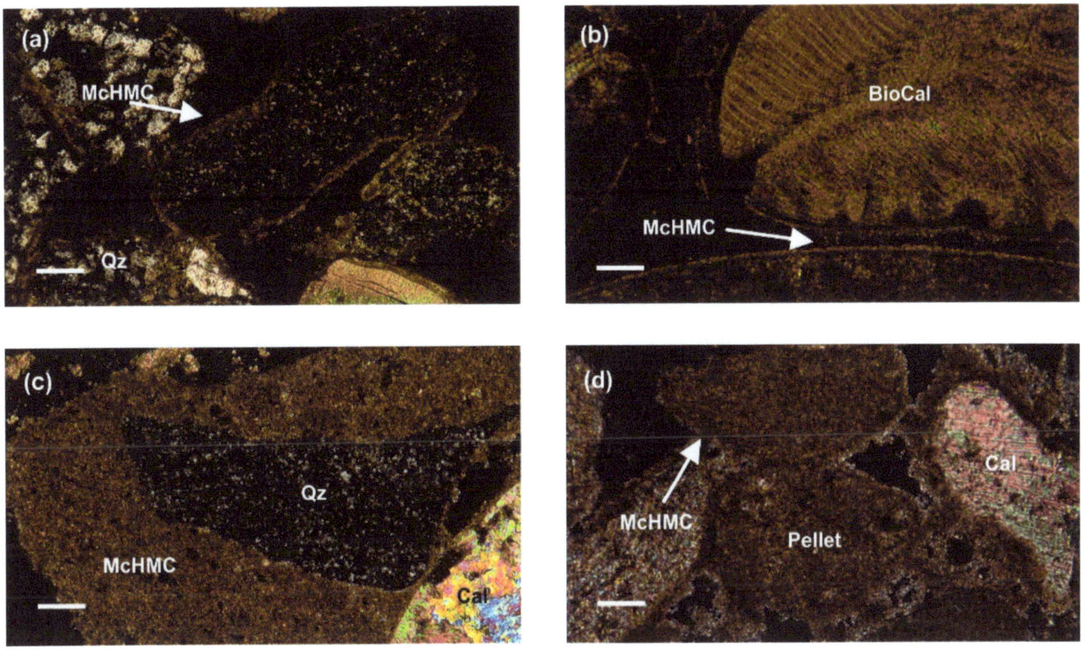

Figure 5. Polarized microscopy image in McHMC: micritic high magnesium calcite, BioCal: biogenic calcite, Cal: calcite, Qz: quartz. (**a**) Sample CoDi1, 10× magnification, scale 20 μm. Micritic HMC cement is coating mineral and lithoclast grains. (**b**) Sample CoDi2, 20× magnification, scale 10 μm. Thin micritic HMC cement is coating calcite grains. (**c**) Sample CoDi3, 20× magnification, scale 10 μm. The micritic HMC cement is fully covering the grains as a pore filling. (**d**) Sample CoDi4, 20× magnification, scale 10 μm. Micritic HMC cement is covering the sediment grains. A pellet concentration is present as a pore filling.

4.3. Sea Level Past Positions

The two samples from the northern and southern part of the beachrock were used as SLIPs. A vertical uncertainty was calculated for each index point derived from the accuracy of height measure and every possible factor associated with sampling and the later analytical processes (i.e., waves that temporally increase the water level) [29,48]. In Table 2, the OSL results of the two samples are provided. The sample DiN1 was collected from 0.17 m and the DiS1 from 0.20 m. Using the OSL dating method, we acquired the ages of 0.381 ka ± 58 years and 1.83 ka ± 14 years, respectively. Both samples were converted into SLIPs (Table 3), showing a sea level at approximately 4 cm and 8 cm, respectively. Table 3 presents all the details of converting the beachrock samples to SLIPs. Both SLIPs were acceptable as OSL dating, and the field measurements did not present methodological or any executive issues.

Table 2. Age determination results of the selected Diolkos samples.

Site	Sample Number	Elevation (m)	Method	No of Aliquots	Equivalent Dose (Gy)	Age BP (ka)
CoDiN	DiN1	0.17	Quartz	12	0.230	0.381
CoDiS	DiS1	0.20	Quartz	15	1190	1.83

Table 3. Past relative sea level calculations.

Beachrock ID	Sample No	Height (m)	Age (ka)	Tidal Range (m)	Measurement Error (m)	Indicative Meaning (m)	RWL (m)	RSL (m)	Error (cm)	Notes
CoDiN	DiN1	0.17	0.381	0.26	0.02	0.26	0.13	0.04	13	Accepted
CoDiS	DiS1	0.20	1.83	0.26	0.02	0.26	0.13	0.08	13	Accepted

5. Discussion

Coastal paleogeographic reconstructions that study sea level fluctuations during the middle–late Holocene in the Mediterranean take advantage of a variety of proxies, which include geomorphological evidence, such as tidal notches [49–52] and beachrocks [12,21,29,31,53], fixed biological indicators [54], and archaeological indicators [49,54–57]. In many studies, different proxies have been used for coastal palaeogeographical reconstructions [12,42,48].

Today, the beachrock of Diolkos is separated at its northern and southern part from the isthmus of Corinth (artificial canal), and it is uplifted. The microscopic examination of the beachrock samples revealed that their formation environment was in the upper intertidal zone [10,12,29,44,58]. The micritic HMC cement with pellet microformations, the high tendency to fill all the beachrock porosity, and the occasional brown color cement are good indicators for using a beachrock as a SLIP [9,10]. The beachrock distribution was quite complex as it had a notable width (105 m maximum), and the human interventions did not facilitate field examination.

The OSL dating was conducted on samples from the most characteristic places. It should be noted that the northern sample (DiN1) derived from the beachrock seaward face and the southern sample (DiS1) derived from the beachrock slab, exactly above the remnants of Diolkos slipway. The northern sample age was estimated at 0.381 ka $\pm$ 57 years (1569 $\pm$ 57 A.D.) and the southern sample at 1.83 ka $\pm$ 14 years (120 $\pm$ 14 A.D). These samples derived from almost the same height. However, their distances from the present-day shoreline are approximately 6 m and 60 m, respectively. Additionally, their conversion to SLIPs showed that the beachrock parts had the same relative water level during the time they were formed. Based on the aforementioned, it seems that the relative sea level did not change between 120 A.D. and 1569 A.D.

Considering the results from the beachrock mineralogical analysis, the sea level reconstruction, and the morphological analysis, it is clear that the two parts of Diolkos beachrock were once a continuous slab.

The area of Corinth isthmus has been uplifting by approximately 0.3 mm/year for the last 200 ka [59–62]. During 1981, an earthquake sequence took place at the area of Loutraki as a result of the fault system of south Alkyonides. That earthquake uplifted Corinth isthmus by 2 cm [63]. The area of Corinth isthmus is a tectonically active area given that, in the last 150 years, two strong seismic events have been described in historical records. In 1858, an earthquake of M = 6.5 destroyed Corinth, and, in 1928, an earthquake M = 6.3 inflicted extensive damage in Corinth and Loutraki [64]. The area is surrounded by a secondary faulting system that is occasionally active. At a distance of 7 km west of Diolkos, the ancient Lechaion port resides. Studies have revealed that local tectonics were responsible for producing tsunamigenic events that affected the ancient port [65,66]. Furthermore, it was noted that repeated phases of uplift and subsidence affected the area of Lechaion even if this situation is in contrast with the general geomorphological and tectonically uplifting regime of the area [66]. According to Morhange et al. [42], a tectonic uplift of 1.2 m was the main reason for the silting of the ancient port of Lechaion. It is clear that the area of Corinth is very tectonically active and the local faulting systems play an important role for the coastal geomorphological changes that affect manmade structures (e.g., ports, slipways, etc.).

Diolkos was an ancient slipway, enabling ships to cross from the Corinthian Gulf to the Saronic sea. Diolkos was constructed during the time of Periander (625–585 B.C.), tyrant of

Corinth [67]. The slipway was functional from at least the 5th to the 1st century B.C. [67]. After 67 A.D., there is no evidence of using Diolkos. At that time, Nero started the first plan to open a canal; however, his efforts failed, and the canal works were abandoned [67]. In Nero's effort to open the canal, Diolkos slipway was partly destroyed. According to Koutsouba and Nakas [41], the frequent usage of Diolkos was already in disuse after the complete destruction of Corinth (146 B.C.). Finally, the canal opening project started in 1882 and was brought to completion in 1893.

Nowadays, a submerged beachrock (hereinafter beachrock A) resides below te Diolkos remnants, which might have been used as a structural background for Diolkos construction [42]. On top of the slipway, a second beachrock exists, uplifted (hereinafter beachrock B). The beachrock A shows similar characteristics and texture with the uplifted one (Figure 6). We can assume that beachrock A had formed within the intertidal zone, similar to beachrock B.

(a) (b)

Figure 6. Southern Diolkos part. (**a**) Inland view. The dashed white line separates the beachrock B from Diolkos remnants. The erosion of Diolkos coastal zone can be observed. (**b**) Contact point of beachrock B with Diolkos remnants. From this site, the sample DiS1 was retrieved.

Our multiproxy analysis, combining historical, archaeological, geomorphological, and mineralogical data along with geochronology, allowed to reconstruct the paleogeographic evolution of Diolkos area. Before Diolkos construction (6th century B.C.), beachrock A was located in the intertidal zone. Assuming that the mean tidal range was almost the same as now, the intertidal zone was ranging between +13 cm to −13 cm. At that time, the Corinthians started constructing Diolkos on top of beachrock A, taking advantage of its slippery surface for pulling the ships. Consequently, the Diolkos slipway was located on the coastline. Beachrock A stopped its development due to the anthropogenic intervention. After the destruction of Corinth during 146 B.C., a co-seismic event must have taken place that submerged beachrock A and part of Diolkos. At that time, human intervention was limited and Diolkos ceased to be used until Nero's canal project in 64 A.D. Considering that the formation of beachrocks requires a calm coastal environment so that sand can accumulate and consequently consolidate, a new beachrock started forming after the abandonment of Nero's canal project. Morhange et al. [42] refer to a co-seismic event around 340 B.C. that might have caused an uplift in the western part of Diolkos. However, this might not be possible as, during that time, Diolkos was functional and beachrock A was still submerged.

Consequently, the first layer of beachrock B, covering the Diolkos remnants, had developed or was developing during the period of 120 ± 14 A.D. The coastal zone had found again its natural equilibrium as the longshore drift currents from Loutraki and Corinth supplied Diolkos with fresh sediment.

Beachrock B started developing and increasing its width and thickness at least up until 1569 ± 57 A.D. After that time, at least one co-seismic event may have occurred, which uplifted the area by ~12 cm. Between 1569 and the Corinth canal opening in 1882, the beachrock may have been still developing (Figure 7). Taking into account the seismic events of 1858 and 1928, we may assume that these events may have uplifted the Diolkos area.

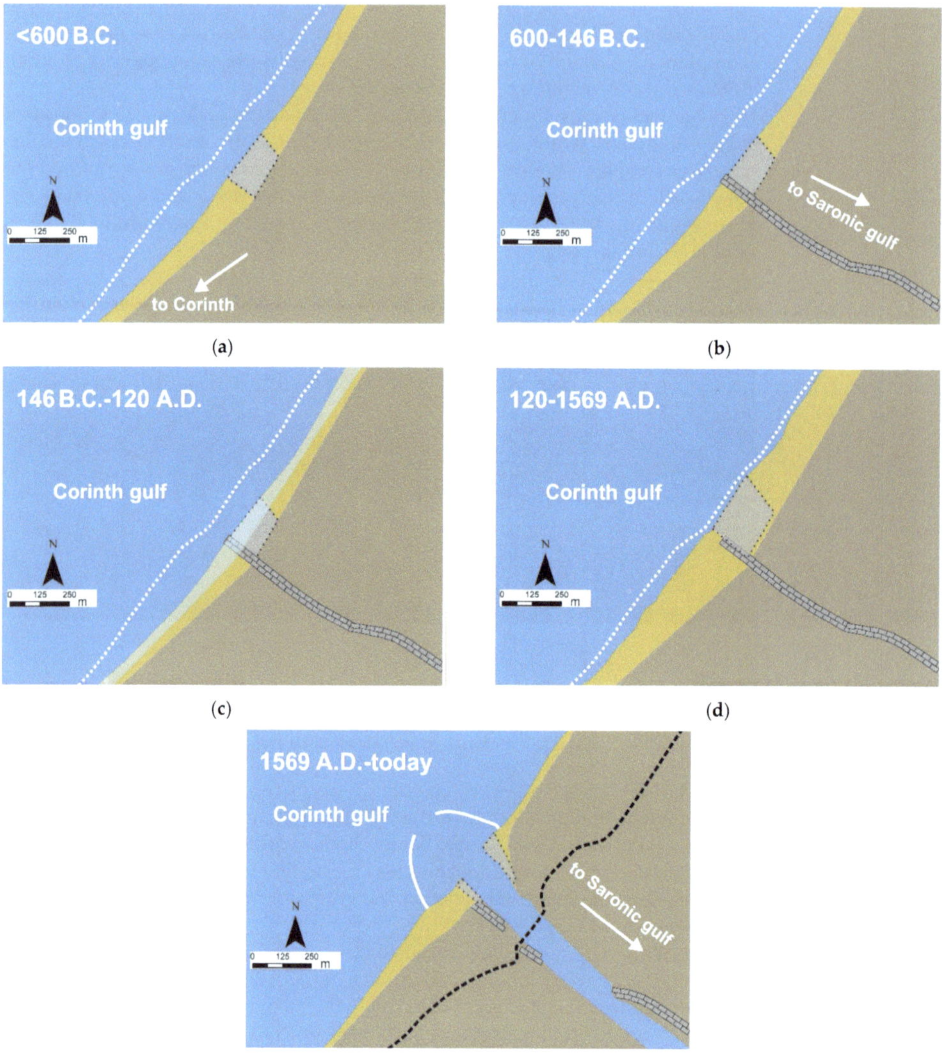

Figure 7. An illustration of the paleogeographic evolution of Diolkos area before the slipway construction until today. The white dashed line indicates the modern coast, the grey rectangle indicates the formed beachrocks, the grey blocks indicate Diolkos slipway, the black dashed line indicates the modern road of Posidonia, and the white lines indicate jetties. (**a**) The coastal zone of Diolkos area before the construction of the slipway (<600 B.C.), indicating the beachrock A. (**b**) Construction of Diolkos slipway on the beachrock A. (**c**) Subsidence of beachrock A and Diolkos slipway. (**d**) The new sediments overlay Diolkos slipway and form the beachrock B. The coast has advanced towards se€. (**e**) The coast reached today's position as the Corinth canal has been opened, and the majority of the beachrock is destroyed.

The functional height of an archaeological remain corresponds to the position (height) of specific architectural parts in relation to the mean sea level at the time of construction. Functional heights set the minimum construction height above the highest local tide [49,68]. There are few studies that refer to the functional height of ancient slipways. In Morhange and Marriner [69], several archaeological sea level indicators were studied along with slipways, which were categorized as meso-tidal functional constructions. Anzidei et al. [70] mentioned that the slipway of Carthage, Tunisia is less constrained as a sea level marker with a functional height of 0.24 ± 0.5 m. Baika [71] investigated a submerged slipway from Kea island, Cyclades, Greece of the Classical/Hellenistic era and identified the relevant RSL of the construction, around −2 to 2.5 ± 0.3 m. Diolkos was constructed by limestone blocks of 35 cm thickness. Taking into account the available literature and the possible Diolkos building material, we can assume that the subsidence of the beachrock A and Diolkos remnants was of the order of ~35 ± 16 cm.

6. Conclusions

In this study, we investigated the evolution of the Diolkos, Corinth, Greece beachrock. Combining data from the geomorphological field study, mineralogical and microscopic analysis, OSL dating, and historical and archaeological records, the paleogeography of the Diolkos area was possible.

- Before the 6th century B.C., a beachrock was lying on the intertidal zone covering the area and reaching at least 65 m from today's coastline.
- Diolkos construction occurred and the beachrock stopped its development until the destruction of Corinth in 146 B.C.
- A subsidence occurred totaling 35 ± 16 cm, and Diolkos along with beachrock A were submerged.
- Sediment accumulation occurred after Nero's intervention at 67 A.D., forming a new beachrock in the intertidal zone in 120 ± 14 A.D.
- After 120 ± 14 A.D., the beachrock continued its development up until 1569 ± 57 A.D.
- After 1569 ± 57 A.D., the beachrock developed another 5 m width until the present-day coastline. The area uplifted by 12 cm owing to one or more co-seismic events.
- The beachrock development stopped when the Corinth canal construction started in 1882.

Author Contributions: Conceptualization, G.S. and N.E.; methodology, G.S., A.K. and N.E.; investigation, G.S.; writing—original draft preparation, G.S. and A.K.; writing—review and editing, G.S., A.K., N.E. and H.M.; visualization, G.S.; supervision, N.E. All authors have read and agreed to the published version of the manuscript.

Funding: This research received no external funding.

Institutional Review Board Statement: Not applicable.

Informed Consent Statement: Not applicable.

Conflicts of Interest: The authors declare no conflict of interest.

References

1. Khan, N.S.; Ashe, E.; Shaw, T.A.; Vacchi, M.; Walker, J.; Peltier, W.R.; Kopp, R.E.; Horton, B.P. Holocene Relative Sea-Level Changes from Near-, Intermediate-, and Far-Field Locations. *Curr. Clim. Chang. Rep.* **2015**, *1*, 247–262. [CrossRef]
2. Rovere, A.; Raymo, M.E.; Vacchi, M.; Lorscheid, T.; Stocchi, P.; Gómez-Pujol, L.; Harris, D.L.; Casella, E.; O'Leary, M.J.; Hearty, P.J. The analysis of Last Interglacial (MIS 5e) relative sea-level indicators: Reconstructing sea-level in a warmer world. *Earth-Sci. Rev.* **2016**, *159*, 404–427. [CrossRef]
3. Engelhart, S.E.; Horton, B.P.; Douglas, B.C.; Peltier, W.R.; Törnqvist, T.E. Spatial variability of late Holocene and 20th century sea-level rise along the Atlantic coast of the United States. *Geology* **2009**, *37*, 1115–1118. [CrossRef]
4. Stattegger, K.; Tjallingii, R.; Saito, Y.; Michelli, M.; Trung Thanh, N.; Wetzel, A. Mid to late Holocene sea-level reconstruction of Southeast Vietnam using beachrock and beach-ridge deposits. *Glob. Planet. Change* **2013**, *110*, 214–222. [CrossRef]

5. Tsanakas, K.; Karymbalis, E.; Cundy, A.; Gaki-Papanastassiou, K.; Papanastasiou, D.; Drinia, H.; Koskeridou, E.; Maroukian, H. Late Holocene geomorphic evolution of the Livadi coastal plain, Gulf of Argostoli, Cephalonia Island, Western Greece. *Geogr. Fis. Din. Quat.* **2019**, *42*, 43–60. [CrossRef]
6. Melis, R.T.; Di Rita, F.; French, C.; Marriner, N.; Montis, F.; Serreli, G.; Sulas, F.; Vacchi, M. 8000 years of coastal changes on a western Mediterranean island: A multiproxy approach from the Posada plain of Sardinia. *Mar. Geol.* **2018**, *403*, 93–108. [CrossRef]
7. Caporizzo, C.; Gracia, F.J.; Aucelli, P.P.C.; Barbero, L.; Martín-Puertas, C.; Lagóstena, L.; Ruiz, J.A.; Alonso, C.; Mattei, G.; Galán-Ruffoni, I.; et al. Late-Holocene evolution of the Northern Bay of Cádiz from geomorphological, stratigraphic and archaeological data. *Quat. Int.* **2021**, *602*, 92–109. [CrossRef]
8. Seeliger, M.; Pint, A.; Frenzel, P.; Marriner, N.; Spada, G.; Vacchi, M.; Başaran, S.; Dan, A.; Seeger, F.; Seeger, K.; et al. Mid- to late-Holocene sea-level evolution of the northeastern Aegean sea. *Holocene* **2021**, *31*, 1621–1634. [CrossRef]
9. Vousdoukas, M.I.; Velegrakis, A.F.; Plomaritis, T.A. Beachrock occurrence, characteristics, formation mechanisms and impacts. *Earth-Sci. Rev.* **2007**, *85*, 23–46. [CrossRef]
10. Mauz, B.; Vacchi, M.; Green, A.; Hoffmann, G.; Cooper, A. Beachrock: A tool for reconstructing relative sea level in the far-field. *Mar. Geol.* **2015**, *362*, 1–16. [CrossRef]
11. Danjo, T.; Kawasaki, S. Characteristics of Beachrocks: A Review. *Geotech. Geol. Eng.* **2014**, *32*, 215–246. [CrossRef]
12. Karkani, A.; Evelpidou, N.; Vacchi, M.; Morhange, C.; Tsukamoto, S.; Frechen, M.; Maroukian, H. Tracking shoreline evolution in central Cyclades (Greece) using beachrocks. *Mar. Geol.* **2017**, *388*, 25–37. [CrossRef]
13. Tucker, M.E.; Bathurst, R.G.C. *Carbonate Diagenesis*; Blackwell Publishing Ltd.: Oxford, UK, 1990.
14. Gischler, E. Beachrock and Intertidal Precipitates. In *Geochemical Sediments and Landscapes*; Blackwell Publishing Ltd.: Hoboken, NJ, USA, 2008; pp. 365–390. ISBN 9781405125192.
15. Alexandersson, T. Intragranular Growth of Marine Aragonite and Mg-Calcite: Evidence of Precipitation from Supersaturated Seawater. *J. Sediment. Res.* **1972**, *42*, 441–460. [CrossRef]
16. Scholle, P.A.; Bebout, D.G.; Moore, C.H. *Carbonate Depositional Environments*; Scholle, P.A., Bebout, D.G., Moore, C.H., Eds.; The American Association of Petroleum Geologists: Tulsa, Oklahoma, 1983; ISBN 0891813101.
17. Kelletat, D. Beachrock as Sea-Level Indicator? Remarks from a Geomorphological Point of View. *J. Coast. Res.* **2006**, *226*, 1558–1564. [CrossRef]
18. Desruelles, S.; Fouache, É.; Ciner, A.; Dalongeville, R.; Pavlopoulos, K.; Kosun, E.; Coquinot, Y.; Potdevin, J.-L. Beachrocks and sea level changes since Middle Holocene: Comparison between the insular group of Mykonos–Delos–Rhenia (Cyclades, Greece) and the southern coast of Turkey. *Glob. Planet. Change* **2009**, *66*, 19–33. [CrossRef]
19. Evelpidou, N.; Kampolis, I.; Pirazzoli, P.A.; Vassilopoulos, A. Global sea-level rise and the disappearance of tidal notches. *Glob. Planet. Change* **2012**, *92–93*, 248–256. [CrossRef]
20. Vacchi, M.; Rovere, A.; Zouros, N.; Desruelles, S.; Caron, V.; Firpo, M. Spatial distribution of sea-level markers on Lesvos Island (NE Aegean Sea): Evidence of differential relative sea-level changes and the neotectonic implications. *Geomorphology* **2012**, *159–160*, 50–62. [CrossRef]
21. Falkenroth, M.; Schneider, B.; Hoffmann, G. Beachrock as sea-level indicator—A case study at the coastline of Oman (Indian Ocean). *Quat. Sci. Rev.* **2019**, *206*, 81–98. [CrossRef]
22. Alexandrakis, G.; Petrakis, S.; Kampanis, N.A. Integrating geomorphological data, geochronology and archaeological evidence for coastal landscape reconstruction, the case of Ammoudara Beach, Crete. *Water* **2021**, *13*, 1269. [CrossRef]
23. Alexandersson, T. Recent littoral and sublittoral high-mg calcite lithification in the mediterranean. *Sedimentology* **1969**, *12*, 47–61. [CrossRef]
24. Neumeier, U. Experimental modelling of beachrock cementation under microbial influence. *Sediment. Geol.* **1999**, *126*, 35–46. [CrossRef]
25. Psomiadis, D.; Tsourlos, P.; Albanakis, K. Electrical resistivity tomography mapping of beachrocks: Application to the island of Thassos (N. Greece). *Environ. Earth Sci.* **2009**, *59*, 233–240. [CrossRef]
26. Nikolakopoulos, K.; Lampropoulou, P.; Fakiris, E.; Sardelianos, D.; Papatheodorou, G. Synergistic Use of UAV and USV Data and Petrographic Analyses for the Investigation of Beachrock Formations: A Case Study from Syros Island, Aegean Sea, Greece. *Minerals* **2018**, *8*, 534. [CrossRef]
27. Hopley, D. Beachrock as a sea-level indicator. In *Sea-Level Research*; van der Plassche, O., Ed.; Springer: Dordrecht, The Netherlands, 1986; pp. 157–173.
28. Evelpidou, N. Geomorphology and Sea Level. In *Encyclopedia of Coastal Science*; Finkl, C., Makowski, C., Eds.; Springer: Cham, Switzerland, 2019; pp. 885–894. ISBN 9783319486574.
29. Polidorou, M.; Saitis, G.; Evelpidou, N. Beachrock development as an indicator of paleogeographic evolution, the case of Akrotiri Peninsula, Cyprus. *Z. Geomorphol.* **2021**, *63*, 3–17. [CrossRef]
30. Erginal, A.E.; Kıyak, N.G.; Öztürk, B. Investigation of Beachrock Using Microanalyses and OSL Dating: A Case Study from Bozcaada Island, Turkey. *J. Coast. Res.* **2010**, *262*, 350–358. [CrossRef]
31. Ozturk, M.Z.; Erginal, A.E.; Kiyak, N.G.; Ozturk, T. Cement fabrics and optical luminescence ages of beachrock, North Cyprus: Implications for Holocene sea-level changes. *Quat. Int.* **2016**, *401*, 132–140. [CrossRef]
32. Tatumi, S.H.; Kowate, E.A.; Gozzi, G.; Kassab, L.R.P.; Suguio, K.; Barreto, A.M.F.; Bezerra, F. Optical dating results of beachrock, eolic dunes and sediments applied to sea-level changes study. *J. Lumin.* **2003**, *102–103*, 562–565. [CrossRef]

33. Barreto, A.M.F.; Bezerra, F.H.R.; Suguio, K.; Tatumi, S.H.; Yee, M.; Paiva, R.P.; Munita, C.S. Late Pleistocene marine terrace deposits in northeastern Brazil: Sea-level change and tectonic implications. *Palaeogeogr. Palaeoclimatol. Palaeoecol.* **2002**, *179*, 57–69. [CrossRef]
34. Thomas, P.J. Luminescence Dating of Beachrock in the Southeast Coast of India—Potential for Holocene Shoreline Reconstruction. *J. Coast. Res.* **2009**, *251*, 1–7. [CrossRef]
35. Papanikolaou, D. Tectonostratigraphic models of the Alpine terranes and subduction history of the Hellenides. *Tectonophysics* **2013**, *595–596*, 1–24. [CrossRef]
36. Higgins, M.D.; Higgins, R.A. A geological companion to Greece and the Aegean. *Choice Rev.* **1993**, 34–3874. [CrossRef]
37. Papanikolaou, D.J.; Royden, L.H. Disruption of the Hellenic arc: Late Miocene extensional detachment faults and steep Pliocene-Quaternary normal faults—Or what happened at Corinth? *Tectonics* **2007**, *26*, TC5003. [CrossRef]
38. Hellenic Navy Hydrographic Service. *Hellenic Navy Hydrographic Service*; Hellenic Navy Hydrographic Service: Athens, Greece, 2015.
39. Verdelis, N. Continuation of Diolkos excavation. *Archaeol. Newsp.* **1956**, *95*, 1–3.
40. Verdelis, N. Diolkos excavation. In *Practicals of the Archaeological Society of Athens*; Archaeological Society at Athens 1960; Archaeological Society of Athens: Athens, Greece, 1960; Volume 160, pp. 136–143.
41. Koutsouba, D.; Nakas, G. Diolkos: A significant engineering structure of Antiquity. In Proceedings of the Corinthia and the Northeast Peloponnesus, Loutraki, Greece, 26–29 March 2009.
42. Morhange, C.; Pirazzoli, P.A.; Evelpidou, N.; Marriner, N. Late Holocene Tectonic Uplift and the Silting Up of Lechaion, the Western Harbor of Ancient Corinth, Greece. *Geoarchaeology* **2012**, *27*, 278–283. [CrossRef]
43. Maroukian, H.; Gaki-Papanastassiou, K.; Papanastassiou, D. Coastal changes in Corinthia, Greece. In *RES MARITIMAE: Cyprus and the Eastern Mediterranean from Prehistory to Late Antiquity*; Swiny, S., Hohlfelder, R.L., WyldeSwiny, H., Eds.; Scholars Press: Atlanta, GA, USA, 1994; pp. 217–226.
44. Saitis, G.; Koutsopoulou, E.; Karkani, A.; Anastasatou, M.; Stamatakis, M.; Gatou, M.-A.; Evelpidou, N. A multi-analytical study of beachrock formation in Naxos and Paros Islands, Aegean Sea, Greece and their palaeoenvironmental significance. *Z. Geomorphol.* **2021**, *63*, 19–32. [CrossRef]
45. Vacchi, M.; Marriner, N.; Morhange, C.; Spada, G.; Fontana, A.; Rovere, A. Multiproxy assessment of Holocene relative sea-level changes in the western Mediterranean: Sea-level variability and improvements in the definition of the isostatic signal. *Earth-Sci. Rev.* **2016**, *155*, 172–197. [CrossRef]
46. Vacchi, M.; Ghilardi, M.; Spada, G.; Currás, A.; Robresco, S. New insights into the sea-level evolution in Corsica (NW Mediterranean) since the late Neolithic. *J. Archaeol. Sci. Rep.* **2017**, *12*, 782–793. [CrossRef]
47. Shennan, I.; Horton, B. Holocene land- and sea-level changes in Great Britain. *J. Quat. Sci.* **2002**, *17*, 511–526. [CrossRef]
48. Dean, S.; Horton, B.P.; Evelpidou, N.; Cahill, N.; Spada, G.; Sivan, D. Can we detect centennial sea-level variations over the last three thousand years in Israeli archaeological records? *Quat. Sci. Rev.* **2019**, *210*, 125–135. [CrossRef]
49. Evelpidou, N.; Pirazzoli, P.; Vassilopoulos, A.; Spada, G.; Ruggieri, G.; Tomasin, A. Late Holocene Sea Level Reconstructions Based on Observations of Roman Fish Tanks, Tyrrhenian Coast of Italy. *Geoarchaeology* **2012**, *27*, 259–277. [CrossRef]
50. Evelpidou, N.; Melini, D.; Pirazzoli, P.A.; Vassilopoulos, A. Evidence of repeated late Holocene rapid subsidence in the SE Cyclades (Greece) deduced from submerged notches. *Int. J. Earth Sci.* **2014**, *103*, 381–395. [CrossRef]
51. Boulton, S.J.; Stewart, I.S. Holocene coastal notches in the Mediterranean region: Indicators of palaeoseismic clustering? *Geomorphology* **2015**, *237*, 29–37. [CrossRef]
52. Sisma-Ventura, G.; Sivan, D.; Shtienberg, G.; Bialik, O.M.; Filin, S.; Greenbaum, N. Last interglacial sea level high-stand deduced from well-preserved abrasive notches exposed on the Galilee coast of northern Israel. *Palaeogeogr. Palaeoclimatol. Palaeoecol.* **2017**, *470*, 1–10. [CrossRef]
53. Damien, A.; Kosmas, P.; Éric, F. Holocene relative sea-level variations and archeological implications, Abu Dhabi western region, United Arab Emirates. *Arab. J. Geosci.* **2020**, *13*, 1–16. [CrossRef]
54. Vacchi, M.; Russo Ermolli, E.; Morhange, C.; Ruello, M.R.; Di Donato, V.; Di Vito, M.A.; Giampaola, D.; Carsana, V.; Liuzza, V.; Cinque, A.; et al. Millennial variability of rates of sea-level rise in the ancient harbour of Naples (Italy, western Mediterranean Sea). *Quat. Res.* **2020**, *93*, 284–298. [CrossRef]
55. Sivan, D.; Widowinski, S.; Lambeck, K.; Galili, E.; Raban, A. Holocene sea-level changes along the Mediterranean coast of Israel, based on archaeological observations and numerical model. *Palaeogeogr. Palaeoclim. Palaeoecol.* **2001**, *167*, 101–117. [CrossRef]
56. Sivan, D.; Lambeck, K.; Toueg, R.; Raban, A.; Porath, Y.; Shirman, B. Ancient coastal wells of Caesarea Maritima, Israel, an indicator for relative sea level changes during the last 2000 years. *Earth Planet. Sci. Lett.* **2004**, *222*, 315–330. [CrossRef]
57. Benjamin, J.; Rovere, A.; Fontana, A.; Furlani, S.; Vacchi, M.; Inglis, R.H.; Galili, E.; Antonioli, F.; Sivan, D.; Miko, S. Late Quaternary sea-level changes and early human societies in the central and eastern Mediterranean Basin: An interdisciplinary review. *Q. Int.* **2017**, *449*, 29–57. [CrossRef]
58. Petropoulos, A.; Baziotis, I.; Anagnostou, C.; Evelpidou, N. Beachrocks cement characteristics and conditions of formation. case study platanias beach, chania, greece. *Bull. Geol. Soc. Greece* **2017**, *50*, 458. [CrossRef]
59. Collier, R.E.L.; Leeder, M.R.; Rowe, P.J.; Atkinson, T.C. Rates of tectonic uplift in the Corinth and Megara Basins, central Greece. *Tectonics* **1992**, *11*, 1159–1167. [CrossRef]

60. Dia, A.N.; Cohen, A.S.; O'Nions, R.K.; Jackson, J.A. Rates of uplift investigated through 230Th dating in the Gulf of Corinth (Greece). *Chem. Geol.* **1997**, *138*, 171–184. [CrossRef]
61. Turner, J.A.; Leeder, M.R.; Andrews, J.E.; Rowe, P.J.; van Calsteren, P.V.; Thomas, L. Testing rival tectonic uplift models for the Lechaion Gulf in the Gulf of Corinth rift. *J. Geol. Soc.* **2010**, *167*, 1237–1250. [CrossRef]
62. Papanikolaou, I.D.; Triantaphyllou, M.; Pallikarakis, A.; Migiros, G. Active faulting at the Corinth Canal based on surface observations, borehole data and paleoenvironmental interpretations. Passive rupture during the 1981 earthquake sequence? *Geomorphology* **2015**, *237*, 65–78. [CrossRef]
63. Mariolakos, I.; Stiros, S.C. Quaternary deformation of the Isthmus and Gulf of Corinthos (Greece). *Geology* **1987**, *15*, 225–228. [CrossRef]
64. Papazachos, B.P.; Papazachou, C.B. *The Earthquakes of Greece*; Ziti Publ.: Thessaloniki, Greece, 1997; p. 304.
65. Hadler, H.; Koster, B.; Mathes-schmidt, M. Lechaion, the Ancient Harbour of Corinth (Peloponnese, Greece) destroyed by tsunamigenic impact. In Proceedings of the 2nd INQUA-IGCP-567 International Workshop on Active Tectonics, Earthquake Geology, Archaeology and Engineeering, Corinth, Greece, 19–24 September 2011; p. 1755.
66. Hadler, H.; Vött, A.; Koster, B.; Mathes-Schmidt, M.; Mattern, T.; Ntageretzis, A.K.; Reicherter, K.; Willershäuser, T. Multiple late-Holocene tsunami landfall in the eastern Gulf of Corinth recorded in the palaeotsunami geo-archive at Lechaion, harbour of ancient Corinth (Peloponnese, Greece). *Z. Geomorphol. Suppl. Issues* **2013**, *57*, 139–180. [CrossRef]
67. Lewis, M.J.T. Railways in the Greek and Roman world in Early Railways. In *Early Railways. A Selection of Papers from the First International Early Railways Conference*; Guy, A., Ed.; Newcomen Society: London, UK, 2001; pp. 8–19.
68. Lambeck, K.; Anzidei, M.; Antonioli, F.; Benini, A.; Esposito, A. Sea level in Roman time in the Central Mediterranean and implications for recent change. *Earth Planet. Sci. Lett.* **2004**, *224*, 563–575. [CrossRef]
69. Morhange, C.; Marriner, N. Archeological and biological relative sea-level indicators. *Handb. Sea-Level Res.* **2015**, 146–156. [CrossRef]
70. Anzidei, M.; Antonioli, F.; Lambeck, K.; Benini, A.; Soussi, M.; Lakhdar, R. New insights on the relative sea level change during Holocene along the coasts of Tunisia and western Libya from archaeological and geomorphological markers. *Quat. Int.* **2011**, *232*, 5–12. [CrossRef]
71. Baika, K. Archaeological indicators of relative sea-level changes in the Attico-Cycladic massif: Preliminary results. *Bull. Geol. Soc. Greece* **2008**, *XLII*, 33–48.

quaternary

MDPI

Article

Digital Deforestation: Comparing Automated Approaches to the Production of Digital Terrain Models (DTMs) in Agisoft Metashape

Matthew D. Howland [1,*], Anthony Tamberino [2], Ioannis Liritzis [3,4,5] and Thomas E. Levy [2]

1 Jacob M. Alkow Department of Archaeology and Ancient Near Eastern Cultures, Tel Aviv University, Tel Aviv-Yafo 6997801, Israel
2 Scripps Center for Marine Archaeology, Department of Anthropology, University of California San Diego, La Jolla, CA 92092, USA; atamberi@ucsd.edu (A.T.); tlevy@ucsd.edu (T.E.L.)
3 Key Research Institute of Yellow River Civilization and Sustainable Development, College of Environment and Planning, University of Henan, Kaifeng 475001, China; liritzis@henu.edu.cn or ioliritzis@gmail.com or ioannis.liritzis@euro-acad.eu or liritzis@aegean.gr
4 Department of Archaeology, School of History, Classics and Archaeology, College of Arts, Humanities & Social Sciences, The University of Edinburgh, Edinburgh EH8 9AG, UK
5 Department of Physics & Electronics, Rhodes University, P.O. Box 94, Makhanda 6140, South Africa
* Correspondence: matth@tauex.tau.ac.il

Abstract: This paper tests the suitability of automated point cloud classification tools provided by the popular image-based modeling (IBM) software package Agisoft Metashape for the generation of digital terrain models (DTMs) at moderately-vegetated archaeological sites. DTMs are often required for various forms of archaeological mapping and analysis. The suite of tools provided by Agisoft are relatively user-friendly as compared to many point cloud classification algorithms and do not require the use of additional software. Based on a case study from the Mycenaean site of Kastrouli, Greece, the mostly-automated, geometric classification tool "Classify Ground Points" provides the best results and produces a quality DTM that is sufficient for mapping and analysis. Each of the methods tested in this paper can likely be improved through manual editing of point cloud classification.

Keywords: photogrammetry; DTM; GIS; IBM

check for updates

Citation: Howland, M.D.; Tamberino, A.; Liritzis, I.; Levy, T.E. Digital Deforestation: Comparing Automated Approaches to the Production of Digital Terrain Models (DTMs) in Agisoft Metashape. *Quaternary* **2022**, 5, 5. https://doi.org/10.3390/quat5010005

Academic Editor: José Javier Baena Preysler

Received: 20 November 2021
Accepted: 4 January 2022
Published: 14 January 2022

1. Introduction

In recent years, aerial laser scanning (ALS; also known as LIDAR) has attracted much attention for its ability to "see through" dense vegetation [1–8]. This technology can record ground surfaces below vegetation as some fraction of the lasers emitted from the scanner strike ground surfaces through small gaps between leaves. This allows for the processing of ALS-derived point clouds to filter out "first returns", representing the canopy, leaving only "last returns", representing the ground surface [8–12]. One end product of such a process is a digital terrain model (DTM; also referred to as a bare-earth digital elevation model), a measure of the elevation of a ground surface that does not include vegetation or structures [13–15]. DTMs are critical datasets for archaeologists, useful for identification of sites and features below dense vegetation but also for performing various kinds of landscape modeling and spatial analyses in GIS [13–17]. Thus, digitally stripping a site of its vegetation through the production of DTMs is critical for many avenues of archaeological inquiry.

Despite the advantages of ALS for DTM production, laser scanning approaches are not always feasible as the technique can be cost-prohibitive [2,18–21]—though costs are likely to come down with the rapid pace of technological development. As such, archaeologists often prefer the combination of low-altitude aerial photography, often from UAVs, and image-based modeling (IBM) for the collection of elevation datasets at sitewide scale [22–30].

Agisoft Metashape (formerly Agisoft Photoscan) stands out as one particularly common software package used for the production of 3D datasets through IBM [25,26,31–33]. Standard IBM procedures produce a digital surface model (DSM; i.e., an elevation dataset including both vegetation and structures) rather than a DTM [34]. DSMs are not suitable for most forms of archaeological analysis and mapping as the elevations in the dataset represent a combination of vegetation, structures, and ground surfaces. In the context of ALS, where point cloud filtering techniques are common and well-known, DSMs are infrequently used by archaeologists [14]. Among IBM practitioners, however, the use of DSMs produced from unfiltered or minimally-filtered point clouds is common [25,35].

Though IBM is often preferred by archaeologists due to cost considerations and ease of use, IBM approaches suffer in comparison to ALS with regard to the inability of the former approach to record data hidden by dense vegetation. As such, a DSM produced through IBM methods does not contain data on archaeological features hidden below trees. Thus, IBM/photogrammetry is often considered inappropriate for application in vegetated areas [20,26,36,37]. However, many—if not most—archaeological sites exist in moderately vegetated areas, often featuring some trees, shrubs, or bushes, but not enough to entirely cover the site and prevent it from being seen from above. Collection of elevation data through methods of low-altitude aerial photography and IBM at sites such as these is an attractive proposition due to the cost-effectiveness and user-friendliness of the techniques [21]. In this context, IBM is viable but conducting additional data processing to produce a DTM is necessary for mapping and spatial analysis.

IBM-derived data can be processed in a number of ways to produce a DTM [38]. Geometric filters normally used for the processing of ALS data are often applied to IBM data as well [39,40]. Examples of this include the use of algorithms such as the fast Fourier transform [41], cloth simulation filtering [42,43], and a TIN-based filtering approach [38,44–46]. These geometric algorithms are frequently applied using specialized point cloud filtration software tools such as LASGround [46], TerraSolid [45], or LP360 [44]. IBM point clouds can also be classified based on the use of RGB and NIR imagery to identify the spectral signature of points, either through the use of an NDVI threshold [38,47,48] or a machine learning-based classification [47]. Machine learning classifiers can also be based on geometric relationships within a scene [49], or both color and geometry [50]. GIS-based filtering of DSMs proves yet another approach to DTM production from IBM data [51]. Most of these tools are suitable for use by specialists in ALS or remote sensing but are beyond the technical expertise of most field archaeologists who nevertheless require DTMs for mapping or analysis. This is problematic as a major appeal of LAAP and IBM-based approaches is their ease of use.

Fortunately, user-friendly alternatives exist for point cloud filtration and DTM production. One prominent example is the point cloud classification toolkit provided by Agisoft Metashape [52]. The set of tools in this program, ranging from fully manual to fully automated, can be used effectively to filter out vegetation and structures from photogrammetric point clouds in an easy and straightforward way [21,28,34,53,54]. The results of these approaches have been validated as accurate and useful for analysis of bare earth surfaces through comparative analysis of derivative DSMs and DTMs [34]. However, studies measuring the accuracy of Agisoft-based point cloud filtration have found that results are not as accurate as those generated from more technically-intensive point cloud filtering algorithms or DTM production through ALS [34,38,55]. Moreover, users of the automated tools in Agisoft have also pointed out the need for manual editing of point cloud classification results [21]. Still, automated approaches for IBM point cloud filtration mean that IBM can be a viable and accurate alternative to ALS even when ground classification is necessary [21]. As such, the availability of user-friendly point cloud classification tools within Agisoft is a boon to archaeologists working at all but the most densely vegetated sites. These tools allow for the use of cost-effective and user-friendly photogrammetric methods to generate GIS-based elevation datasets that can facilitate both sophisticated spatial analyses and aid more straightforward goals of contour generation and mapping.

This paper explores the point cloud classification and DTM production functionality offered by Agisoft Metashape Professional [52,56]. The approaches offered by Agisoft are largely automated and user-friendly and as such, provide a useful alternative to more technical approaches requiring specialized knowledge or additional software. The utility of the approaches offered by Agisoft has not been adequately tested despite the ubiquity of use of the software by archaeologists. Each of the three automated approaches to point cloud classification provided by Agisoft Metashape is tested based on IBM data recorded at the Late Helladic period site Kastrouli, in central Greece. The generation of derivative contour data for both an unmodified DSM and a DTM using standard GIS techniques is also presented in order to discuss the utility of DTM production using these techniques.

Kastrouli is located in the Phokis region of central Greece, on the Desfina Peninsula [57]. The site lies at the end of a ridgeline at the top of a hill, the top of which is encircled by a fortification wall. The slopes of the hill on which the site lies are rocky with sparse vegetation, while the top of the hill is partially covered by trees, large bushes, and shrubs, as well as several low field walls. The site's earliest and most notable remains are, in addition to the fortification wall, three tombs that date to the Late Helladic period, one of which (Tomb A) was dated more precisely to the LH IIIB 2 [58,59], though the tomb was re-assigned to the LH IIIA 2 or slightly later based on excavations in 2016 and optically stimulated luminescence dating [60,61]. The tomb was also re-used through the LH IIIC and again later in the Middle Geometric period, before being partially looted in the 20th century [58,59]. The fortification wall was also constructed in the Late Helladic period and was reinforced in later periods [59,62,63]. Excavations in 2017 and scientific analysis of finds and skeletal remains have shed additional light on the site's occupation in the Late Helladic period [62–67]. The current case study at Kastrouli was part of the 2016 excavation campaign at Kastrouli, which excavated Tomb A at the site and conducted two small wall sample probes in other parts of the site. The campaign also featured a 3D and spatial recording program designed to comprehensively and intensively document the site and the progress of the excavation through spatial and 3D recording [58–60,68].

2. Materials and Methods

The data collected to create a sitewide model of Kastrouli was collected using a helium balloon with an attached, custom frame and a Canon EOS 50D DSLR triggered by an interval timer (updated from the LAAP system described in [25]). The balloon was tethered to an operator on the ground, who maneuvered the balloon around the site with the goal of collecting images in transects across the site and with a great deal of (ca. 90%+) overlap between consecutive images. The balloon, though less predictable in flight than a typical small UAV, carries a higher resolution camera and allows for longer flight time. As such, the balloon system generates comparable results to a UAV system for photogrammetric purposes if coverage is adequate, though recent technological development in UAVs has rendered them the gold standard of LAAP recording. In total, sidewide LAAP photography at Kastrouli captured 790 images of the site, a sufficient number to generate a detailed model of the site.

Once these images were captured, they were processed using a straightforward and standard workflow in Agisoft Metashape Pro. The ways in which these stages work within Agisoft is discussed in further detail elsewhere [25,27]. At Kastrouli, the model was georeferenced using nine control points established at the site using differential GPS. The model overall had a horizontal spatial error—as reported by Agisoft—of 8.23 cm, which we regard as acceptable for a sitewide model, especially given the acquisition of smaller, more precise models at key parts of the site. The processed and georeferenced model was sufficient to generate a high-resolution (2.5 cm) orthophotograph suitable for the project's mapping goals (Figure 1) as well as a DSM of the site. As Kastrouli is moderately vegetated, a DSM includes elevations of vegetation and architecture at the site, meaning that the dataset is inadequate for mapping and spatial analysis. Thus, producing a DTM was necessary. As discussed above, many methods for producing a DTM from IBM datasets exist, though most require a relatively high level of technical expertise. Those provided by Agisoft Metashape, however, are relatively user-friendly, though varied in degree of automation and effectiveness.

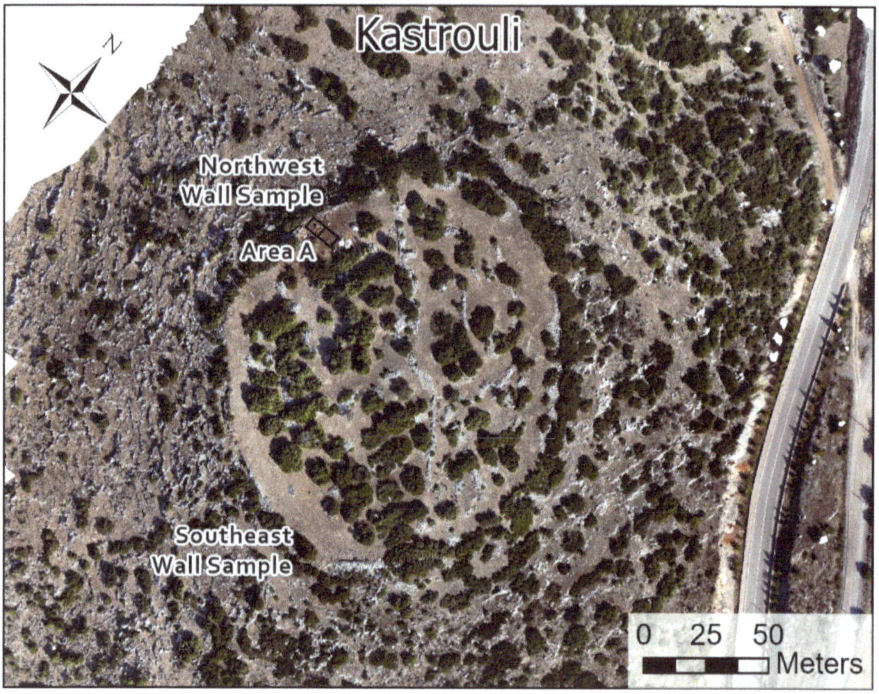

Figure 1. An orthophotograph of the site of Kastrouli, Greece. Note the vegetation obscuring the ground surface across the site.

Agisoft Metashape provides a suite of point cloud classification tools that facilitate the production of DTMs from a subset of points in a model (Table 1) [52]. These include four options to classify points at varying levels of automation. First, a nearly fully automated classification (Classify Points) is available, applying machine learning techniques to sort points into standard LIDAR classes. Users have the option to select which classes will be used and a confidence parameter for point classification. At 0.00 confidence, this tool will classify every point in a scene based on Agisoft's proprietary classification algorithm. This tool is the easiest to apply as it requires only the input of one parameter.

Table 1. Methods of point cloud classification in Agisoft Metashape Professional and the corresponding user parameters used to produce optimal DTMs at Kastrouli.

Point Cloud Classification Method	User Parameters
Classify Points	Confidence: 0.00
Classify Ground Points	Max angle (deg): 15 Max distance (m): 0.05 Cell size (m): 10
Select Points by Color	Color: #b69b8a Tolerance: 15 Channels: red, green, blue, hue, saturation, value
Assign Class	Fully manual

A second, mostly automated tool (Classify Ground Points) requires users to set three geometric parameters, which are used to distinguish points representing the bare earth from those representing other features based on the angle and slope from adjacent points [34,38]. This tool uses a TIN-based algorithm [21]. The Classify Ground Points tool in Agisoft Metashape is very easy to use but requires an iterative process in which the three geometric metrics used to differentiate points are manually adjusted and the selection tool re-run in order to find a quality result [34]. As such, it is difficult to ensure that the best possible result is attained.

A third tool provided by Agisoft (Select Points by Color) allows for the classification of points by their color values, with an option to set a tolerance range. This tool can be applied to select and classify points into the various categories one at a time. At Kastrouli, a medium-light shade of orange (hex code #b69b8a) generally representing the color of the surface of the site in the acquired imagery was selected for identifying ground points. This tool also requires an iterative process in which colors and tolerance are adjusted in order to find a visually optimal result.

The final tool (Assign Class) is fully manual, allowing users to select points by hand and sort them into classes. The first three automated approaches are tested and presented here. In each case, the threshold for user-provided values was chosen based on the best appearance of the resulting classification (Table 1). In order to test the reliability of these approaches, results are presented below without any manual editing of point cloud classification, which should normally be applied for best results [21]. These values will naturally vary by site and data collection methods.

Producing a DTM from a subset of a dense point cloud in Agisoft Metashape is a straightforward task and only requires the selection of relevant point classes during DEM production [52]. A DTM was produced from the results of each point cloud classification process described above. The results of these processes are presented below, along with smoothed contours generated from the DTMs/DSM for comparative purposes.

3. Results

The methods described above resulted in four elevation models of 2.5 cm spatial resolution: an unmodified DSM (Figure 2a) and three DTMs produced through classifying points geometrically (Figure 2b), through color selection (Figure 2c), and using the fully automatic procedure (Figure 2d). Each DTM is intended to represent the bare earth surface of the site, stripped of vegetation and architecture, though results vary by method. In each method, in areas of the site where the points were classified as ground, the cloud of points was used by Agisoft to generate a continuous raster surface. However, in areas covered by vegetation or architecture in which no ground classified points existed due to the limitations of photography and photogrammetry discussed above, the program interpolated a surface from the ground points on each side of the open space. In other words, the elevation measurements in areas covered by vegetation or stone walls serve as estimates based on the nearest ground points [21]. Ultimately the entire DTM is generated through an interpolative process, though this interpolation becomes more speculative in areas with fewer relevant points—i.e., those with dense vegetation and a complex ground surface [21]. It is difficult to quantify the error in elevation induced by the interpolation of vegetated areas, given a lack of a control dataset of points below vegetation. Testing of the quality of the DTM-generation process was not a primary goal of the field project, and so the data needed to conduct this type of accuracy testing was not collected in the field. However, visual inspection of the resulting DTMs is insightful for the relative utility of the results and subsequent research can investigate the accuracy of each classification method [38].

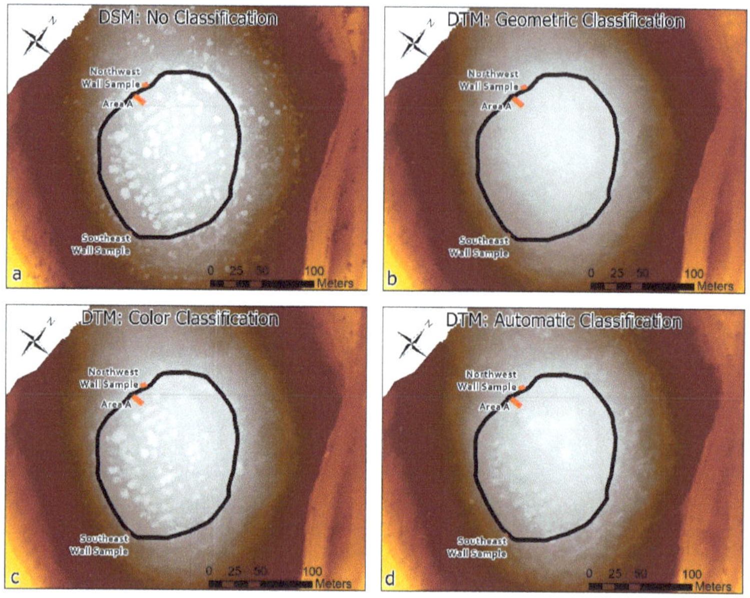

Figure 2. Comparison between an unmodified DEM of Kastrouli produced by (**a**) LAAP and photogrammetry and DTMs of the site generated by LAAP, photogrammetry, and point cloud classification through: (**b**) Classify Ground Points, (**c**) Select Points by Color, and (**d**) Classify Points. The site wall and excavation areas are shown for context. DEMs (**a**,**b**) are also viewable in interactive format here: https://storymaps.arcgis.com/stories/39280437536b4c458817fa4aa5c0b541 (accessed on 19 November 2021).

4. Discussion

Visual inspection of each resulting DEM is useful for understanding the extent to which each point cloud classification method was successful in identifying ground points, and therefore removing vegetation and architecture at the site from the elevation model. The DSM produced without classifying points (Figure 2a) serves as a control for comparison. On this DSM, the vegetation across the site is clearly visible as roughly circular high points across the site, especially its summit. Architecture is also apparent on closer inspection.

Of the other elevation models, the DTM produced by the geometric classification tool Classify Ground Points is clearly superior. The model is free of any elements of elevation representing large vegetation and architecture at the site is not apparent in the dataset. Previous testing of this point cloud classification technique for DTM production has reported that quantitative error results for the approach are highly competitive, but qualitative results are less so [43]. In the present case study, visual inspection of the results shows that the qualitative results in terms of the algorithm's ability to differentiate vegetation from ground points are superior to other methods provided within Agisoft. However, further testing can examine the extent to which quantitative error rates compare to other approaches.

Other methods provided by Agisoft were less successful in differentiating ground points from vegetation. A color-based approach (Figure 2b), despite classifying the lowest number of points out of the three classification techniques, was the least successful in removing vegetation from the DEM. The DEM produced through this method clearly illustrates that points on top of vegetation were classified as ground points, even despite the iterative process and fine tuning of the color and tolerance parameters. Despite the flawed result, this tool still has potential for point cloud classification as it can be used in combination with other tools, including manual identification, to produce a more refined result. However, at Kastrouli, this tool was not sufficient to produce a DTM without manual editing.

The final approach tested here is the fully automated tool, Classify Points, which applies Agisoft's proprietary algorithm, developed through machine learning, to sort points in the model into classes. Points identified as "ground" were subsequently used to produce the DEM. Unfortunately, though easy to use, this approach was not effective in classifying points at Kastrouli. This automated approach classified many points on the surface of the site as "building" (orange points in Figure 3), despite the lack of buildings in the scene. Low stone walls across the site were correctly classified as building; however, bedrock outcrops and much of the bare earth at the site was incorrectly grouped as such. Only a fraction of the actual ground surface was classified as "ground" (brown points in Figure 3) as well, with this class also featuring enough low-lying vegetation to disrupt the quality of the DTM. However, much of the vegetation at the site was classified correctly, suggesting potential for more nuanced use of this approach to classify and remove vegetation rather than rely on the identification of ground points. Overall, this tool was not accurate at Kastrouli, though it may be effective in other contexts and sites.

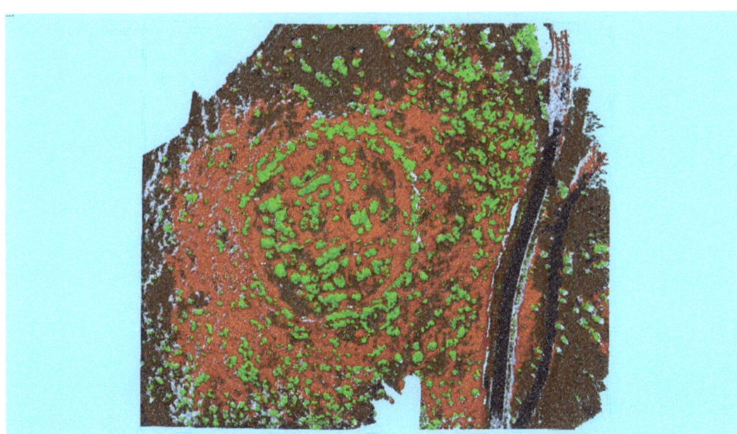

Figure 3. The dense point cloud from Kastrouli, as classified by the fully automated Classify Points algorithm. Green points are classified as "high vegetation", orange points are classified as "building", and brown points are classified as "ground". Much of the bare earth at the site is incorrectly classified as "building."

Though these approaches were of varying effectiveness, in total they suggest that the use of multiple methods, combined with some manual editing of point cloud classification, will provide effective results. However, at Kastrouli, the geometric procedure Classify Ground Points provided quality results without even a need for manual editing. However, producing a highly accurate DTM would likely require manual editing of automated classifications [21], which in turn requires many hours of additional labor in classification by a trained eye able to differentiate rocky outcrops from stone walls and ground-truthing. These tasks are important for best results, but they exponentially complicate and limit the efficiency of the overall process. As such, projects producing DTMs from photogrammetric datasets should consider the balance between efficiency of data processing and the ultimate accuracy of a final DTM.

In this case study, the geometric approach produced a DTM that generally reflects the topography of Kastrouli, especially on the top area of the site. The elevation dataset clearly has removed the artificially high measurements that represent the highest trees and bushes. This is a valuable accomplishment, as the current vegetation on the site most likely does not reflect ancient vegetation, and these elements also disrupt mapping and analysis efforts.

The smoother surface of the DTM on the top of the site more accurately reflects the ground surface in that area, as it is largely flat and gently sloped within the area encircled by the fortification wall. The vast majority of the archaeological remains at Kastrouli are also found in this area. That the DTM generation method performed most strongly in this area is an encouraging result for study and analysis of the site's anthropogenic component and therefore the project's goals. This is evidenced by the contours generated from the DTM, which—by contrast with contours generated from an unmodified DEM of the site—provide a much simpler and more intuitive representation of the site's elevation (Figure 4). The DTM-generated contours serve as a strong basis for mapping of key site features and excavation areas. Meanwhile, many of the contours generated from the DSM appear to represent the topography of the vegetation at the site rather than the variations in ground level elevations. Ultimately, the DTM produced through these methods provides a more useful basis for understanding ancient occupation of the site as it more closely reflects the ancient occupation surface rather than more recent vegetation. Future investigation at the site will be able to make use of this dataset to examine the patterns of occupation across the site.

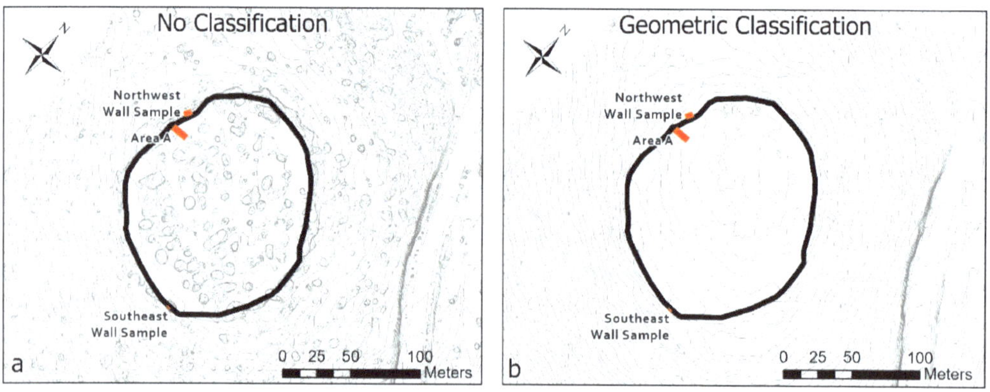

Figure 4. (**a**) Contour map produced from the unmodified DSM. Note the contours representing vegetation. (**b**) Contour map produced from the Classify Ground Points-derived, geometrically filtered DTM. In both cases, contour lines below 5 m have been removed and contours have been smoothed. These maps are also viewable in interactive format here: https://storymaps.arcgis.com/stories/39280437536b4c458817fa4aa5c0b541 (accessed on 19 November 2021).

One of the main benefits to the workflow described here is the relative simplicity and efficiency of the approach. Combined LAAP-IBM methods have become relatively standard applications at archaeological sites, with Agisoft also commonly being applied for its integrated photogrammetric workflow [27,33]. The steps for generating a DTM are also efficient in terms of the manual labor required, with the most time-consuming steps being field collection of data and testing different point selection and classification methods and standards. Refining the classification of the point cloud to reduce false negatives and false positives can also be manually intensive, depending on the standards needing to be met. The work described here produced quality results without manual editing in order to demonstrate the efficiency of that process, though review and manual editing of point cloud classification would likely improve outputs [21]. Processing photographs into 3D models, and subsequently DTMs, is another potential limiting factor on the efficiency of the process, depending on the quality of the computational resources available. In general, however, these LAAP, photogrammetric processing, and point cloud classification methods serve as an effective and efficient method for DTM generation and in many cases may complement already practiced workflows on archaeological projects. These methods also allow for the generation of accurate and simple contours that are important for cartography at archaeological sites. The approach presented here is applicable to all moderately-vegetated sites recorded with IBM, though the quality of point cloud classification methods may vary according to the topography and vegetation at the site. Though more sophisticated methods may also provide better results, the methods tested here are extremely user-friendly and can be improved by manual editing of point cloud classification. As such, this approach provides an easy and straightforward method of DTM production.

Drawbacks of the methods described above are its applicability only to areas with limited vegetation or architecture and its dependence on automated methods that may not capture the complexity of an archaeological site. In areas with no vegetation or architecture, the DTM does not depend on interpolation, which means that the dataset can achieve high fidelity to the ground surface given the potential for photogrammetric methods to achieve high levels of accuracy [35,69]. However, any part of a site where the ground surface is obscured by vegetation or architecture relies on interpolation from the nearest ground surface, which increases the amount of estimation at the expense of actual measurement. Thus, more vegetated or built-up sites will suffer in accuracy compared to more sparsely-covered sites. This trade-off should be factored in when choosing this method for DTM generation. The combined approach described above also should be applied with the understanding that automated methods lack the sophistication provided by expertise and a trained eye. A strictly automated approach may be unable to distinguish between a low stone wall and a rocky outcrop, for example. However, tools for manual point cloud classification allow archaeologists to consider the extent to which they would like to modify or replace the results of an automated system. Ultimately, maps and spatial analyses are subjective enterprises that can be facilitated by objective methods, so a combination of automated and manual approaches seems appropriate for this type of study. Looking forward, future studies should address the extent to which an interpolative method of creating a DTM reflects the reality of the ground surface of the site below vegetation and architecture by comparing the results generated using the method described above to measurements derived from other means [38,55]. This verification will help to demonstrate the accuracy of an IBM-based point cloud classification approach to DTM generation.

5. Conclusions

A combined LAAP-photogrammetry-point cloud classification approach is an effective and efficient workflow for generating a DTM of an archaeological site sufficient for cartography and spatial analysis. Such a methodology bests LiDAR on cost and may integrate more fully into already existing archaeological practices. The popular IBM software package Agisoft Metashape provides a number of mostly-automated tools for point cloud classification that can be applied to DTM production. Of these, the Classify Ground Points tool appears to be the most reliable based on a case study from Kastrouli, Greece. However, it is likely that automated methods can be improved by manual editing of results, at the expense of efficiency. Overall, the use of automated point cloud classification tools in Agisoft provides a streamlined and user-friendly process for the production of DTMs. This approach then represents a useful addition to the toolbox of archaeological projects interested in mapping and spatial analysis at moderately-vegetated or moderately-built up sites.

Author Contributions: Conceptualization, M.D.H.; Methodology, M.D.H. and A.T.; Investigation, M.D.H.; Writing—original draft preparation, M.D.H.; Writing—review and editing, M.D.H., A.T., I.L. and T.E.L.; Visualization, M.D.H.; Project administration, I.L. and T.E.L.; Funding acquisition, T.E.L. All authors have read and agreed to the published version of the manuscript.

Funding: Some graduate student support was provided by the University of California Office of the President through a Research Catalyst Grant for At-Risk Cultural Heritage and the Digital Humanities (Grant ID: CA16-376911; Lead PI: Thomas Levy).

Institutional Review Board Statement: Not applicable.

Informed Consent Statement: Not applicable.

Data Availability Statement: GIS data described in this article are available in a brief StoryMap: https://storymaps.arcgis.com/stories/39280437536b4c458817fa4aa5c0b541 (accessed on 19 November 2021). 3D datasets are available on request.

Acknowledgments: The authors would like to thank the Ministry of Culture for granting permission to dig and A. Tsaroucha, member and representative of the Ephoreia of Antiquities at Phokis (EAPh) in Delphi, for excellent collaboration. We also thank Malcom Wiener for his thoughts about the Tomb A assemblage, as well as Fotis Dasios for discussing with us the Sykia chamber tomb and its finds. We are grateful to the villagers of Kastrouli for hosting the 2016 expedition in their homes and tavernas with warmth and good will. The Kastrouli municipality was very helpful throughout the project and we are grateful for their help and support. We thank also the president of Desfina municipality community, Babis Kaliakoudas, for excellent support prior, during, and after the 2016 fieldwork; Panagiota Karamani, conservator, for her assistance; Fotini Koukou, the guardian, for excellent cooperation; and Alina Levy for her care with logistics and the whole project. Tom Levy is thankful to the Kershaw Family Foundation for their support, and we are grateful to Liz Anne and Phokion Potamianos of La Jolla, California, for their generous support of this project given to Tom Levy. All archaeological works and surveys were supervised by the EAPh's delegate, Anthoula Tsaroucha, and made possible through the constant support of the EAPh and Nancy Psalti, its director.

Conflicts of Interest: The authors declare no conflict of interest.

References

1. Opitz, R.S.; Cowley, D.C. *Interpreting Archaeological Topography. AIRBORNE Laser Scanning, 3D Data and Ground Observation*; Oxbow Books (Occasional Publication of the Aerial Archaeology Research Group, 5): Oxford, UK, 2013.
2. Casana, J.; Laugier, E.J.; Hill, A.C.; Reese, K.M.; Ferwerda, C.; McCoy, M.D.; Ladefoged, T. 2021 Exploring archaeological landscapes using drone-acquired lidar: Case studies from Hawai'i, Colorado, and New Hampshire, USA. *J. Archaeol. Sci. Rep.* **2021**, *39*, 103133.
3. Chase, A.F.; Chase, D.Z.; Weishampel, J.F.; Drake, J.B.; Shrestha, R.L.; Slatton, K.C.; Awe, J.J.; Carter, W.E. Airborne LiDAR, archaeology, and the ancient Maya landscape at Caracol, Belize. *J. Archaeol. Sci.* **2011**, *38*, 387–398. [CrossRef]
4. Devereux, B.J.; Amable, G.S.; Crow, P.; Cliff, A.D. The potential of airborne lidar for detection of archaeological features under woodland canopies. *Antiquity* **2005**, *79*, 648–660. [CrossRef]
5. Doneus, M.; Briese, C.; Fera, M.; Janner, M. Archaeological prospection of forested areas using full-waveform airborne laser scanning. *J. Archaeol. Sci.* **2008**, *35*, 882–893. [CrossRef]
6. Johnson, K.M.; Ouimet, W.B. Rediscovering the lost archaeological landscape of southern New England using airborne Light Detection and Ranging (LiDAR). *J. Archaeol. Sci.* **2014**, *43*, 9–20. [CrossRef]
7. Inomata, T.; Triadan, D.; Pinzón, F.; Burham, M.; Ranchos, J.L.; Aoyama, K.; Haraguchi, T. Archaeological application of airborne LiDAR to examine social changes in the Ceibal region of the Maya lowlands. *PLoS ONE* **2018**, *13*, e0191619.
8. Doneus, M.; Briese, C. Full-waveform airborne laser scanning as a tool for archaeological reconnaissance. *BAR Int. Ser.* **2006**, *1568*, 99.
9. Meng, X.; Currit, N.; Zhao, K. Ground filtering algorithms for airborne LiDAR data: A review of critical issues. *Remote Sens.* **2010**, *2*, 833–860. [CrossRef]
10. Montealegre, A.L.; Lamelas, M.T.; de la Riva, J. A comparison of open-source LiDAR filtering algorithms in a Mediterranean forest environment. *IEEE J. Sel. Top. Appl. Earth Obs. Remote Sens.* **2015**, *8*, 4072–4085. [CrossRef]
11. Opitz, R.S. An overview of airborne and terrestrial laser scanning in archaeology. In *Interpreting Archaeological Topography: 3D Data, Visualization and Observation*; Opitz, R.C., Cowley, D.C., Eds.; Oxbow Books: Oxford, UK, 2013; pp. 13–31.
12. Sithole, G.; Vosselman, G. Experimental comparison of filter algorithms for bare-Earth extraction from airborne laser scanning point clouds. *ISPRS J. Photogramm. Remote Sens.* **2004**, *59*, 85–101. [CrossRef]
13. Mlekuž, D. 2013 Skin deep: LiDAR and good practice of landscape archaeology. In *Good Practice in Archaeological Diagnostics*; Corsi, C., Slapšak, B., Vermeulen, F., Eds.; Springer: Cham, Switzerland, 2013; pp. 113–129.
14. Štular, B.; Eichert, S.; Lozić, E. Airborne LiDAR Point Cloud Processing for Archaeology. Pipeline and QGIS Toolbox. *Remote Sens.* **2021**, *13*, 3225. [CrossRef]
15. White, D.A. LiDAR, point clouds, and their archaeological applications. In *Mapping Archaeological Landscapes from Space*; Comer, D.C., Harrower, M.J., Eds.; Springer: New York, NY, USA, 2013; pp. 175–186.
16. Evans, D.H.; Fletcher, R.J.; Pottier, C.; Chevance, J.-B.; Soutif, D.; Tan, B.S.; Im, S.; Ea, D.; Tin, T.; Kim, S.; et al. Uncovering archaeological landscapes at Angkor using lidar. *Proc. Natl. Acad. Sci. USA* **2013**, *110*, 12595–12600. [CrossRef]
17. Masini, N.; Lasaponara, R. Airborne lidar in archaeology: Overview and a case study. In *Computational Science and Its Applications–ICCSA 2013, Proceedings of the International Conference on Computational Science and Its Applications, Ho Chi Minh City, Vietnam, 24–27 June 2013*; Springer: Berlin/Heidelberg, Germany, 2013; pp. 663–676.
18. Fernandez-Diaz, J.C.; Carter, W.E.; Shrestha, R.L.; Glennie, C.L. Now you see it . . . now you don't: Understanding airborne mapping LiDAR collection and data product generation for archaeological research in Mesoamerica. *Remote Sens.* **2014**, *6*, 9951–10001. [CrossRef]

19. Howland, M.D. 3D Recording in the Field: Style Without Substance? In *Cyber-Archaeology and Grand Narratives*; Jones, I.W.N., Levy, T.E., Eds.; Springer: Cham, Switzerland, 2018; pp. 19–33.
20. Vilbig, J.M.; Sagan, V.; Bodine, C. Archaeological surveying with airborne LiDAR and UAV photogrammetry: A comparative analysis at Cahokia Mounds. *J. Archaeol. Sci. Rep.* **2020**, *33*, 102509. [CrossRef]
21. Jiménez-Jiménez, S.I.; Ojeda-Bustamante, W.; Marcial-Pablo, M.d.J.; Enciso, J. Digital Terrain Models Generated with Low-Cost UAV Photogrammetry: Methodology and Accuracy. *ISPRS Int. J. Geo-Inf.* **2021**, *10*, 285. [CrossRef]
22. Campana, S. Drones in archaeology. State-of-the-art and future perspectives. *Archaeol. Prospect.* **2017**, *24*, 275–296. [CrossRef]
23. Hill, A.C.; Rowan, Y. Droning on in the Badia: UAVs and site documentation at Wadi al-Qattafi. *Near East. Archaeol.* **2017**, *80*, 114–123. [CrossRef]
24. Hill, A.C. Economical drone mapping for archaeology: Comparisons of efficiency and accuracy. *J. Archaeol. Sci. Rep.* **2019**, *24*, 80–91. [CrossRef]
25. Howland, M.D.; Kuester, F.; Levy, T.E. Photogrammetry in the field: Documenting, recording, and presenting archaeology. *Mediterr. Archaeol. Archaeom.* **2014**, *14*, 101–108.
26. Magnani, M.; Douglass, M.; Schroder, W.; Reeves, J.; Braun, D.R. The digital revolution to come: Photogrammetry in archaeological practice. *Am. Antiq.* **2020**, *85*, 737–760. [CrossRef]
27. Olson, B.R.; Placchetti, R.A.; Quartermaine, J.; Killebrew, A.E. The Tel Akko Total Archaeology Project (Akko, Israel): Assessing the suitability of multi-scale 3D field recording in archaeology. *J. Field Archaeol.* **2013**, *38*, 244–262. [CrossRef]
28. Themistocleous, K. 2020 The use of UAVs for cultural heritage and archaeology. In *Remote Sensing for Archaeology and Cultural Landscapes*; Hadjimitsis, D.G., Themistocleous, K., Cuca, B., Agapiou, A., Lysandrou, V., Lasaponara, R., Masini, N., Schreier, G., Eds.; Springer: Cham, Switzerland, 2020; pp. 241–269.
29. Waagen, J. New technology and archaeological practice. Improving the primary archaeological recording process in excavation by means of UAS photogrammetry. *J. Archaeol. Sci.* **2019**, *101*, 11–20.
30. Wernke, S.A.; Adams, J.A.; Hooten, E.R. Capturing Complexity: Toward an Integrated Low-Altitude Photogrammetry and Mobile Geographic Information System Archaeological Registry System. *Adv. Archaeol. Pract.* **2014**, *2*, 147–163. [CrossRef]
31. Hill, A.C.; Rowan, Y.; Kersel, M.M. Mapping with aerial photographs: Recording the past, the present, and the invisible at Marj Rabba, Israel. *Near East. Archaeol.* **2014**, *77*, 182–186. [CrossRef]
32. Jones, C.A.; Church, E. Photogrammetry is for everyone: Structure-from-motion software user experiences in archaeology. *J. Archaeol. Sci. Rep.* **2020**, *30*, 102261. [CrossRef]
33. Verhoeven, G. Taking Computer Vision Aloft–Archaeological Three-Dimensional Reconstructions from Aerial Photographs with Photoscan. *Archaeol. Prospect.* **2011**, *18*, 67–73. [CrossRef]
34. Dubbini, M.; Curzio, L.I.; Campedelli, A. Digital elevation models from unmanned aerial vehicle surveys for archaeological interpretation of terrain anomalies: Case study of the Roman castrum of Burnum (Croatia). *J. Archaeol. Sci. Rep.* **2016**, *8*, 121–134. [CrossRef]
35. Uysal, M.; Toprak, A.S.; Polat, N. DEM generation with UAV Photogrammetry and accuracy analysis in Sahitler hill. *Measurement* **2015**, *73*, 539–543. [CrossRef]
36. O'Driscoll, J. Landscape applications of photogrammetry using unmanned aerial vehicles. *J. Archaeol. Sci. Rep.* **2018**, *22*, 32–44. [CrossRef]
37. Sapirstein, P.; Murray, S. Establishing best practices for photogrammetric recording during archaeological fieldwork. *J. Field Archaeol.* **2017**, *42*, 337–350. [CrossRef]
38. Anders, N.; Valente, J.; Masselink, R.; Keesstra, S. Comparing filtering techniques for removing vegetation from UAV-based photogrammetric point clouds. *Drones* **2019**, *3*, 61. [CrossRef]
39. Zeybek, M.; Şanlıoğlu, İ. Point cloud filtering on UAV based point cloud. *Measurement* **2019**, *133*, 99–111. [CrossRef]
40. Serifoglu Yilmaz, C.; Gungor, O. Comparison of the performances of ground filtering algorithms and DTM generation from a UAV-based point cloud. *Geocarto Int.* **2016**, *33*, 1–16. [CrossRef]
41. Fernández-Lozano, J.; Gutiérrez-Alonso, G. Improving archaeological prospection using localized UAVs assisted photogrammetry: An example from the Roman Gold District of the Eria River Valley (NW Spain). *J. Archaeol. Sci. Rep.* **2016**, *5*, 509–520. [CrossRef]
42. Polat, N.; Uysal, M. DTM generation with UAV based photogrammetric point cloud. *ISPRS* **2017**, *XLII-4/W6*, 77–79. [CrossRef]
43. Zhang, W.; Qi, J.; Wan, P.; Wang, H.; Xie, D.; Wang, X.; Yan, G. An easy-to-use airborne lidar data filtering method based on cloth simulation. *Remote Sens.* **2016**, *8*, 501. [CrossRef]
44. Jensen, J.L.R.; Mathews, A.J. Assessment of image-based point cloud products to generate a bare earth surface and estimate canopy heights in a woodland ecosystem. *Remote Sens.* **2016**, *8*, 50. [CrossRef]
45. Rahmayudi, A.; Rizaldy, A. Comparison of Semi Automatic DTM from Image Matching with DTM from LIDAR. *ISPRS* **2016**, *41*, 373–380.
46. Zhang, Z.; Gerke, M.; Vosselman, G.; Yang, M.Y. Filtering Photogrammetric Point Clouds using Standard LIDAR Filters Towards DTM Generation. *ISPRS Ann. Photogramm. Remote Sens. Spat. Inf. Sci.* **2018**, *4*, 319–326. [CrossRef]
47. Durupt, M.; Flamanc, D.; le Bris, A.; Iovan, C.; Champion, N. Evaluation of the potential of Pleiades system for 3D city models production: Building, vegetation and DTM extraction. In Proceedings of the ISPRS Commission I Symposium, Karlsruhe, Germany, 9–12 October 2016.

48. Skarlatos, D.; Vlachos, M. Vegetation removal from UAV derived DSMS, using combination of RGB and NIR imagery. In Proceedings of the ISPRS Annals of Photogrammetry, Remote Sensing and Spatial Information Sciences, Riva del Garda, Italy, 4–7 June 2018; Volume IV-2, pp. 255–262.
49. Özdemir, E.; Remondino, F.; Golkar, A. Aerial point cloud classification with deep learning and machine learning algorithms. *ISPRS* **2019**, *42*, 843–849. [CrossRef]
50. Becker, C.; Rosinskaya, E.; Häni, N.; d'Angelo, E.; Strecha, C. Classification of aerial photogrammetric 3D point clouds. *Photogramm. Eng. Remote Sens.* **2018**, *84*, 287–295. [CrossRef]
51. Sammartano, G.; Spanò, A. DEM Generation based on UAV Photogrammetry Data in Critical Areas. In Proceedings of the GISTAM, Rome, Italy, 26–27 April 2016; pp. 92–98.
52. Dense Cloud Classification. Available online: https://agisoft.freshdesk.com/support/solutions/articles/31000148866-dense-cloud-classification#Automatic-Classify-Ground-Points%C2%A0 (accessed on 31 October 2021).
53. Hatzopoulos, J.N.; Stefanakis, D.; Georgopoulos, A.; Tapinaki, S.; Pantelis, V.; Liritzis, I. Use of Various Surveying Techonologies to 3D Digital Mapping and Modelling of Cultural Heritage Structures for Maintenance and Restoration Purposes: The Tholos in Delphi, Greece. *Mediterr. Archaeol. Archaeom.* **2017**, *17*, 311–336.
54. Sonnemann, T.F.; Hung, J.U.; Hofman, C.L. 2016 Mapping indigenous settlement topography in the Caribbean using drones. *Remote Sens.* **2016**, *8*, 791. [CrossRef]
55. Salach, A.; Bakuła, K.; Pilarska, M.; Ostrowski, W.; Górski, K.; Kurczyński, Z. 2018 Accuracy assessment of point clouds from LiDAR and dense image matching acquired using the UAV platform for DTM creation. *ISPRS Int. J. Geo-Inf.* **2018**, *7*, 342. [CrossRef]
56. Tutorial (Intermediate Level): Dense Cloud Classification and DTM Generation with Agisoft PhotoScan Pro 1.1. Available online: https://www.agisoft.com/pdf/PS_1.1%20-Tutorial%20(IL)%20-%20Classification%20and%20DTM.pdf (accessed on 31 October 2021).
57. Liritzis, I. Kastrouli fortified settlement (desfina, phokis, Greece): A chronicle of research. *Sci. Cult.* **2021**, *7*, 17–32.
58. Raptopoulos, S. Phokis (Φωκίδα). *Arkheologiko Deltio* **2005**, *60*, 463–464.
59. Raptopoulos, S. 2012 Mycenaean tholos tomb in Desfina of Phokis (Μυκηναϊκός θολωτός τάφος στη Δεσφίνα Φωκίδος). *Arkheologiko Ergo Thessal. Kai Ster. Elladas* **2012**, *3*, 1071–1078.
60. Liritzis, I.; Polymeris, G.S.; Vafiadou, A.; Sideris, A.; Levy, T.E. Luminescence dating of stone wall, tomb and ceramics of Kastrouli (Phokis, Greece) Late Helladic settlement: Case study. *J. Cult. Herit.* **2019**, *35*, 76–85. [CrossRef]
61. Sideris, A.; Liritzis, I.; Liss, B.; Howland, M.D.; Levy, T.E. At-Risk Cultural Heritage: New Excavations and Finds from the Mycenaean Site of Kastrouli, Phokis, Greece. *Mediterr. Archaeol. Archaeom.* **2017**, *17*, 271–285.
62. Liritzis, I.; Sideris, A. The Mycenaean Site of Kastrouli, Phokis, Greece: Second Excavation Season, July 2017. *Mediterr. Archaeol. Archaeom.* **2018**, *18*, 209–224.
63. Baziotis, I.; Xydous, S.; Manimanaki, S.; Liritzis, I. An integrated method for ceramic characterization: A case study from the newly excavated Kastrouli site (Late Helladic). *J. Cult. Herit.* **2020**, *42*, 274–279. [CrossRef]
64. Chovalopoulou, M.-E.; Bertsatos, A.; Manolis, S.K. Identification of skeletal remains from a Mycenaean burial in Kastrouli-Desfina, Greece. *Mediterr. Archaeol. Archaeom.* **2017**, *17*, 265–269.
65. Koh, A.J.; Birney, K.J.; Roy, I.M.; Liritzis, I. 2020 The Mycenaean citadel and environs of Desfina-Kastrouli: A transdisciplinary approach to southern Phokis. *Mycen. Archaeol. Archaeom.* **2020**, *20*, 47–73.
66. Kontopoulos, I.; Penkman, K.; Liritzis, I.; Collins, M.J. 2019 Bone diagenesis in a Mycenaean secondary burial (Kastrouli, Greece). *Archaeol. Anthropol. Sci.* **2019**, *11*, 5213–5230. [CrossRef]
67. Liritzis, I.; Xanthopoulou, V.; Palamara, E.; Papageorgiou, I.; Iliopoulos, I.; Zacharias, N.; Vafiadou, A.; Karydas, A.G. Characterization and provenance of ceramic artifacts and local clays from Late Mycenaean Kastrouli (Greece) by means of p-XRF screening and statistical analysis. *J. Cult. Herit.* **2020**, *46*, 61–81. [CrossRef]
68. Levy, T.E.; Sideris, T.; Howland, M.; Liss, B.; Tsokas, G.; Stambolidis, A.; Fikos, E.; Vargemezis, G.; Tsourlos, P.; Georgopoulos, A.; et al. At-risk world heritage, cyber, and marine archaeology: The Kastrouli–Antikyra Bay land and sea project, Phokis, Greece. In *Cyber-Archaeology and Grand Narratives*; Jones, I.W.N., Levy, T.E., Eds.; Springer: Cham, Switzerland, 2018; pp. 143–234.
69. Doneus, M.; Verhoeven, G.; Fera, M.; Briese, C.; Kucera, M.; Neubauer, W. From deposit to point cloud – a study of low-cost computer vision approaches for the straightforward documentation of archaeological excavations. *Geoinf. FCE CTU* **2011**, *6*, 81–88. [CrossRef]

quaternary

MDPI

Article

Characterization of the Obsidian Used in the Chipped Stone Industry in Kendale Hecala

Üftade Muşkara [1,*] and Ayşin Konak [2]

[1] Department of Conservation and Restoration of Cultural Properties, Faculty of Fine Arts, Kocaeli University, İzmit 41380, Turkey

[2] Department of Archaeology, Faculty of Science and Letters, Kocaeli University, İzmit 41380, Turkey; aysink@kocaeli.edu.tr

* Correspondence: uftade.muskara@kocaeli.edu.tr

Abstract: Kendale Hecala is located on the Ambar River in the Upper Tigris Basin, province of Diyarbakır in Southeast Anatolia. Various raw materials, including obsidian, radiolarite, chert, jasper, chalcedony, and quartzite, were used in the lithic industry. Obsidian artefacts constitute an average of 64% of the chipped stone assemblage. Technological analysis reveals that obsidian was brought to the settlement as nodules and chipped into various tools at the settlement. Understanding the operational sequence of the lithic industry, chaîne opératoire, including the distribution of raw material from source to site, is important to demonstrate the socio-cultural organization of the settlement in Southeastern Anatolia during the Ubaid period. In order to identify source varieties, the obsidian artefacts uncovered from Ubaid layers of Kendale Hecala were analyzed by macro-observations, and the characterization of archaeological samples was performed using a handheld XRF. Multivariate analysis of the data indicates the use of obsidian from different resources at the settlement, including Nemrut Dağ, Bingöl B, and Group 3d.

Keywords: Upper Tigris; obsidian sourcing; pXRF; Nemrut Dağ; Group 3d; Ubaid

check for updates

Citation: Muşkara, Ü.; Konak, A. Characterization of the Obsidian Used in the Chipped Stone Industry in Kendale Hecala. *Quaternary* **2022**, 5, 3. https://doi.org/10.3390/quat5010003

Academic Editor: Ioannis Liritzis

Received: 20 November 2021
Accepted: 4 January 2022
Published: 7 January 2022

1. Introduction

Ubaid, which originally referred to a pottery style, characterizes the material culture that shared similar tools, architectural forms, and practices for a specific period. The distribution of the Ubaid pottery and consequently the extension of culture depending on the mutual relation between different societies approximately reached an area from the Strait of Hormuz to Northern Mesopotamia and Upper Tigris, and the Mediterranean shores [1]. The Ubaid culture spread from southern Mesopotamia to the north across Mesopotamia in the later Ubaid 3 and 4 phases, which correspond to the Northern Ubaid period, dating from about 5300 to 4500 BC [2]. The interaction within the regions is evident, especially from pottery tradition, while the obsidian circulation could also reflect the communication pattern.

The characterization studies elucidate the variety of obsidian sources reached to a specific site and obsidian procurement strategies. Determination of compositional groups will also reveal the preference for specific raw material properties for particular obsidian artefact production. The lithic chaîne opératoire was defined by Lemonnier in four stages, including extraction, reduction, production, and transformation [3,4]. Identifying the source could explain the extraction phase comprising the selection of raw material and the transportation system of obsidian [5]. On the other hand, the sequences of technical processes corresponding to the places of activity for core preparation, flake, and blade production from the core and tool kits received less recognition. The inadequate sampling strategy in provenance studies could not provide relating information.

The recent studies focused on a more systematic sampling strategy that spatially and chronologically represents the site, allowing for the detection of any possible shift in the obsidian procurement over time. This strategy also provides insights for reduction, production, and tool finishing processes; therefore, we could obtain a more extensive perspective for (1) the lithic industry at the settlement, (2) the social and cultural structure of the society, (3) the role of the settlement in obsidian distribution, (4) the interaction between societies through trade, and (5) the spread in skill and knowledge.

In Anatolia, two major groups of geological sources are defined as Central Anatolia, including Acıgöl, Göllüdağ, Nenezidağ, Hasandağ, and Eastern Anatolia, comprising the Upper Euphrates, Lake Van, and Northeastern Anatolia regions. The previous provenance studies indicated that Bingöl B obsidians and peralkaline obsidians from Bingöl A and Nemrut Dağ were the primary sources in Southeastern Anatolia during the PPNA and PPNB (Hasankeyf, Çayönü, Hallan Çemi, and Gusir Höyük). More thorough studies that examine the geochemical data to describe obsidian selection systems and circulation networks for a period from PPN to Ubaid periods in the area detected changes in the proportions of the various geological sources over time [6–8]. Although the change could be related to the sample size, the shift in obsidian procurement and diversity of obsidian sources may suggest a modification in the taste for the physical appearances of obsidian materials, such as color, in the technology or exchange system. The effects of a shift in trade networks and centralization are reflected in the place of activity, i.e., on-site domestic production or off-site standardized blade industry [5].

In the Upper Tigris region, Kenan Tepe obsidian assemblage from Ubaid levels, for instance, suggests a difference in the production stage of obsidian tools based on their raw material [9]. Due to the proximity, the obsidian raw material diversity and *chaîne opératoire* at Kenan Tepe could bear some similarities to Kendale Hecala. However, we also acknowledge that the obsidian procurement could show regional varieties. Obsidian sources arrived at Kenan Tepe mainly include Bingöl B, Bingöl A, Nemrut Dağ, Muş, and the compositional group defined as Group 3d by Renfrew et al. [10]. The elemental composition and the physical appearance of Group 3d obsidian at the site are well described in a recent study by Campbell et al. [11]. Due to a similar proportion of Group 3d artefacts to Bingöl and Nemrut Dağ sources, the authors suggest a region between Bingöl and Nemrut Dağ for the location of Group 3d. The obsidian raw material diversity at Tell Zeidan in Northern Mesopotamia was studied by Khaldi et al. [12]. The results suggested that obsidian obtained from the Sıcaksu sub-source of Nemrut Dağ, Bingöl B, and Bingöl A were available at the site for Ubaid period contexts.

The obsidian source characterization study at Gre Fılla was conducted on fifty samples comprising various debitage and tools in 2021. The assigned origins of obsidian samples revealed that Gre Fılla was located in the distribution area of Bingöl B, Bingöl A (Solhan), and Nemrut Dağ sources [13]. While obsidian from Bingöl B constituted the majority, we could follow the temporal change in the proportions of different sources within the layers of PPNB. The archaeological evidence and C^{14} dating indicate an occupation at Gre Fılla dating from the PPNA to the late PPNB (9300–8200 cal. BC) [14]. On the other hand, the C^{14} dating at Kendale Hecala suggests a cultural sequence between 6700 and 5500 BC, including the Pottery Neolithic (PN) and the Ubaid periods. Gre Fılla and Kendale Hecala sites by the Ambar Çayı, one of the tributaries of the Upper Tigris basin, are located only 800 m. apart (Figure 1). Therefore, characterization studies at two mounds will provide valuable data to understand the obsidian procurement and exchange patterns in the region during the Neolithic and Chalcolithic periods. This study presents the elemental compositions of twenty-one obsidian artefacts from Kendale Hecala obtained from Ubaid layers using a handheld, portable X-ray fluorescence (pXRF) spectrometer. We aimed to establish the primary obsidian sources that arrived at Kendale Hecala during the Ubaid period to assign the site in the obsidian distribution system in the Upper Tigris region and define whether Group 3d obsidian was available at the site.

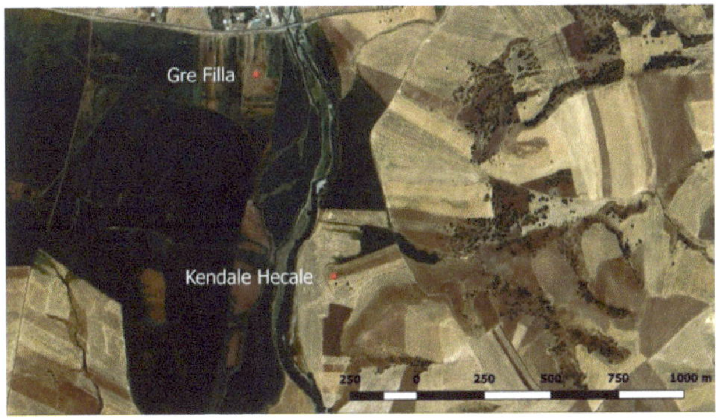

Figure 1. The aerial photography showing the locations of Gre Fılla and Kendale Hecala in the Ambar Çayı valley (Google Earth image).

2. The Archaeological Site

Ambar Dam is proposed to be constructed in the northern valley of Ambar Çayı, located in the upper Tigris valley. The salvage excavations in the region, including three mounds, Ambar Höyük, Gre Fılla, and Kendale Hecala, were initiated in 2018 by the Diyarbakır Archaeological Museum under the supervision of Prof. Dr. A. Tuba Ökse. The archaeological evidence indicates that Gre Fılla was settled from PPNA to PPNB. The site was then abandoned and used as a cemetery during the Late Antiquity and early centuries of the Middle Age. Kendale Hecala, situated at 800 m. south of Gre Fılla, is approximately 3–4 m in height, and about 0.65 ha in extent (Figures 1 and 2). According to C^{14} dating and the material remains, including pottery, lithic, and architecture, the settlement had been used in three periods spanning 6660–4540 cal. BC (Period KH III-II) and AD 500–1500 (Period KH I) (Table 1) [14,15]. Following the Early Chalcolithic settlement, the site was abandoned and reoccupied after a period of ca. 5300 years during the medieval period. The excavation at the site was sustained at ten trenches at the northern operation and four at the southern operation in the 2019 season.

Figure 2. The aerial photography of the trenches at Kendale Hecala (excavation archive).

Table 1. The cultural sequences at Kendale Hecala.

Date (BC)	Period	Pottery	Lithics	Architecture
6660–5500	KH III-PN [1]			
5500–4550	KH II-Ubaid 3–4			
500–1500 [2]	KH I-Middle Age			

[1] Pottery Neolithic. [2] AD.

3. The Lithic Industry at Kendale Hecala

The various raw materials of lithic materials consisted mainly of obsidian (65%) and other knappable siliceous rocks (35%), commonly known as flint, chert, and radiolite. Macro observations presented that the color of the obsidian materials ranges from opaque black, red-dotted black-gray, translucent green, and gray. The debitage constitutes 60% of lithic assemblage, and tools form 40%, while debris is also present. The percentage of obsidian cores (26%) is considerably lower than flint cores (74%). Bladelets, blades, and flakes comprise the major portion of the obsidian assemblage, while the exhausted obsidian cores were mainly used to produce blades and flakes based on the scares on the debitage planes. Cores with pyramidal shapes (Figure 3) or polyhedral with opposite platforms and shapeless core fragments are also defined. The lesser amount of obsidian cores and the reduction strategies (exhausted cores) suggest that obsidian was less readily available at the settlement, contrary to local knappable siliceous materials. The reason for the preference for obsidian is not apparent yet; however, it could be derived from conventional practices in the society, social aspects of the organization, or transferred skills with technology. The nodules and flakes exhibiting the primary cortex on the surfaces, along with the amount of debris, point to on-site production activity.

Figure 3. Unipolar pyramidal blade core sample.

The length of the blades is less than 10 cm. The higher percentage of the obsidian blade and flakes, as opposed to the percentage of flint blades and flakes, may suggest that obsidian was the choice material for the production of blades and flakes. Typological classification of obsidian tools found at Kendale includes retouched blades and flakes, scrapers, splintered tools, scrapers, bores, denticulated tools, burins, notched tools, backed blades, retouched blades, and microliths (Figures 4 and 5). Furthermore, obsidian materials found at the site reflect an ad hoc flake and blade industry.

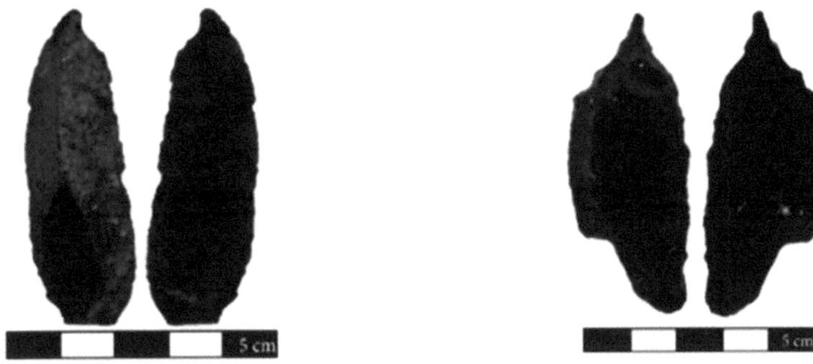

Figure 4. Retouched blade and bore samples from Kendale Hecala.

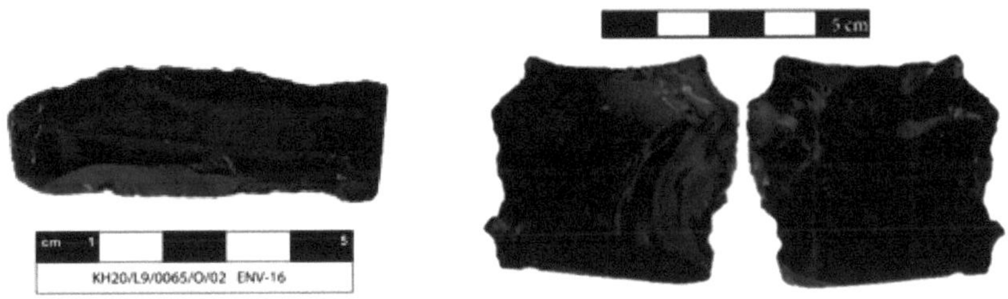

Figure 5. A retouched blade and splintered tools from Kendale Hecala.

4. Materials and Methods

4.1. Sampling

The obsidian samples were collected among 2018–2019 finds for element analysis to present the obsidian geological source variety at Kendale Hecala. All twenty-one samples are assigned to the Ubaid period and include flakes, cores, and tools such as retouched materials and scrapers (Table 2). We selected the samples to represent spatial differences at the site and followed the protocols for dimension and thickness recommended for XRF measurements [16,17]. Obsidian artefacts of various colors observed through macro analysis were selected in order to ensure our sampling strategy would reflect the raw material variety at the site. The sample morphology is an important parameter to obtain accurate and precise data when using a handheld XRF instrument. The surface roughness of the samples is one of the major limitations of archaeological studies. Irregular surfaces of archaeological samples may create an air gap between the instrument probe, resulting in a significant change in the analytical signal [18,19]. Therefore, we excluded obsidian artefacts which do not meet this requirement.

Among the sources in Eastern Anatolia, only Nemrut Dağ and Bingöl A sources were classified as peralkaline obsidians (Figure 6). Due to the proximity to Kendale Hecala, geological samples were collected from Bingöl A (Solhan) as a possible source of peralkaline obsidian artefacts found at the site. We used the geochemical data obtained by different studies using various analytical methods to compare with our dataset and to relate obsidian assemblage of Kendale Hecala to the calcalkaline obsidian sources of Bingöl B, Muş, and Süphan Dağ.

Table 2. Contextual and technological information for the analyzed Kendale Hecala II artefacts dated to cal. 5600–5000 B.C.

Sample Nr	Stratum	Trench	Color	Assemblage
KH-O-001	KH II	K8	Black	Retouched
KH-O-002	KH II	L8	Gray	Retouched
KH-O-003	KH II	L8	Black	Retouched
KH-O-004	KH II	L8	Gray	Retouched
KH-O-005	KH II	L8	Black	Retouched
KH-O-006	KH II	K8	Green	Retouched
KH-O-007	KH II	K8	Green	Flake
KH-O-008	KH II	K8	Green	Core
KH-O-009	KH II	K8	Green	Splitter
KH-O-010	KH II	K8	Green	Retouched
KH-O-011	KH II	K8	Green	Retouched
KH-O-012	KH II	K8	Green	Flake
KH-O-013	KH II	K8	Smokey gray	Core
KH-O-014	KH II	K8	Smokey gray	Retouched
KH-O-015	KH II	K8	Green	Core
KH-O-016	KH II	K8	Green	Flake
KH-O-017	KH II	L8	Green-gray	Flake
KH-O-018	KH II	L8	Green-gray	Retouched
KH-O-019	KH II	L8	Green-gray	Splitter
KH-O-020	KH II	L8	Black	Retouched
KH-O-021	KH II	L8	Green-gray	Retouched

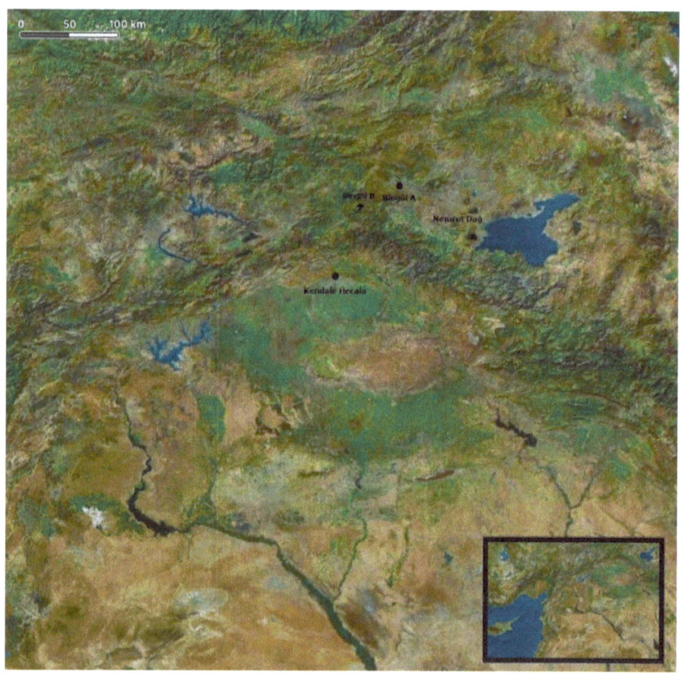

Figure 6. Map showing the locations of Kendale Hecala and obsidian sources at Bingöl A, Bingöl B and Nemrut Dağ (Google Earth image).

4.2. Analytical Procedure

Chemical analysis of the obsidian samples was conducted with a handheld EDXRF; Hitachi X-Met8000 Expert. The apparatus was equipped with a silicon-drift detector (SDD) and an excitation source of X-ray tube Rh target. We use pXRF for the determination of the major elements Fe, K, Zr, Ca, and Ti, as well as the trace elements Ba, Mn, Rb, Zn, Sr, Pb, and As.

Archaeological and geological samples were cleaned with distilled water in an ultrasonic bath for 15 min. The instrument was set to 40 kV for 90 s as an optimal time. We also give at least 5 min intervals between each measurement to allow the instrument to cool and stabilize as suggested by Steiner et al. [20]. The measurements were performed according to the calibration method mentioned in Muşkara and Konak [13]. The matrixed matching strategy to check the accuracy of the calibration method requires using international standard materials (SRM or certified reference material CRM), which are generally in powdered form to ensure homogeneity. The best way to ensure appropriate validation is using the sample preparation method in the same way of reference material [21], since the comparison of the analytical signals obtained by different sample preparation methods may not always be proper (for further discussion on the issue, see [19,22]. We used East Göllüdağ obsidian samples as secondary standards to confirm the analyzer accuracy and stability before and during the analysis of the samples to ensure the validity of the calibration method for intra-laboratory comparison.

The flattest spots on the surface were selected for measurements to minimize the air gap between the sample and probe. The analysis spot size of the instrument was 5 mm, and the positioning of the analyzer was controlled by an integrated camera. All samples were analyzed in duplicate and some in triplicate to ensure the reliability of the measurements.

4.3. Data Analysis

The accuracy and reliability of obsidian provenance studies or source assignment of obsidian artefacts by pXRF have been discussed over the last decade. Various authors suggested that the data obtained by pXRF measurements can indeed be used to discriminate various geological sources. The obsidian procurement patterns at many sites in Anatolia and Near East have been usefully defined using pXRF and other portable or non-destructive devices such as SEM-EDX (scanning electron microscope-energy dispersive X-ray) PIXE (proton-induced X-ray emission), EMPA (electron micro probe analyzer). However, the concerns on the validity of pXRF measurements have not been fully addressed. The validity of pXRF has been described by Nazaroff et al. [23], referring to its ability to distinguish different geological sources, although a systematic error introduced by pXRF was observed when the results were compared with the data obtained by a lab-based XRF instrument. Two sample *t*-tests were applied between the datasets to check the accuracy of the results acquired by pXRF. K-means cluster analysis, then, was used to calculate the analytical capacity of the pXRF instrument to assign the archaeological sample to the specific source. On the other hand, a similar approach was followed by Frahm to test the validity of pXRF, although he did not apply the appropriate calibration producers, even on the artefacts that did not have the desired morphology for the measurements [24]. He concluded that "Even with systematic error from the (lack of) calibration and random error due to problematic artefacts, the compositions of these Near Eastern obsidians, as measured using HHpXRF, have greater inter-source than intra-source variations" [24] (p. 1091), expect for the peralkaline sources Nemrut Dağ and Bingöl A. However, of course, the validation of pXRF using reference materials is still required to establish accuracy and precision.

While multiple bivariate scatterplots are generally used for obsidian-source assignment, principal component analysis (PCA) has been suggested to identify compositional groups in a dataset [25–31]. In this study, we applied PCA to characterize archaeological samples and define compositional groups. The data of archaeological samples and geological samples from Bingöl A were transformed to base-10 logarithms before PCA to reduce the differences in magnitudes of the concentrations recorded and address the skewness of the original data [32,33]. In contrast, other data-transformation methods include Z score standardization [31], multivariate Box–Cox transformation [28], and a minimum/maximum normalization [34]. PCA was performed using a correlation matrix with no rotation axis. A ratio plot of Fe/Mn vs. Rb/Zr and a ternary graph of Zr, Rb, and Zn are also used to distinguish peralkaline and calc-alkaline obsidian artefacts. In this study, multivariate statistical analysis is carried out with the help of SPSS software 28.0.1.0 (142) (IBM 2021).

5. Results

The elemental composition of geological samples originated from Bingöl A, and East Göllüdağ were reported in Muşkara and Konak [13]. Since the Ba and Sr values of Bingöl A samples acquired by pXRF was below the detection limits of the instrument, the results for the same elements originated from East Göllü Dağ obsidian samples provides insight for the lower limits of the linear dynamic range of our instrument for Ba and Sr. The element compositions of Kendale Hecala obsidian artefacts are presented in Table 3.

Table 3. Element composition of artefacts from Kendale Hecala obtained by pXRF.

Sample	Fe	Ca	Zr	Ti	Ba	Mn	Rb	Zn	Sr	Pb	As	Source
KH-O-001	12,132	4666	268	1168	317	305	192	36	38	25	5	Bingöl B
KH-O-002	12,860	7271	286	1288	318	296	208	33	39	29	3	Bingöl B
KH-O-003	15,396	7204	291	1186	419	310	203	36	38	25	6	Bingöl B
KH-O-004	12,537	8398	281	1180	373	275	194	39	37	29	1	Bingöl B
KH-O-005	12,673	5371	295	1228	431	323	205	39	38	29	3	Bingöl B
KH-O-006	28,952	2145	1148	1173	n.d. [1]	541	216	168	n.d.	34	23	Nemrut D.
KH-O-007	26,570	1791	1051	1067	n.d.	486	194	140	n.d.	31	20	Nemrut D.
KH-O-008	28,504	2342	1148	916	n.d.	500	210	167	n.d.	36	18	Nemrut D.
KH-O-009	24,063	2170	1022	899	n.d.	504	202	145	n.d.	39	22	Nemrut D.
KH-O-010	27,422	1922	1078	1149	n.d.	490	196	162	n.d.	33	17	Nemrut D.
KH-O-011	27,345	1859	892	1157	n.d.	584	194	136	n.d.	33	15	Nemrut D.
KH-O-012	24,812	1871	1042	1008	n.d.	477	206	156	n.d.	38	25	Nemrut D.
KH-O-013	10,206	5486	170	368	n.d.	331	394	74	n.d.	63	15	Group 3d
KH-O-014	11,825	3571	189	449	n.d.	430	448	84	n.d.	67	14	Group 3d
KH-O-015	11,553	4442	273	1079	356	284	198	33	34	28	4	Bingöl B
KH-O-016	30,038	2532	1260	1230	n.d.	554	229	170	n.d.	40	22	Nemrut D.
KH-O-017	18,320	2379	1082	759	n.d.	392	201	130	n.d.	31	14	Nemrut D.
KH-O-018	29,772	2807	1019	1294	n.d.	631	223	151	n.d.	42	16	Nemrut D.
KH-O-019	18,665	2587	1080	794	n.d.	408	202	142	n.d.	33	14	Nemrut D.
KH-O-020	12,158	4280	196	467	317	383	476	89	11	70	11	Group 3d
KH-O-021	33,332	2215	1098	1212	318	734	235	163	38	41	20	Nemrut D.

[1] n.d. not detected.

The log-transformed data for Fe, Zr, Ca, Ti, Mn, Rb, Zn, Pb, and As of archaeological and geological samples were used for PCA in order to distinguish various compositional groups among the Kendale artefacts. Using the scree plot and eigenvalues, two components were found significant. Examination of biplots of element vectors and the data of obsidian samples against the first and second principal components indicates four compositional groups in our dataset (Figures 7 and 8). Group 1, representing Bingöl A, is characterized by higher amounts of As and Zn. None of the archaeological samples are assigned to this group. On the other hand, Ba and Sr values found below the detection limits of the instrument and relatively higher contents of Zr and Fe identify 12 artefacts designated to Group 2 that are derived from peralkaline obsidian of Nemrut Dağ.

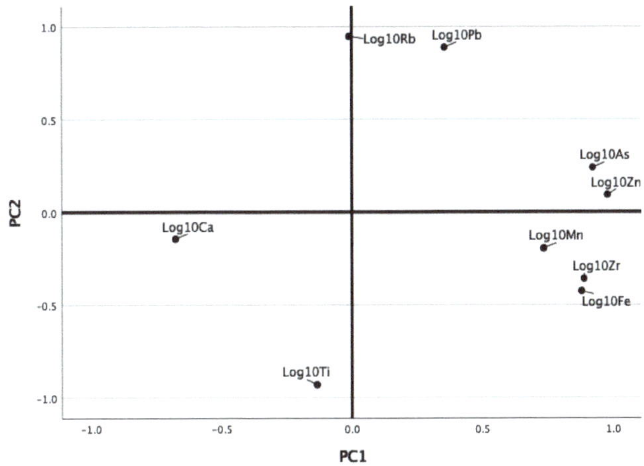

Figure 7. Bivariate plot of loading PC1 and PC2.

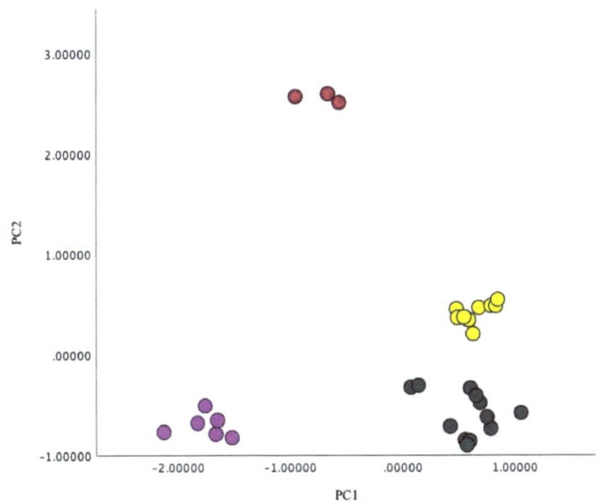

Figure 8. Bivariate plot showing obsidian artefacts and Bingöl A geological samples on PC1 and PC2 for the four compositional groups (yellow circles coding Bingöl A).

The second principal component distinctively separates Group 3, including three artefacts. The discriminating positive loading of Pb and Rb elements is interesting since it suggests that these artefacts were produced from Group 3d obsidian. In a Rb vs. Pb scatterplot, which was recommended for assigning Group 3d obsidian [11], three artefacts (KH-O-013, 014, and 020) exhibit significantly higher values of Rb and Pb (Figure 9). The Pb content with a mean of 66 $\pm$ 4.54 ppm (mg/kg) obtained by Campbell et al. [11] for the artefacts defined as Group 3d is consistent with our data. The mean concentration of Rb is 439 $\pm$ 42 ppm corresponds to values that have previously been published [11,35,36], while the Rb content of KH-O-013 is relatively lower. However, it is compatible with the data of four artefacts from Yarım Tepe II obtained by Francaviglia using XRF [37] (see also [35]), and one sample from Eridu reported by Renfrew [10].

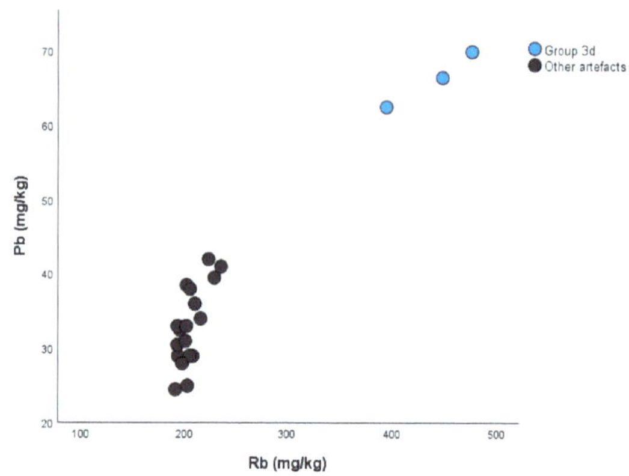

Figure 9. Rb vs. Pb scatterplot showing obsidian artefacts and Bingöl A geological samples.

Group 4, consisting of six artefacts, is separated for its higher Ca and Ti values in Figure 6. In the Ba vs. Zr scatter plot, our data for Group 4 are compared to published values in order to define the geological source of the samples (Figure 10). Group 4 artefacts have a Ba content ranging from 317 to 431 ppm, while their Zr content is between 268 and 291 ppm. Therefore, the element composition of this group corresponds to Bingöl B source.

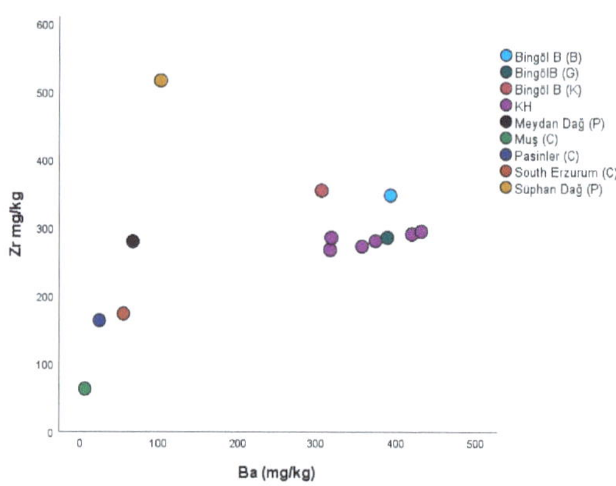

Figure 10. Ba vs. Zr scatterplot showing Kendale Hecala obsidian artefacts (KH-O-01, 02, 03, 04, 05 and 15) and obsidian sources of Eastern Anatolia: Bingöl B (G) [38].; Bingöl B (B) [39]; Bingöl B (K) [40], Süphan Dağ (P) [41].

The Fe/Mn vs. Rb/Zr scatterplot of the samples discriminates three geochemical groups (Figure 11). One group closer to Bingöl A reference samples represents the peralkaline obsidians at Kendale Hecala; however, we understand that there is an overlapping between Nemrut Dağ obsidian, especially between Sıcaksu and Bingöl A [13]. On the contrary, the scatterplot of principal component analysis is more efficient to discriminate between two peralkaline sources. The artefacts assigned to Bingöl B are separated from peralkaline obsidian, while three artefacts identified as Group 3d appear on the upper left corner of the plot, is notably differentiated from the other artefacts. This pattern is related to a significantly higher concentration of Rb.

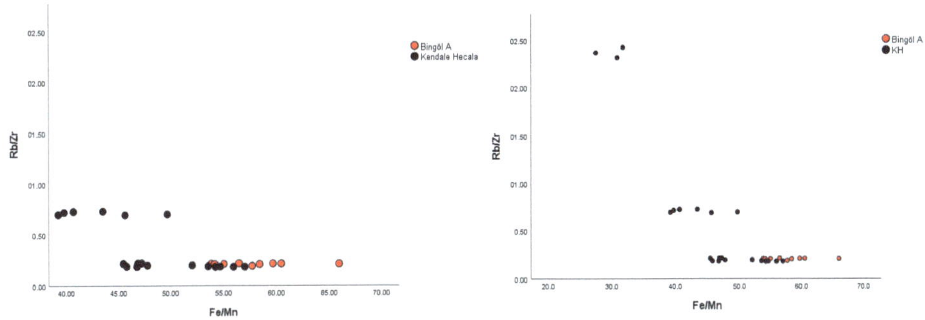

Figure 11. Bivariate plot of Fe/Mn vs. Rb/Zr showing obsidian artefacts and Bingöl A geological samples.

The compositional groups defined at Kendale Hecala are also distinguished by plotting the values of Zr, Rb, and Zn in a ternary graph (Figure 12). Most obsidian artefacts show consistencies with geological samples collected from Solhan, representing the peralkaline obsidian source. The higher Rb concentration separates three artefacts (KH-O-013, 014, and 020) and the lower values of Zn assign six artefacts (KH-O-001-005, and 015) to calcalkaline obsidians.

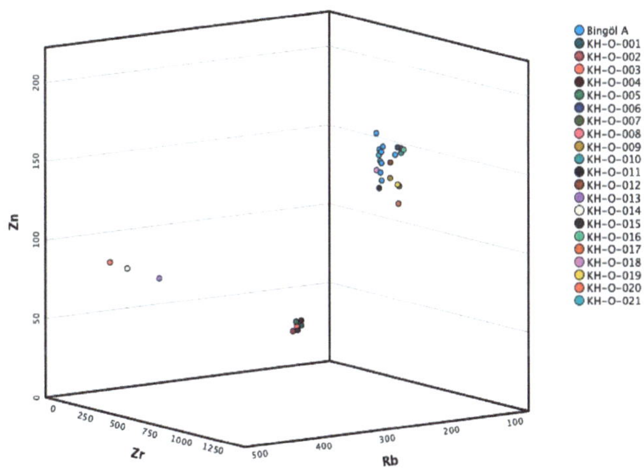

Figure 12. Ternary plot of Zr, Rb and Zn showing obsidian artefacts and Bingöl A geological samples.

6. Discussion

The geological source of 21 obsidian artefacts, including cores, flakes, retouched materials, and scrapers, were determined as Bingöl B, Nemrut Dağ, and Group 3d. Since our geological sampling represents only Bingöl A source obtained from Solhan, we applied a principal component analysis for the discrimination of different compositional groups. The total variance explained by the first and second components is 82%. The first component is positively correlated with Zn and As, while also less correlated with other components (Table 4). The second component is positively correlated with Rb and negatively correlated with Ti.

Table 4. The component matrix of PCA.

	1	2	3
Log10Fe	0.878	−0.426	0.165
Log10Zr	0.887	−0.359	−0.188
Log10Ca	−0.675	−0.145	0.644
Log10Ti	−0.134	−0.931	0.251
Log10Mn	0.733	−0.196	0.596
Log10Rb	−0.011	0.948	0.269
Log10Zn	0.979	0.093	0.000
Log10Pb	0.351	0.888	0.223
Log10As	0.921	0.240	−0.025

The PC1 separates Group 1, Group 2 (peralkaline), and Group 4 (calc-alkaline) obsidians, and PC2 discerns Group 3. On the other hand, discriminating Bingöl A and Nemrut Dağ compositional group has been difficult, especially when the data were acquired by pXRF. Various binary diagrams, including La vs. Nb and Zr vs. Nb, were applied by Chataigner [42]. Although Bingöl A and Nemrut Dağ sources are represented homogeneously, she suggests that flow dating could provide better results when archaeological samples are not allocated to either source precisely. Other quantitative analysis methods of the data have been applied, such as ratio plot of Nb/Pb vs. Y/Nb [30], Fe vs. Zr [12], or Al_2O_3 vs. FeO(T) vs. Zr [43]. Characterization studies have also identified sub-sources in Nemrut Dağ [29,42–44]. Among the defined outcrops, Sıcaksu obsidian has the most similar geochemical compositional to Böngöl A, while it was reported as the most available source at settlements in Northern Mesopotamia from the Late Neolithic to Chalcolithic period [12,29]. However, Frahm [43] argued that other than Sıcaksu, three other Nemrut sub-sources occurred at Körtik Tepe, Domuztepe, and Tell Mozan. Our data for Gre Fılla artefacts also suggest that obsidian raw material from various outcrops of Nemrut Dağ (especially Sıcaksu and Kayacık) arrived at the settlement [13].

The scatterplot of PC1 and PC2 in this study identifies peralkaline obsidians; however, archaeological samples differentiate from Bingöl A geological samples due to similar behavior of Zn-As (positive loading for Bingöl A) and Fe-Zr elements (positive loading for archaeological samples). Due to the clear separation between the groups, Group 2 artefacts are assigned to Nemrut Dağ, although no overlapping regions may have resulted from the number of archaeological samples analyzed. On the other hand, the representation of the sample on the bivariate plot of Fe/Mn vs. Rb/Zr supports the allocation of obsidian artefacts to Nemrut Dağ. Future studies will include obsidian samples from Sıcaksu, Kayacık, and other outcrops of Nemrut Dağ in order to test the efficiency of the variable set used in the principal component analysis. While Fe, Zr, and Zn have been used for identifying peralkaline obsidians, As appears to be among the set of discriminating variables.

The Group 3d obsidian compositional group was initially defined by Renfrew et al. [10]. Various obsidian sources, including Central Anatolia, Eastern Anatolia, and Armenia, were analyzed in this study using OES (optical emission spectroscopy) instrument. After appointing four distinct compositional groups, they specified the chemical characteristics of Group 3d as: "Two specimens that fall in group 3c on the barium-zirconium graph (nos. 171, Ras Shamra, and 235, Dahran) have been distinguished as Group 3d, on the basis of an exceptionally high content of rubidium and lithium. It is not yet clear whether these are from an otherwise undocumented source, possibly in Armenia, or are anomalous analyses from known sources" [10] (p. 33). The most recent study on Group 3d is published by Campbell et al. [11]. Although the main focus was the artefacts recovered from Kenan Tepe, they provide a comprehensive description of the physical and geochemical properties of Group 3d obsidian. They suggest a scatter plot of a Rb vs. Pb is the most accurate method for identification of this type of raw material since the concentrations of these elements are significantly higher. The previous studies and Kenan Tepe data provide consistent result for the elements that define for Group 3d. The mean concentration of Rb is reported to be higher than 450 ppm, and the Pb content is around 60 ppm. Other than Rb and Pb, the authors mentioned that the concentrations of Li, B, and Cs could be distinctive for this group. In our dataset, we assigned three artefacts, KH-O-013 (core), KH-O-014 (retouched), and KH-O-020 (retouched), to Group 3d.

Kenan Tepe is the only settlement so far that the large-scale use of this compositional group has been detected [11]. Due to the intensive use of Group 3d and on-site production, the authors proposed a possible location of Group 3d as more accessible to Kenan Tepe, probably between Nemrut Dağ and Bingöl area. When we compare the proportion of Group 3d artefact among the other sources, the ratio of Group 3d artefacts is 14% (Figure 13). On the other hand, obsidian assemblage at the site indicates on-site production, regardless of obsidian type. However, the number of obsidian artefacts was limited in our study since the sampling was undertaken at the earlier stages of the excavation. Future studies will produce data to better understand the obsidian chaîne opératoire at Kendale Hecala during the Ubaid period.

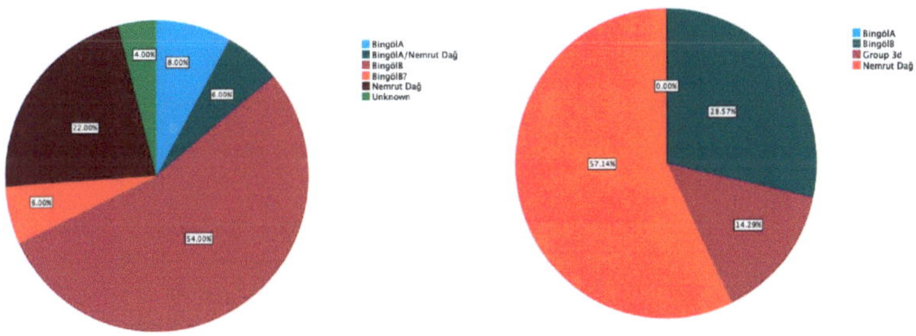

Figure 13. Proportion of main sources at Gre Fılla (PPNB) and Kendale Hecala II (Ubaid period).

Other than Group 3d, the primary sources at Kendale Hecala during the Ubaid period were Bingöl B and Nemrut Dağ. We follow a dramatic shift in obsidian diversity when the available sources at Gre Fılla during PPNB are compared to Kendale Hecala during the Ubaid period. On the other hand, the relative proportion of each source also varied between the two settlements.

7. Conclusions

The lithic industry at Kendale Hecala is significantly dominated by obsidian during the Ubaid period. The proportion of obsidian in the lithic industry is almost twice as much as other knappable siliceous rocks. The lithic analysis reveals obsidian nodules, primary flakes, and production debris, indicating a local work organization, including the reduction and production stages of chaîne opératoire. The three core samples in our study assigned to three different sources may also indicate that the debitage was produced on site, regardless of the raw material type. Therefore, we assume that Kendale Hecala was not connected to a central lithic distribution system. The obsidian as a raw material probably arrived to the site by the local groups (for instance, by transhumance routes) from relatively short distances. Although an exchange and transportation system from a settlement closer to the obsidian source to the vicinity can be expected [45], obsidian artefacts at Kendale Hecala suggest direct contact with raw material prior to the initial stages of obsidian preparation. Except for Group 3d, the distances of the primary sources to Kendale Hecala are approximately 150 km south from Bingöl B and 250 km southwest from Nemrut Dağ [46]. The production activity is in accordance with Kenan Tepe, where the on-site reduction is apparent, although obsidian material constitutes around 18 to 20% of the lithic industry at Kenan Tepe. On the other hand, the long-distance transportation system of obsidian from eastern Anatolia to northwestern Iran, between Lake Urmia and the Lake Van regions, during the Late Neolithic and Chalcolithic periods is also evident from obsidian-sourcing studies [47].

The previous study at Gre Fılla demonstrates a diachronic change in obsidian use in the region. Bingöl A has so far not been detected at Kendale Hecala; therefore, we may assume that Nemrut Dağ was the main supplier of peralkaline obsidian. The relative dominance of Bingöl B in PPNB at Gre Fılla disappeared at Kendale Hecala, as illustrated in Figure 11. Due to the proximity of the two sites, the continuous chronological sequences, and material culture, it is safe to argue that technological knowledge and cultural information were shared between Gre Fılla (PPNA-PNNB periods) and Kendale Hecala (PN-Ubaid periods). However, future studies will provide more data for identifying the dynamics that affected obsidian procurement at Ambar Çay Valley in the Upper Tigris region between the PPNA and Ubaid periods. On the other hand, the shift in dominant sources and changes in the diversity of raw material between Gre Fılla (PPNB) and Kendale Hecala (Ubaid period) suggest that the distance of the source from the settlement was not the primary factor that influenced the obsidian consumption strategies.

Author Contributions: Conceptualization, Ü.M. and A.K.; methodology, Ü.M. and A.K.; software, Ü.M.; validation, Ü.M.; formal analysis, Ü.M. and A.K.; investigation, Ü.M. and A.K.; resources, Ü.M. and A.K.; data curation, Ü.M.; writing—original draft preparation, Ü.M.; writing—review and editing, Ü.M. and A.K.; visualization, Ü.M. and A.K. All authors have read and agreed to the published version of the manuscript.

Funding: This research received no external funding.

Data Availability Statement: The data presented in this study are available on request from the corresponding author.

Acknowledgments: We want to thank A. Tuba Ökse, Kocaeli University, for providing the obsidian artefacts from Kendale Hecala. The project was conducted with permission from Diyarbakır Archaeological Museum 2019). Although the work was not financially supported, we are grateful to Troy-Met and Tuna Eroğlu for kindly providing Hitachi XMET 8000 handheld system during the study. We would also like to thank Şakir Can, Özlem Ekinbaş Can, and the Kendale Hecala archaeological team for all their assistance, as well as the museum staff, especially Mevlut Çolak and Azad Gül, for their help.

Conflicts of Interest: The authors declare no conflict of interest.

References

1. Carter, R.A.; Philip, G. *Beyond the Ubaid: Transformation and Integration in the Late Prehistoric Societies of the Middle East*; The University of Chicago: Chicago, IL, USA, 2010; pp. 1–23.
2. Porada, E.; Hansen, D.P.; Dunham, S.; Babcock, S.H. The Chronology of Mesopotamia, ca. 7000–1600 BC. In *Chronologies in Old World Archaeology*; Ehrick, R.W., Ed.; University of Chicago Press: Chicago, IL, USA, 1992; pp. 77–121.
3. Lemonnier, P. La description des chaînes opératoires: Contribution à l'analyse des systèmes techniques. *Techniques et Culture* **1976**, *1*, 100–151.
4. Lemonnier, P. *Elements for an Anthropology of Technology*; The University of Michigan: Ann Arbor, MI, USA, 1992; pp. 25–32.
5. Thomalsky, J. Lithic industries of the Ubaid and Post-Ubaid period in northern Mesopotamia. *Publ. L'Institut Français D'Études Anatol.* **2012**, *27*, 417–439.
6. Batist, Z. Obsidian Circulation Networks in Southwest Asia an Anatolia (12,000–5700 B.P.): A Comparative Approach. Master's Thesis, McMaster University, Hamilton, ON, Canada, 2014.
7. Campbell, S.; Healey, E. Diversity in obsidian use in the prehistoric and early historic Middle East. *Quat. Int.* **2018**, *468*, 141–154. [CrossRef]
8. Barge, O.; Kharanaghi, H.A.; Biglari, F.; Moradi, B.; Mashkour, M.; Tengberg, M.; Chataigner, C. Diffusion of Anatolian and Caucasian obsidian in the Zagros Mountains and the highlands of Iran: Elements of explanation in 'least cost path' models. *Quat. Int.* **2018**, *467*, 297–322. [CrossRef]
9. Campbell, S.; Healey, E. Multiple sources: The pXRF analysis of obsidian from Kenan Tepe, SE Turkey. *JAS Rep.* **2016**, *10*, 377–389. [CrossRef]
10. Renfrew, C.; Dixon, J.E.; Cann, J.R. Obsidian and early cultural contact in the Near East. *East. Proc. Prehist. Soc.* **1966**, *32*, 30–37. [CrossRef]
11. Campbell, S.; Healey, E.; Maeda, O. Profiling an unlocated source: Group 3D obsidian in prehistoric and early historic near East. *JAS Rep.* **2020**, *33*, 102533. [CrossRef]
12. Khalidi, L.; Gratuze, B.; Stein, G.; McMahon, A.; Al-Quntar, S.; Carter, R.; Cuttler, R.; Dreshchler, P.; Healey, E.; Inizan, M.L.; et al. The growth of early social networks: New geochemical results of obsidian from the Ubaid to Chalcolithic Period in Syria, Iraq and the Gulf. *JAS Rep.* **2016**, *9*, 743–757. [CrossRef]
13. Muşkara, Ü.; Konak, A. Obsidian source identification at Gre Fılla, Turkey. *JAS Rep.* **2021**, *38*, 103003. [CrossRef]

14. Ökse, A.T. Ambar Dam Salvage Excavations in 2018–2020: Ambar Höyük, Gre Filla and Kendale Hecala. In *Archaeology of Anatolia Volume IV: Recent Discoveries (2018–2020)*; Sharon, R.S., McMahon, G., Eds.; Cambridge Scholars Press: Newcastle upon Tyne, UK, 2021; pp. 4–20.
15. Ökse, A.T. New Data on the Late Neolithic Pottery from the Northern Upper Tigris Region: Ambar Dam Reservoir. In *Neolithic Pottery from the Near East: Production, Distribution and Use, Proceedings of the 2019 Third International Workshop (Antalya, Turkey)*; Özbal, R., Erdalkıran, M., Tonoike, Y., Eds.; Koç University Press: Istanbul, Turkey, 2020; Volume 259, pp. 301–322.
16. Davis, M.K.; Jackson, T.L.; Shackley, M.S.; Teague, T.; Hampel, J.H. Factors affecting the energy-dispersive X-ray fluorescence (EDXRF) analysis of archaeological obsidian. In *Archaeological Obsidian Studies: Method and Theory*; Shackley, M.S., Ed.; Plenum Press: New York, NY, USA, 1998; pp. 159–180.
17. Shackley, M.S. An introduction to X-ray fluorescence (XRF) analysis in archaeology. In *X-ray Fluorescence Spectrometry (XRF) in Geoarchaeology*; Shackley, M.S., Ed.; Springer: New York, NY, USA, 2011; pp. 7–44.
18. Gauvin, R.; Lifshin, E. Simulation of X-ray emission from rough surfaces. *Microchim. Acta* **2000**, *132*, 201–204. [CrossRef]
19. Liritsiz, I.; Zacharias, N. Portable XRF of Archaeological Artifacts: Current Research, Potentials and Limitations. In *X-ray Fluorescence Spectrometry (XRF) in Geoarchaeology*; Shackley, M.S., Ed.; Springer: New York, NY, USA, 2011; pp. 1009–1142.
20. Steiner, A.E.; Conrey, R.M.; Wolff, J.A. PXRF calibrations for volcanic rocks and the application of in-field analysis to the geosciences. *Chem. Geol.* **2017**, *453*, 35–54. [CrossRef]
21. Hunt, A.M.; Speakman, R.J. Portable XRF analysis of archaeological sediments and ceramics. *JAS* **2015**, *53*, 626–638. [CrossRef]
22. Ferguson, J.R. *X-ray Fluorescence of Obsidian: Approaches to Calibration and the Analysis of Small Samples. Handheld XRF for Art and Archaeology*; Leuven University Press: Leuven, Belgium, 2012; pp. 401–422.
23. Nazaroff, A.J.; Prufer, K.M.; Drake, B.L. Assessing the applicability of portable X-ray fluorescence spectrometry for obsidian provenance research in the Maya lowlands. *JAS* **2010**, *37*, 885–895. [CrossRef]
24. Frahm, E. Validity of "off-the-shelf" handheld portable XRF for sourcing Near Eastern obsidian chip debris. *JAS* **2013**, *40*, 1080–1092. [CrossRef]
25. Glascock, M.D.; Neff, H. Neutron activation analysis and provenance research in archaeology. *Meas. Sci. Technol.* **2003**, *14*, 1516–1526. [CrossRef]
26. Grave, P.; Kealhofer, L.; Marsh, B.; Gates, M.H. Using neutron activation analysis to identify scales of interaction at Kinet Höyük, Turkey. *JAS* **2008**, *35*, 1974–1992. [CrossRef]
27. Forster, N.; Grave, P. Non-destructive PXRF analysis of museum-curated obsidian from the Near East. *JAS* **2012**, *35*, 1974–1992. [CrossRef]
28. Prokeš, L.; Galiová, M.V.; Hušková, S.; Vaculovič, T.; Hrdlička, A.; Mason, A.Z.; Neff, H.; Přichystal, A.; Kanický, V. Laser microsampling and multivariate methods in provenance studies of obsidian artefacts. *Chem. Pap.* **2015**, *69*, 761–778. [CrossRef]
29. Robin, A.K.; Mouralis, D.; Akköprü, E.; Gratuze, B.; Kuzucuoğlu, C.; Nomade, S.; Pereira, A.; Doğu, A.F.; Erturaç, K.; Khalidi, L. Identification and characterization of two new obsidian sub-sources in the Nemrut volcano (Eastern Anatolia, Turkey): The Sıcaksu and Kayacık obsidian. *JAS Rep.* **2016**, *9*, 705–717. [CrossRef]
30. Orange, M.; Carter, T.; Le Bourdonnec, F.X. Sourcing obsidian from Tell Aswad and Qdeir 1 (Syria) by SEM-EDS and EDXRF: Methodological implications. *Comptes Rendus Palevol* **2013**, *12*, 173–180. [CrossRef]
31. Baxter, M.J.; Buck, C.E. Data Handling and Statistical Analysis. In *Modern Analytical Methods in Art and Archaeology*; Ciliberto, E., Spoto, G., Eds.; John Wiley and Sons: New York, NY, USA, 2000; pp. 681–746.
32. Glascock, D.M. A systematic approach to geochemical sourcing of obsidian artifacts. *Sci. Cult.* **2020**, *2*, 35–47.
33. Liritzis, I.; Xanthopoulou, V.; Palamara, E.; Papageorgiou, I.; Iliopoulos, I.; Zacharias, N.; Vafiadou, A.; Karydas, A.G. Characterization and provenance of ceramic artifacts and local clays from Late Mycenaean Kastrouli (Greece) by means of p-XRF screening and statistical analysis. *J. Cult. Herit.* **2020**, *46*, 61–81. [CrossRef]
34. López-García, P.A.; Vidal-Aldana, C.I.; Gómez-Ambríz, E.A.; Argote, D.L. The obsidian of la ferrería site: Local consumption and long-distance interactions in north and northwestern Mexico. *JAS Rep.* **2021**, *9*, 103081. [CrossRef]
35. Chataigner, C.; Poidevin, J.L.; Arnaud, N.O. Turkish occurrences of obsidian and use by prehistoric peoples in the Near East from 14,000 to 6000 BP. *J. Volcanol. Geotherm. Res.* **1998**, *85*, 517–537. [CrossRef]
36. Chabot, J.; Poidevin, J.L.; Chataigner, C.; Fortin, M. Caracterisation et provenance des artefacts en obsidienne de Tell'Atij et de Tell Gudeda (III millenaire, Syrie). *Cahiers d'Acheologie Du CELAT* **2001**, *10*, 241–256.
37. Francaviglia, V.M. L'origine des outils en obsidienne de Tell Magzalia, Tell Sotto, Yarim Tepe et Kül Tepe, Iraq. *Paléorient* **1994**, *20*, 18–31. [CrossRef]
38. Gratuze, B. Obsidian characterization by laser ablation ICP-MS and its application to prehistoric trade in the Mediterranean and the Near East: Sources and distribution of obsidian within the Aegean and Anatolia. *JAS* **1999**, *26*, 869–881. [CrossRef]
39. Bressy, C.; Poupeau, G.; Yener, K.A. Cultural interactions during the Ubaid and Halaf periods: Tell Kurdu (Amuq Valley, Turkey) obsidian sourcing. *JAS* **2005**, *32*, 1560–1565. [CrossRef]
40. Khalidi, L.; Gratuze, B.; Boucetta, S. Provenance of obsidian excavated from Late Chalcolithic levels at the sites of Tell Hamoukar and Tell Brak, Syria. *Archaeometry* **2009**, *51*, 879–893. [CrossRef]
41. Poupeau, G.; Le Bourdonnec, F.X.; Carter, T.; Delerue, S.; Shackley, M.S.; Barrat, J.A.; Dubernet, S.; Moretto, P.; Calligaro, T.; Milić, M.; et al. The use of SEM-EDS, PIXE and EDXRF for obsidian provenance studies in the Near East: A case study from Neolithic Çatalhöyük (central Anatolia). *JAS* **2010**, *37*, 2705–2720. [CrossRef]

42. Chataigner, C. Les propriétés géochimiques des obsidiennes et la distinction des sources de Bingöl et du Nemrut Dag. *Paléorient* **1994**, *20*, 9–17. [CrossRef]
43. Frahm, E. Distinguishing Nemrut Dağ and Bingöl A obsidians: Geochemical and landscape differences and the archaeological implications. *JAS* **2012**, *39*, 1436–1444. [CrossRef]
44. Frahm, E. Variation in Nemrut Dağ obsidian at Pre-Pottery Neolithic to Late Bronze Age sites (or: All that's Nemrut Dağ obsidian isn't the Sıcaksu source). *JAS* **2020**, *32*, 102438. [CrossRef]
45. Abedi, A.; Vosough, B.; Razani, M.; Kasiri, M.B.; Steimiger, D.; Ebrahimi, G. Obsidian Deposits from Northwestern Iran and First Analytical Results: Implications for Prehistoric Production and Trade. *Mediterr. Archaeol. Archaeom.* **2018**, *18*, 107–118.
46. Ökse, A.T. Yukarı Dicle Havzasında Ambar Çayı Vadisi Yerleşim Tarihi. *Olba* **2020**, *28*, 1–34.
47. Abedi, A.; Varoutsikos, B.; Chataigner, C. Provenance of obsidian artifacts from the Chalcolithic site of Dava Göz in NW IRAN using portable XRF. *JAS Rep.* **2018**, *20*, 756–767. [CrossRef]

quaternary

Review

Direct Dating of Chinese Immovable Cultural Heritage

Robert G. Bednarik

College of History and Culture, Hebei Normal University, Shijiazhuang 050016, China;
ifrao@hotmail.com; Tel.: +86-163-9523-0549

Abstract: The most extensive corpus of ancient immovable cultural heritage is that of global rock art. Estimating its age has traditionally been challenging, rendering it difficult to integrate archaeological evidence of early cultural traditions. The dating of Chinese rock art by 'direct methods' began in the late 1990s in Qinghai Province. Since then, China has acquired the largest body of direct dating information about the rock art of any country. The establishment of the International Centre for Rock Art Dating at Hebei Normal University has been the driving force in this development, with its researchers accounting for most of the results. This centre has set the highest standards in rock art age estimation. Its principal method, microerosion analysis, secured the largest number of determinations, but it has also applied other methods. Its work with uranium-thorium analysis of carbonate precipitates in caves is of particular significance because it tested this widely used method. The implications of this work are wide-ranging. Most direct-dating of rock art has now become available from Henan, but results have also been reported from Heilongjiang, Inner Mongolia, Ningxia, Jiangsu, Hubei, Guangxi, Yunnan, Qinghai, Tibet, and Xinjiang. Intensive work by several teams is continuing and is expected to result in a significantly better understanding of China's early immovable cultural heritage.

Keywords: rock art; petroglyph; microerosion dating; radiocarbon dating; uranium–thorium dating; China

check for updates

Citation: Bednarik, R.G. Direct Dating of Chinese Immovable Cultural Heritage. *Quaternary* **2021**, *4*, 42. https://doi.org/10.3390/quat4040042

Academic Editor: Ioannis Liritzis

Received: 25 October 2021
Accepted: 24 November 2021
Published: 25 November 2021

1. Introduction

Immovable cultural heritage occurs throughout the world and in many forms, of which rock art is the most numerous of manifestations. In the case of China, the connection between rock art and other such heritage is particularly important because features such as statues, religious and secular structures or rock inscriptions of known ages have been used extensively to calibrate the direct dating of rock art. Estimating the ages of rock art is one of the most challenging tasks of archaeology and is riddled with controversies [1]. Many approaches have been tried, and it has become evident that the methodology of 'direct' dating is the most dependable of them. It is characterized by a direct physical relationship between the rock art in question and the dating criterion, and the falsifiability of the propositions concerning that relationship.

A wide range of potential 'dating criteria' has been appraised, but there are difficulties with many of them. Most importantly, the demand for falsifiability renders it is essential that the analysis should be repeatable: another researcher must be able to test the claim by repeating the experiment. Such replication is not possible with many methods proposed or already used because they involve the removal of physical samples that are sacrificed in the process of analysis. Such methods may also be challenged on ethical grounds by arguing that these interventions damage the integrity of the rock art or its relationship with contiguous features, such as mineral accretions. Examples include extracting carbon-bearing substances contained in rock art paint residues, cations present in rock varnishes covering petroglyphs, or determining the nature of uranium and thorium components of reprecipitated carbonates. Many of these applications are severely hampered by the significant variations of the concentrations of the dating criteria elements in coeval mineral skins on a millimetre-scale, which may be well above 100% [2,3]. For instance, the method of cation-ratio dating of iron and manganese-rich mineral accretions has long been discredited [4], and the uranium-series analysis of speleothem skins is currently under intensive review (see below).

The topic of rock art age estimation in China was first reviewed over three decades ago [5], and given the significant progress made in this field since then, it is worthwhile assessing how much change there has been. It is notable that the first-ever academic report about Chinese rock art in a Western language only appeared in 1984 [6]. Since 1991, when a sizeable Australian delegation attended a rock art conference in Yinchuan, Ningxia Province, the collaboration between Chinese and Australian rock art researchers developed and eventually flourished. The status of rock art dating in China in 1991 was that such practices were then limited almost wholly to archaeological or indirect means, such as "presumed association with a dated sediment deposit, perceived stylistic connection, spatial association and similar" [5]. Much rock art 'dating' derived from the pareidolic 'identification' of presumed animal species depicted or from correlation with ancient documents, such as the *jia gu wen* (writing on tortoise shells or bones). Other indirect approaches were the perceived degree of weathering, presumably depicted activity themes, and alleged styles [7]. Only two examples of direct rock art dating were then known in China: radiocarbon dating of stalactitic deposit physically related to a rock painting at the massive Huashan site in Guangxi Zhuang Autonomous Region [8]; and the ^{14}C content of flowstone laminae and pollen in the underlying paint layer were determined at one of the rock painting sites at Cangyuan, Yunnan Province [6]. The Huashan motif appears to date from between 2370 and 2115 BP, but more recent analytical work at the site has suggested a somewhat younger age. The Cangyuan image seems to be approximately 3000 years old, an estimate that has recently been confirmed [9]. A careful assessment of the 57 U–Th and four AMS radiocarbon results secured from four Cangyuan rock paintings has suggested that the paintings seem to be between 3800 and 2700 years old. Excavation results from the sites corroborate this conclusion.

2. Introduction of Direct Rock Art Dating in China

These first two direct dating attempts of Chinese rock art refer to endeavors that were not testable by replicating the experiments on which they were based. The subsequent results were derived from Tang Huisheng, who, in 1997–1998, introduced the use of microerosion analysis in Qinghai Province [10]. He collected microerosion calibration data from three petroglyph sites: Shuixia, Lebogou and Kexiaotu. These were then used to place petroglyphs from three more sites chronologically: Lushan, Lumanggou and Yeniugou. These were found to be approximately E2000, E2300 and E3200 years old, respectively (the 'E' prefix indicates that the age estimate was derived from erosion data). Since these measurements are repeatable, they fully comply with the requirements of direct rock art dating (Figure 1). Tang then secured age estimates from three cupules at the Jiangjunya site at Lianyungang City, Jiangsu Province, ranging from E4300 to c. E11,000 years BP, using calibration obtained from a Buddhist inscription at nearby Kongwang Hill, dating from April 61 CE [11].

All methods currently used to estimate the ages of rock art are experimental, and that includes microerosion analysis [1]. However, that method offers significant advantages, such as full replicability and lack of physical intervention. Microerosion-derived age estimates of petroglyphs can only be approximate because precipitation can vary as a function of time. Therefore, in the present report, only approximate estimates are given (for proposed tolerances, see individual publications cited). Nevertheless, the results of seven 'blind tests' conducted in Russia, Portugal, Italy, Bolivia, Australia (2) and China matched archaeological expectations very well [12–18]. In terms of their magnitude, results from this method are fully reliable. Radiocarbon analysis, by contrast, can provide very precise results, but when obtained from rock art these may be entirely false. Those obtained from paint residues can only be accepted if the substance analyzed has been identified and separated, be it at the molecular or at the object level [19].

Figure 1. Tang Huisheng conducting the first replicable direct dating of rock art in China in 1997 at the Lushan petroglyph site in Qinghai Province (photograph by Gao Zhiwei, with permission).

The discovery of a major rock art concentration in Henan Province [20] prompted a very successful rock art dating expedition in that region and Ningxia and Jiangsu Provinces during June and July 2014 [21]. It utilized China's wealth of rock surfaces suitable for microerosion calibration, especially soundly dated rock inscriptions. Several calibration curves, as well as twenty-seven age estimates from petroglyphs, were secured. This work included testing previous archaeological age predictions for some of the best-known Chinese petroglyph complexes, such as those of Helanshan and Jiangjunya. For instance, there had been numerous age estimates for the famous Helanshan petroglyphs near Yinchuan, ranging from the Pleistocene to recent centuries, and based on various methods. The 2014 campaign furnished reliable estimates of E2000 to E2330 years BP.

On the other hand, the seven dates secured from the Jiangjunya site, 14 km west of Lianyungang, ranged widely, from about E369 to E5380 years BP, demonstrating the use of the site from Neolithic to recent historical times. Some of the site's many petroglyphs have been demonstrated to have been retouched after their initial creation. For example, Petroglyph 1 was made about E2210 years ago but was retouched some E360 years ago. Such reworking of a petroglyph cannot be readily identified by any other dating method. All age estimates of this campaign were again obtained by microerosion analysis and are thus repeatable. Now or centuries into the future, any researcher can locate the dated motif and even the specific micro-wane and re-measure it. This adds the benefit that future erosion rates can be determined.

The microerosion method endeavors to ascertain when crystals in the grooves of percussion grooves were fractured by impact during petroglyph production. At that time, the edges of these fractures were totally sharp, but erosion gradually rounds them at the microscopic level in a quantifiable process that is a function of time. The resulting micro-wanes reflect the time since the fractures occurred [12,13]. In contrast to most other known direct dating methods, it refers to criteria that are functions of actual age rather than minimum or maximum ages. It is also non-invasive and involves no contact with the rock art and there are no contaminating factors. The method even allows age determinations in the field. However, it also entails several disadvantages: it has so far only been applied to two minerals (quartz and feldspar); it requires minimum grain sizes of about 1.5 mm with fractures of about 90° between the cleavage surfaces, orientated so that the micro-wane faces the microscope; and the rock surface must have been exposed to precipitation ever since the petroglyph was created. Microerosion analysis provides very reliable but imprecise age estimates, with tolerances often in the order of 20–25%. The significant differences in rainfall in different environments can be accounted for by calibration against the microerosion of surfaces of known ages. In recent years a universal calibration has been created that is based on relative regional precipitation and can be applied where local calibration is not possible [22,23]. The only minerals calibrated so far are quartz and feldspar and the former is thought to have a range of up to maximal 50 ka.

In June and October 2015, rock art dating missions were undertaken in the Xinjiang, Inner Mongolia, Ningxia, Guangxi, and Henan provinces [18]. Although new age estimates were only secured from two of these regions, the expeditions provided an essential overview of the scale of Chinese rock art and the logistics of developing comprehensive approaches to its dating, as well as several other scientific data. More than any such work previously undertaken in China, these journeys impressed the need to develop flexible approaches utilizing multiple methods, with geomorphological procedures forming the most reliable core. Nevertheless, a series of five other microerosion estimates were secured from the Henan sites Xuanluoling, Taibailing and Paomaling, all forming part of the extensive Mt Juci complex. Of particular interest was determining the age of a mask/face petroglyph at Xiao Fengshan site in Inner Mongolia. This motif on rhyolite turned out to be late Neolithic at about E4730 +1400/−810 years of age. This was subsequently found to confirm the archaeological expectation that the area's face/mask images are of that period. However, there is no proof that this age estimate can be applied to all face/mask motifs of the region.

Another microerosion-based rock art dating program was undertaken in 2016 [24]. Five sites were investigated in Xianju County of eastern Zhejiang Province. They yielded fifteen microerosion results, including calibration from three surfaces of the Wufubei site complex. Several motifs at the Xiaofangyan and Songlongshan sites provided seven quite consistent dates ranging from E1200 to E1360 at the second site and slightly earlier results from the first.

The effects of the 2014 rock art dating campaign have led to developments beyond the provision of more credible rock art dates than any previous project. Most importantly, it persuaded Tang to establish the International Centre of Rock Art Dating and Conservation at the College of History and Culture, Hebei Normal University, Shijiazhuang.

3. The International Centre of Rock Art Dating (ICRAD)

This agency of Hebei Normal University was formally established on 16 June 2016 [25]. That university was chosen because it already possessed facilities for AMS radiocarbon, uranium–thorium and OSL analyses, three of the methods used in rock art dating work. The Centre will establish a comprehensive archive for global information on all direct rock art dating projects and results in the world since the early 1980s, and it will conduct its own research in age estimation of rock art in China. The ICRAD established a simple ground rule to ascertain the scientific integrity of records: they must be presented so that another researcher can try to duplicate (or refute) the reported results, be it by the same or another method. Therefore, the dating criterion must be described so that the second researcher can re-locate the criterion reliably. ICRAD also emphasizes the need to establish protocols that would stand the test of time and will not need to be significantly modified in the future.

To facilitate the implementation of these protocols, ICRAD has established a system of numbering each rock art age determination attempt with a unique code, much in the way radiocarbon dating results are identified. Without such a system, the growing mass of uncollated and incompatible data would eventually become unmanageable. ICRAD's direct dating register will eventually be made available publicly to facilitate its use globally.

Since the establishment of ICRAD, the efforts of direct-dating Chinese have continued unabated—in fact, they appear to be accelerating. In 2017, a large team conducted the first rock art dating program undertaken in Hubei Province, focusing on a mountainous area east of Tongbai [26]. Huai River rock art corpus includes numerous sites that generally resemble the Henan rock art to the north. Eight of them yielded age estimates, which in all cases derive from cupules. They all fall under 1270 years, ranging down to about 650 years, indicating that the extensive rock art complex is relatively young. The results of this work were interpreted according to the recently established universal calibration curve (UCC) [22,23].

A second team revisited petroglyph sites in the granite region of Fangcheng in Henan Province and secured a series of nine age estimates from the sites Fangshan 2 and 4, Zhangzhuang, Wushigou 2 and 4 [27]. The first site produced the earliest results from two zoomorphs that are between 4000 and 5000 years old. This was soon followed by the team's assessment of several sites at Lianyungang in Jiangsu Province during two field seasons [28]. The authors provided 14 microerosion age estimates from eight sites, including three results from two of the Jiangjunya sites. They ranged from E710 to E2020 years BP, broadly confirming that the area's petroglyphs cover a considerable period, but the majority is 1000 to 2000 years old. Calibration was secured from a rock inscription at Xioaxishan 1 site, which was also confirmed by the UCC recently established. This paper also introduced a local phenomenon, standing stones bearing petroglyphs that are very common in northern Chinese regions but rare further south. In Lianyungang, they are called *shiganma* (stone mother) and bear anthropomorphous petroglyphs. Jin and Chao studied eleven of them, managing to secure age estimates from four. These corresponded well with ancient literature and inscriptions.

Jin and Chao then presented the first rock art dating results from Liaoning Province in north-eastern China [29]. They investigated three site complexes featuring eight sites near Anshan City. The authors provided age estimates of three cupules from the site Bafen'gou and one each from cupules at Wangjiayu 1 and 2, ranging from E1140 to E2030 years. Of interest is their detection of KEM (kinetic energy metamorphosis [30]), which they had also reported previously from the granite of Wushigou 1 at Fangsheng [27]. KEM was only discovered in recent years but has since been investigated intensively [31]. It has been recognized as an essential tribological variable in the study of petroglyphs.

Most recently, the focus of rock art dating has turned to the Tibetan Plateau, specifically to Garze Tibet Autonomous Prefecture in Sichuan Province and Yushu Tibetan Autonomous Prefecture in Qinghai Province [32]. Twelve petroglyph sites were investigated in that area, featuring vast numbers of zoomorphs and more recent Buddhist rock inscriptions. Despite strenuous endeavors, only one dating could be extracted from the petroglyphs. It is from a geometric design at the Kewa site that was E2089 +218/−295 years old, i.e., most probably of the Han Dynasty.

Although most direct rock art dates from China were secured by microerosion analysis, it would be wrong to assume that no other methods were used or at least tried. For example, a project investigating cave art in Guangxi Region that found a tradition rich in feline depictions used ^{14}C analysis to estimate the ages of two types of material: charcoal applied in rock painting and the wax of a small beehive superimposed over paint residues [33]. The charcoal flakes in the white paint of a feline provided a date of 250 $\pm$ 30 years BP, and the beehive yielded 80 $\pm$ 30 years BP. This relatively recent tradition has been explained in terms of available ethnographic information provided by the Zhuang people of the region.

Another analytical method much used for rock art age estimation in China is one of the uranium-series techniques, determining the ^{230}Th/^{234}U ratio. It has been used predominantly in two regions, Yunnan, and Heilongjiang Provinces, but was recently also applied in Tibet. The method demands that the initial ratio of ^{230}Th/^{234}U at the sample formation must be known or determined. Thorium is not soluble in water under naturally occurring conditions, whereas uranium is, and the optimistic assumption is made that freshly formed carbonate precipitates are free of Th. The method was first introduced in rock art age estimation in 1981 [34] when it was found sometimes to provide significantly misleading results. In China, the method was first used at the Baiyunwan rock art site in Yunnan, yielding only inconsistent results, and the uncertainties were attributed to U depletion, detrital ^{230}Th occurrence and the presence of 'dead' carbon [35]. Tang collected samples from Jinshajiang sites in Yunnan, some of which were subjected to analysis by two different laboratories. Significant were the ^{230}Th/^{234}U results of the June 2017 ICRAD expedition to Heilongjiang Province, conducting the first scientific rock art research in

China's northernmost region [2]. Its results led to a fundamental reassessment of a method that had been the subject of scientific controversy for many years.

4. The Trouble with U–Th Dating of Rock Art

The first application of U–Th analysis to estimate the age of rock art relates to petroglyphs on the ceiling of Malangine Cave in South Australia [34]. A speleothem lamina covers one generation of them that in turn bears another tradition of petroglyphs, thus providing a minimum date for one and maximum date for the other. Its radiocarbon age was 5550 ± 55 years bp, but the sample's U–Th date was five times greater, 28.0 ± 2.0 ka. All subsequently dated similar carbonate speleothems subjected to both tests showed a similar pattern: the U–Th results were always older and, in most cases, significantly older than ^{14}C or archaeological estimates (Figure 2) [36–42]. Indeed, in two cases, both from China, the U–Th dates were more than one hundred times as old. A reprecipitated carbonate film at Yilin in Heilongjiang that can only be a few centuries old at most has provided a U–Th raw age of 134.6 ka, i.e., hundreds of times its realistic age [2]. An international team recently discovered a few hand and foot impressions of juveniles in a hardened travertine deposit at the Quesang Hot Spring site in Tibet. They correctly proposed that the age of these prints should approximate the rock's age, which must have been soft and still forming at the time they were produced. They secured U–Th 'dates' from the travertine that would place the age of the formation between 169 ka and 226 ka. On that basis, they claimed to have found the oldest known rock art globally, probably made by Denisovans [43].

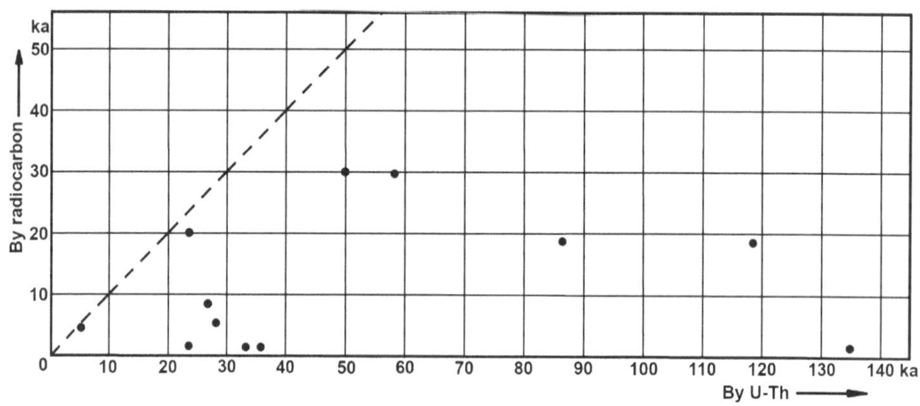

Figure 2. U–Th age determinations of speleothems compared with archaeologically realistic or radiocarbon ages of these same deposits.

This follows similarly spectacular claims from several cave sites in Spain, also based on U–Th data, that paintings thought to be of the late Upper Paleolithic were much older and were made by Neanderthals [44–46]. Due to these many concerns about the credibility of U–Th dates from non-crystalline reprecipitated carbonates, an intensive debate of the method when applied to thin or porous carbonates has developed over the last decade [47–56]. The primary cause of the excessive ages attributed to reprecipitated carbonate deposits is the depletion of U by moisture. Solution may also remove detrital Th, there may be a transformation of aragonite to calcite, or samples may be contaminated by components of the support rock [57–59].

Two other factors are of great concern. One issue needing more attention is the significant variation of U concentrations in coeval calcite skins demonstrated to occur on a millimetre-scale that may be greater than 100% [2,3]. The second concern stems from 'blind tests' we conducted due to the grossly incongruous U–Th results from Heilongjiang sites Mohe and Yilin 2 [2]. We split four samples from Yunnan Jinshajiang sites and submitted the two sets to two different U–Th laboratories [60]. Not only did this yield two entirely different sets of results, but the reporting protocols also differed profoundly. Moreover, three results produced negative values, probably attributable to significant leaching of U and other contaminating factors (Table 1). The stochastic distribution of the dates in Figure 2 suggests that the distortion is not systematic but seems to be a random function of taphonomic processes distorting the U–Th ratios. Most notably, the water-soluble U can be readily mobilized when the deposit is subjected to moisture. This frequently occurs with speleothems and even more so with travertine that is fully exposed to precipitation. Travertines are not dense crystalline formations like stalagmites; they have varying degrees of porosity which assists the reaction with carbonic acid to revert to their soluble (bicarbonate) phase.

Table 1. Comparison of the raw U–Th ages of four split samples provided by two laboratories: all ages in ka.

Sample	MR-1	HY-1	YDG-1	YDG-2
Laboratory 1	1.359 ± 0.179	2.362 ± 2.573	4.674 ± 5.118	20.077 ± 2.742
Laboratory 2	-7 $+21/-26$	-20 $+26/-35$	-14 $+33/-45$	0.4 ± 7.7

There are also a few more minor issues related to extraordinary claims of this nature about rock art. Although we have no reliable information on soft tissue dimensions of any robust humans, especially not on Denisovans, we assume that Neanderthals had thicker fingers than moderns, and we know that their feet differed from those of gracile humans [61,62]. The footprints at Quesang were made by 'modern' humans, as the authors correctly note, suggesting that they are much younger than proposed. Moreover, the earliest rock art currently known is not, as suggested, in Sulawesi: there are several earlier candidates in India, France, Spain, South Africa, even Australia [63]. Moreover, the age of the Sulawesi rock art was also determined by the unreliable U–Th method.

Considerable efforts have been made to date petroglyphs by the uranium–thorium (U-Th) disequilibrium method and optically stimulated luminescence (OSL) surface dating. For the former, the chronology of early human fossil remains overlain by precipitated calcite (stalagmites) also requires a plea for caution with regards to contamination, as described above. Associated methodological corrections are reported for Petralona Cave in Greece [64–66]. For the latter, Liritzis' surface luminescence dating of sun-exposed archaeological stone, masonry surfaces has been introduced [67] which has been extended and applied to rock art cases of exfoliated engraved fragments [68–70].

5. Summary and Outlook

It needs to be emphasized that U–Th results of the Holocene, especially the second half of that period, seem to match ^{14}C dates from the same deposits frequently. It is only as we approach the Pleistocene that the results of the two methods diverge. By the time 30,000 carbon years is reached, the corresponding U–Th ages are around 50,000 years—and this also appears to apply to fossil bone [71]. Nevertheless, the $^{230}Th/^{234}U$ method has been widely used to date carbonate speleothems, and when it produces extraordinary results, its advocates reject the need for checking these with another method [54]. One of the most consequential outcomes of the work by the International Centre for Rock Art Dating (ICRAD) is that it has found a path to test the results of U–Th analysis and thereby help resolve the deadlock between the opposing parties. First, it has begun to take multiple samples of coeval carbonate skins, confirming dramatic differences [2]. Second, the processing of split samples by multiple laboratories has shown no correspondence whatsoever, be it in actual dates or reporting protocols [60]. Direct rock art dating results that cannot be verified are questionable, and if different laboratories deliver wildly diverging dates of split samples, there is no basis for even the most rudimentary comparison. The refusal of the advocates of exclusive use of U–Th dating to consider applying a second method [54] also deprives the discipline of the most crucial attribute of good science—the facility of testability.

The method of microerosion analysis has become the most intensively used by ICRAD researchers, despite its lack of high precision. It offers reliability instead, simplicity of application, unlimited repeatability, the benefit of obtaining target dates rather than maximum or minimum ages, and its lack of physical intervention. In China, with so many historical sources, rock inscriptions and archaeological sources of dating information, the method has already been widely applied. Its results have, in many cases, been verified independently by archaeologically derived information of several types. By comparison, the exclusive application of U–Th analysis, especially in presumed Pleistocene contexts, has universally provided ages that are archaeologically far too great, and the reasons for this are well understood.

Funding: This research received no external funding.

Acknowledgments: The author thanks the International Centre for Rock Art Dating for its outstanding support, particularly its Director, Tang Huisheng. He also thanks the many participants of numerous rock art expeditions in many parts of China.

Conflicts of Interest: The author declares no conflict of interest.

References

1. Bednarik, R.G. The dating of rock art: A critique. *J. Archaeol. Sci.* **2002**, *29*, 1213–1233. [CrossRef]
2. Tang, H.; Kumar, G.; Jin, A.; Bednarik, R.G. Rock art of Heilongjiang Province, China. *J. Archaeol. Sci. Rep.* **2020**, *31*, 102348. [CrossRef]
3. Hoffmann, D.L.; Spötl, C.; Mangini, A. Micromill and in situ laser ablation sampling techniques for high spatial resolution MC-ICPMS U-Th dating of carbonates. *Chem. Geol.* **2009**, *259*, 253–261. [CrossRef]
4. Watchman, A. Investigating the cation-ratio calibration curve: Evidence from South Australia. *Rock Art Res.* **1992**, *9*, 106–110.
5. Bednarik, R.G.; Li, F. Rock art dating in China: Past and future. *Artefact* **1991**, *14*, 25–33.
6. Wang, N. An introduction to rock paintings in Yunnan Province, People's Republic of China. *Rock Art Res.* **1984**, *1*, 75–84.
7. Tang, H. Theory and methods in Chinese rock art studies. *Rock Art Res.* **1993**, *10*, 83–90.
8. Qin, S.; Qin, T.; Lu, M.; Yü, J. *The Investigation and Research of the Cliff and Mural Paintings of the Zuojiang River Valley in Guangxi*; Guangxi National Printing House: Nanning, China, 1987.
9. Shao, Q.; Wu, Y.; Pons-Branchu, E.; Zhu, Q.; Dapoigny, A.; Jiang, T. U-series dating of carbonate accretions reveals late Neolithic age for the rock paintings in Cangyuan, southwestern China. *Quat. Geochronol.* **2021**, *61*, 101127. [CrossRef]
10. Tang, H.; Gao, Z. Dating analysis of rock art in the Qinghai-Tibetan Plateau. *Rock Art Res.* **2004**, *21*, 161–172.
11. Tang, H.; Mei, Y. Dating and some other issues on the prehistoric site at Jiangjunya. *Southeast Cult.* **2008**, *202*, 11–23.
12. Bednarik, R.G. A new method to date petroglyphs. *Archaeometry* **1992**, *34*, 279–291. [CrossRef]
13. Bednarik, R.G. Geoarchaeological dating of petroglyphs at Lake Onega, Russia. *Geoarchaeology* **1993**, *8*, 443–463. [CrossRef]
14. Bednarik, R.G. The age of the Coa valley petroglyphs in Portugal. *Rock Art Res.* **1995**, *12*, 86–103.
15. Bednarik, R.G. Microerosion analysis of petroglyphs in Valtellina, Italy. *Origini* **1997**, *21*, 7–22.
16. Bednarik, R.G. Age estimates for the petroglyph sequence of Inca Huasi, Mizque, Bolivia. *Andean Past* **2000**, *6*, 277–287.
17. Bednarik, R.G. About the age of Pilbara rock art. *Anthropos* **2002**, *97*, 201–215.
18. Tang, H.; Kumar, G.; Jin, A.; Wu, J.; Liu, W.; Bednarik, R.G. The 2015 rock art missions in China. *Rock Art Res.* **2018**, *35*, 25–34.
19. Bednarik, R.G. *Yanhua Kexue—Yuangu Yishu de Kexue Yanjiu (Rock Art Science: The Scientific Study of Palaeoart)*; Jin, A., Translator; Shaanxi Xinhua Publishing & Media Group: Xi'an, China, 2020.
20. Tang, H. New discovery of rock art and megalithic sites in the Central Plain of China. *Rock Art Res.* **2012**, *29*, 157–170.
21. Tang, H.; Kumar, G.; Liu, W.; Xiao, B.; Yang, H.; Zhang, J.; Lu, X.H.; Yue, J.; Li, Y.; Gao, W.; et al. The 2014 microerosion dating project in China. *Rock Art Res.* **2017**, *34*, 40–54.
22. Beaumont, P.B.; Bednarik, R.G. Concerning a cupule sequence on the edge of the Kalahari Desert in South Africa. *Rock Art Res.* **2015**, *32*, 162–177.
23. Bednarik, R.G. Advances in microerosion analysis. *Rock Art Res.* **2019**, *36*, 43–48.
24. Jin, A.; Zhang, J.; Xiao, B.; Tang, H. Microerosion dating of Xianju petroglyphs, Zhejiang Province, China. *Rock Art Res.* **2016**, *33*, 3–7.
25. Bednarik, R.G. The International Centre of Rock Art Dating and Conservation (ICRAD). *Rock Art Res.* **2016**, *33*, 111–112.
26. Tang, H.; Jin, A.; Li, M.; Fan, Z.; Liu, W.; Kumar, G.; Bednarik, R.G. The 2017 rock art mission in Hubei Province, China. *Rock Art Res.* **2020**, *37*, 67–73.
27. Jin, A.; Chao, G. The 2018 expedition to Fangcheng cupule sites in central China. *Rock Art Res.* **2019**, *36*, 157–163.
28. Jin, A.; Chao, G. The 2018 and 2019 rock art expeditions to Lianyungang, east China. *Rock Art Res.* **2020**, *37*, 74–81.

29. Jin, A.; Chao, G. The 2018 expedition to Anshan cupule sites, northeast China. *Rock Art Res.* **2021**, *38*, 3–9.
30. Bednarik, R.G. The tribology of cupules. *Geol. Mag.* **2015**, *152*, 758–765. [CrossRef]
31. Bednarik, R.G. *Tribology in Geology and Archaeology*; Nova Science Publishers: New York, NY, USA, 2019.
32. Li, M.; Lari, J.; Tang, H.; Li, Y.; Bednarik, R.G. The 2019 survey of petroglyphs in the Qinghai-Tibet Plateau, western China. *Rock Art Res.* **2022**, *39*. in press.
33. Li, M.; Shi, L.; Wu, X.; Tang, H. Discovery of new type of cave rock paintings in Guangxi Zhuang Autonomous Region, China. *Rock Art Res.* **2020**, *37*, 5–18.
34. Bednarik, R.G. Die Bedeutung der paläolithischen Fingerlinientradition. *Anthropologie* **1984**, *23*, 73–79.
35. Taçon, P.S.C.; Aubert, M.; Gang, L.; Yang, D.; Liu, H.; May, S.K.; Fallon, S.; Ji, X.; Curnoe, D.; Herries, A.I.R. Uranium-series age estimates for rock art in southwest China. *J. Archaeol. Sci.* **2012**, *39*, 492–499. [CrossRef]
36. Bard, E.; Hamelin, B.; Fairbanks, R.G.; Zindler, A. Calibration of the ^{14}C timescale over the past 30,000 years using mass spectrometric U–Th ages from Barbados corals. *Nature* **1990**, *345*, 405–410. [CrossRef]
37. Holmgren, K.; Lauritzen, S.-E.; Possnert, G. ^{230}Th/^{234}U and ^{14}C dating of a late Pleistocene stalagmite in Lobatse II cave, Botswana. *Quat. Sci. Rev.* **1994**, *13*, 111–119. [CrossRef]
38. Labonne, M.; Hillaire-Marcel, C.; Ghaleb, B.; Goy, J.L. Multi-isotopic age assessment of dirty speleothem calcite: An example from Altamira Cave, Spain. *Quat. Sci. Rev.* **2002**, *21*, 1099–1110. [CrossRef]
39. Plagnes, V.; Causse, C.; Fontugne, M.; Valladas, H.; Chazine, J.-M.; Fage, L.-H. Cross dating (Th/U-^{14}C) of calcite covering prehistoric paintings in Borneo. *Quat. Res.* **2003**, *60*, 172–179. [CrossRef]
40. Quiles, A.; Fritz, C.; Medina, M.A.; Pons-Branchu, E.; Sanchidrián, J.L.; Tosello, G.; Valladas, H. Chronologies croisées (C-14 et U/Th) pour l'étude de l'art préhistorique dans la grotte de Nerja: Méthodologie. In *Sobre Rocas y Huesos: Las Sociedades Prehistóricas y Sus Manifestaciones Plásticas*; Medina-Alcaide, M.A., Romero Alonso, A., Ruiz-Márquez, R.M., Sanchidrián Torti, J.L., Eds.; Fundación Cueva de Nerja: Córdoba, Spain, 2014; pp. 420–427.
41. Sanchidrián, J.L.; Valladas, H.; Medina-Alcaide, M.A.; Pons-Branchu, E.; Quiles, A. New perspectives for ^{14}C dating of parietal markings using $CaCO_3$ thin layers: An example in Nerja Cave (Spain). *J. Archaeol. Sci. Rep.* **2017**, *12*, 74–80. [CrossRef]
42. Valladas, H.; Pons-Branchu, E.; Dumoulin, J.P.; Quiles, A.; Sanchidrián, J.L.; Medina-Alcaide, M.A. U/Th and ^{14}C crossdating of parietal calcite deposits: Application to Nerja Cave (Andalusia, Spain) and future perspectives. *Radiocarbon* **2017**, *59*, 1955–1967. [CrossRef]
43. Zhang, D.D.; Bennett, M.R.; Cheng, H.; Wang, L.; Zhang, H.; Reynolds, S.C.; Zhang, S.; Wang, X.; Li, T.; Urban, T.; et al. Earliest parietal art: Hominin hand and foot traces from the middle Pleistocene of Tibet. *Sci. Bull.* **2021**, in press. [CrossRef]
44. Hoffmann, D.L.; Standish, C.D.; García-Diez, M.; Pettitt, P.B.; Milton, J.A.; Zilhã, J.; Alcolea-Gonzalez, J.J.; Cantalejo-Duarte, P.; Collado, H.; De Balbin, R.; et al. U–Th dating of carbonate crusts reveal Neanderthal origin of Iberian cave art. *Science* **2018**, *359*, 912–915. [CrossRef] [PubMed]
45. Hoffmann, D.L.; Standish, C.D.; García-Diez, M.; Pettitt, P.B.; Milton, J.A.; Zilhão, J.; Alcolea-González, J.J.; Cantalejo-Duarte, P.; Collado, H.; De Balbín, R.; et al. Response to Comment on 'U-Th dating of carbonate crusts reveals Neandertal origin of Iberian cave art'. *Science* **2018**, *362*, eaau1736. [CrossRef] [PubMed]
46. Hoffmann, D.L.; Standish, C.D.; Pike, A.W.; García-Diez, M.; Pettitt, P.B.; Angelucci, D.E.; Villaverde, V.; Zapata, J.; Milton, J.A.; Alcolea-González, J.; et al. Dates for Neanderthal art and symbolic behaviour are reliable. *Nat. Ecol. Evol.* **2018**, *2*, 1044–1045. [CrossRef] [PubMed]
47. Bednarik, R.G. U–Th analysis and rock art: A response to Pike et al. *Rock Art Res.* **2012**, *29*, 244–246.
48. Clottes, J. U-series dating, evolution and Neandertal. *Int. Newsl. Rock Art* **2012**, *64*, 1–6.
49. Pike, A.W.G.; Hoffmann, D.L.; García-Diez, M.; Pettitt, P.B.; Alcolea, J.; De Balbin, R.; González-Sainz, C.; De Las Heras, C.; Lasheras, J.-A.; Montes, R.; et al. U-series dating of Paleolithic art in 11 caves in Spain. *Science* **2012**, *336*, 1409–1413. [CrossRef]
50. Pons-Branchu, E.; Bourrillon, R.; Conkey, M.W.; Fontugne, M.; Fritz, C.; Gárate, D.; Quiles, A.; Rivero, O.; Sauvet, G.; Tosello, G.; et al. Uranium-series dating of carbonate formations overlying Paleolithic art: Interest and limitations. *Bull. Soc. Préh. Franç.* **2014**, *111*, 211–224.
51. Sauvet, G.; Bourrillon, R.; Conkey, M.; Fritz, C.; Garate-Maidagan, D.; Rivero Vila, O.; Tosello, G.; White, R. Answer to 'Comment on uranium-thorium dating method and Palaeolithic rock art' by Pons-Branchu et al. *Quat. Int.* **2017**, *432*, 96–97. [CrossRef]
52. Hoffmann, D.L.; Utrilla, P.; Bea, M.; Pike, A.W.G.; García-Diez, M.; Zilhão, J.; Domingo, R. U-series dating of Palaeolithic rock art at Fuente del Trucho (Aragón, Spain). *Quat. Int.* **2016**, *432*, 50–58. [CrossRef]
53. Hoffmann, D.L.; Pike, A.W.G.; García-Diez, M.; Pettitt, P.B. Methods for U-series dating of $CaCO_3$ crusts associated with Palaeolithic cave art and application to Iberian sites. *Quat. Geochron.* **2016**, *36*, 104–116. [CrossRef]
54. Pike, A.W.G.; Hoffmann, D.L.; Pettitt, P.B.; García-Diez, M.; Zilhão, J. Dating Palaeolithic cave art: Why U–Th is the way to go. *Quat. Int.* **2017**, *432*, 41–49. [CrossRef]
55. Aubert, M.; Brumm, A.; Huntley, J. Early dates for 'Neanderthal cave art' may be wrong. *J. Hum. Evol.* **2018**, *125*, 215–217. [CrossRef]
56. White, R.; Bosinski, G.; Bourrillon, R.; Clottes, J.; Conkey, M.W.; Corchón Rodriguez, S.; Cortés-Sánchez, M.; de la Rasilla Vives, M.; Delluc, B.; Delluc, G.; et al. Still no archaeological evidence that Neanderthals created Iberian cave art. *J. Hum. Evol.* **2019**, *144*, 102640. [CrossRef]

57. Lachniet, M.S.; Bernal, J.P.; Asmerom, Y.; Polyal, V. Uranium loss and aragonite-calcite age discordance in a calcitized aragonite stalagmite. *Quat. Geochron.* **2012**, *14*, 26–37. [CrossRef]
58. Bajo, P.; Hellstrom, J.; Frisia, S.; Drysdale, R.; Black, J.; Woodhead, J.; Borsato, A.; Zanchetta, G.; Wallace, M.W.; Regattieri, E.; et al. 'Cryptic' diagenesis and its implications for speleothem geochronologies. *Quat. Sci. Rev.* **2016**, *148*, 17–28. [CrossRef]
59. Fontugne, M.; Shao, Q.; Frank, N.; Thil, F.; Guidon, N.; Boeda, E. Cross dating (Th/U-14C) of calcite covering prehistoric paintings at Serra da Capivara National Park, Piauí, Brazil. *Radiocarbon* **2013**, *55*, 1191–1198. [CrossRef]
60. Tang, H.; Bednarik, R.G. Rock art dating by ^{230}Th/^{234}U analysis: An appraisal of Chinese case studies. *Archaeol. Anthrop. Sci.* **2021**, *13*, 19. [CrossRef]
61. Facorellis, Y.; Kiparissi-Apostolika, N.; Maniatis, Y. The cave of Theopetra, Kalambaka: Radiocarbon evidence for 50,000 years of human presence. *Radiocarbon* **2001**, *43*, 1029–1048. [CrossRef]
62. Bednarik, R.G. Antiquity and authorship of the Chauvet Cave rock art. *Rock Art Res.* **2007**, *24*, 21–34.
63. Bednarik, R.G. *Palaeoart of the Ice Age*; Cambridge Scholars Publishing: Newcastle upon Tyne, UK, 2017.
64. Liritzis, I.; Vafiadou, A.; Zacharias, N.; Polymeris, G.; Bednarik, R.G. Advances in surface luminescence dating: Some new data from three selected Mediterranean sites. *Medit. Archaeol. Archaeom.* **2013**, *13*, 105–115.
65. Liritzis, I.; Bednarik, R.G.; Kumar, G.; Polymeris, G.; Iliopoulos, I.; Xanthopoulou, V.; Zacharias, N.; Vafiadou, A.; Bratitsi, M. Daraki-Chattan rock art constrained OSL chronology and multianalytical techniques: A first pilot investigation. *J. Cult. Herit.* **2018**, *37*, 29–43. [CrossRef]
66. Liritzis, I.; Panou, E.; Exarhos, M. Novel approaches in surface luminescence dating of rock art: A brief review. *Medit. Archaeol. Archaeom.* **2017**, *17*, 89–102.
67. Liritzis, I. A new dating method by thermoluminescence of carved megalithic stone building. *Comptes Rendus Acad. Sci. Ser. 2 Sci. Terre Planetes* **1994**, *319*, 603–610.
68. Poulianos, A.N. Petralona Cave dating controversy. *Nature* **1982**, *299*, 280–281. [CrossRef]
69. Liritzis, Y. A critical dating revaluation of Petralona hominid: A caution for patience. *Athens Ann. Archaeol.* **1984**, *15*, 285–296.
70. Liritzis, Y.; Galloway, R.B. The Th230/U234 disequilibrium dating of cave travertines. *Nucl. Instr. Methods* **1982**, *201*, 507–510. [CrossRef]
71. Bednarik, R.G. The dating of rock art and bone by the uranium–thorium method. *Rock Art Res.* **2022**, *39*. in press.

MDPI AG
Grosspeteranlage 5
4052 Basel
Switzerland
Tel.: +41 61 683 77 34

Quaternary Editorial Office
E-mail: quaternary@mdpi.com
www.mdpi.com/journal/quaternary

www.ingramcontent.com/pod-product-compliance
Lightning Source LLC
Chambersburg PA
CBHW041355210926
51850CB00013B/3084
* 9 7 8 3 7 2 5 8 3 0 2 0 6 *